4 Progress in Molecular and Subcellular Biology

With Contributions by
St. Bram · P. Chandra · U. Ebener · H. Laube
W. Lotz · D. Oesterhelt · R. A. Olsson
R. E. Patterson · L. K. Steel · M. Sussman
G. Will · H. J. Witmer · M. Woltersdorf

With 88 Figures

Springer-Verlag Berlin Heidelberg New York 1976

ISBN 3-540-07487-2 Springer-Verlag Berlin Heidelberg New York
ISBN 0-387-07487-2 Springer-Verlag New York Heidelberg Berlin

Printed in Germany.

Offsetprinting and bookbinding: Konrad Triltsch, Graphischer Betrieb, Würzburg.

Progress in Molecular and Subcellular Biology 4

Twenty Five Years of Molecular Biology

Fred E. Hahn

Molecular Biology has apparently come of age. The name was introduced into the modern scientific literature a quarter of a century ago (ASTBURY, 1950). The Journal of Molecular Biology began its publication in 1959, followed by Molecular Pharmacology, Molekularnaia Biologiia, Molecular Photochemistry, Journal of Molecular and Cellular Biochemistry, Clinical Science and Molecular Medicine, Molecular and General Genetics, Molecular Etiology and probably other periodicals dedicated to molecular aspects of the life sciences whose titles may have escaped our cursory search. If one adds to this the numerous sets of monographs and review collections, including this series, the source and secondary literatures in molecular biology comprise, by now, a sizeable library.

Molecular biology is, no doubt, a chemical science. However, during the past two decades it has become fashionable to substitute for the traditional adjective, biochemical, the more alluring "molecular". One author (KOSOWER, 1962) even published a treatise, entitled Molecular Biochemistry whose redundancy in title was commented upon by writing, "In the past few years, a new research area has emerged from the application of the physical-organic approach to the problem in chemical transformation found in biochemistry. We choose to call this area molecular biochemistry, and delimit it as the study of the detailed chemical mechanisms of the chemical transformations in biology, usually as they are described in biochemistry".

Hence, underneath the molecular vogue in communicating biochemical research, there lives an expectation that the mechanics of life can be resolved into their component processes and molecular entities with the distant goal of being able to reassemble this resolved totality conceptually or even physically into a working resemblance of the original living object. "Having pulled apart the chemical continuity of the living organism, we are challenged to reintegrate the scattered pieces into a whole" (LIPMANN, 1971).

In one of his satirical writings, CHARGAFF (1963) recognized clearly that the term molecular biology, in superceding biochemistry, has a programmatic connotation and signals the resolutionistic position in the life sciences. However, resolutionism has not halted at the molecular level of biological organization. We point to the existence of quantum biology whose early entries into the literature, after SCHRÖDINGER's prophesy (1945), were SZENT-GYÖRGYI's Introduction to a Submolecular Biology (1960) and PULLMAN and PULLMAN's Quantum Biochemistry (1962). Among the ideas introduced were that (1) biopolymers possess certain solid-state physical properties which cannot be derived or extrapolated from the physical properties of their component molecules so that the elctronic states of the polymers depart from those of the individual subunits; and (2) the consideration of biological functions of molecules must penetrate below the classical aspects of structural organic chemistry to consider "the essential importance of molecular systems with mobile electrons, and therefore of electronic delocalization" (PULLMAN and PULLMAN, 1962).

We recognize that resolutionistic, i.e. molecular and submolecular biology with its built-in ambition of logical and/or physical reassembly of components into a biologically functional whole is related to positivism. Ultimately it strives, therefore, to "explain" the phenomenon of the living state by uncovering its underlying mechanics. LIPMANN (1971) writes about "molecular technology". Molecular biology, hence, is a contemporary form of mechanistic, i.e. deterministic biology. It has derived its impetus and philosophical justification from the successful development of a physical theory of heredity, i.e. from the resolution of classical genetics into molecular genetics: its symbol has become the genetic code.

The ascendancy of molecular biology has, therefore, occurred for deeper reasons than the mere opportunism of its adepts, although some of its adversaries, e.g. Erwin CHARGAFF (1963, 1968) keep pointing, not entirely without justification, to opportunistic tendencies at work. It would be surprising if the intellectual hybris with which molecular biologists deal with the objects of their study would not, to some extent, find its counterpart in the manner(s) in which they deal with their fellow scientists and with science as a society.

An Editorial for this Progress series, finally, should at least ask the question if the attainment of the ultimate conceptual aim of molecular biology, *viz.*, resolution and reassembly of a living "system" (the term itself being mechanistic), can explain such a system. Physicists are more cautious and modest. Mathematical formulations which dualistically describe wave and particle functions of light and matter are accepted side by side not as "explanations" but as models. Modern physics does not aspire to explain the physical universe but to develop a unified mathematical imagery which deals with phenomena not in the positivistic sense of what they are but in the functional sense of how they interact and behave. The relationship of such models to physical reality lies in their predictive value. Molecular biology will truly come of age when it begins to examine its underlying philosophical premises.

References

ASTBURY, W.T.: Adventures in molecular biology. Harvey Lect. 46, 3 (1950-51).

CHARGAFF, E.: Amphisbaena, In: Essays on Nucleic Acids. Amsterdam: Elsevier 1963.
CHARGAFF, E.: A quick climb up Mount Olympus. Science 159, 1448 (1968).
KOSOWER, E.M.: Molecular Biochemistry. New York: McGraw Hill 1962.
LIPMANN, F.: Wanderings of a Biochemist. New York: Wiley Interscience 1971.
PULLMAN, B., PULLMAN, A.: Quantum Biochemistry. New York: Interscience 1963.
SCHRÖDINGER, E.: What is Life? New York: Cambridge Univ. Press 1945.
SZENT-GYÖRGYI, A.: Introduction to a Submolecular Biology. New York: Academic Press 1960.

Contents

List of Contributors

STANLEY BRAM, Départment de Biologie Moléculaire, Institut Pasteur, F-75000 Paris 15

PRAKASH CHANDRA, Gustav-Embden-Zentrum der Biologischen Chemie, 6 Frankfurt am Main 70, Theodor-Stern-Kai 7

U. EBENER, Gustav-Embden-Zentrum der Biologischen Chemie, 6 Frankfurt am Main 70, Theodor-Stern-Kai 7

H. LAUBE, Gustav-Embden-Zentrum der Biologischen Chemie, 6 Frankfurt am Main 70, Theodor-Stern-Kai 7

WOLFGANG LOTZ, Institute for Enzyme Research, University of Wisconsin, Madison, Wisconsin 53706, USA

DIETER OESTERHELT, Institut für Biochemie der Universität Würzburg, 87 Würzburg, Röntgenring 11

RAY A. OLSSON, COL MC, Division of Medicine, Walter Reed Army Instiute of Research, Walter Reed Army Medical Center, Washington D.C. 20012, USA

RANDOLPH E. PATTERSON, COL MC, Division of Medicine, Walter Reed Army Institute of Research, Walter Reed Army Medical Center, Washington D.C. 20012, USA

LINDA K. STEEL, Gustav-Embden-Zentrum der Biologischen Chemie, 6 Frankfurt am Main 70, Theodor-Stern-Kai 7

MAURICE SUSSMAN, Hebrew University, Section of Developmental and Molecular Biology, Migrash Harussin, Jerusalem, Israel

G. WILL, Gustav-Embden-Zentrum der Biologischen Chemie, 6 Frankfurt am Main 70, Theodor-Stern-Kai 7

HEMAN J. WITMER, Department of Biological Sciences, University of Illinois of Chicago Circle, P.O.Box 4348, Chicago, Illinois 60680, USA

M. WOLTERSDORF, Gustav-Embden-Zentrum der Biologischen Chemie, 6 Frankfurt am Main 70, Theodor-Stern-Kai 7

The Polymorphism of DNA

Stanley Bram

I. Introduction

The salts of deoxyribonucleic acid (DNA) are in many ways living molecules whose morphological characteristics are dependent on their genetic information and on their past and present environment. However, until very recently, DNA was not considered to possess variable structures – in short, it was not thought to be polymorphic. There were three so-called canonical forms having an almost inorganic regularity. Three invariant forms are not consistent with the diversity of DNA functions or its specific recognition. As recently pointed out by CRICK (1971), double-stranded DNA in the invariant classical states cannot provide enough detail for the very specific recognition of DNA by proteins. Recognition schemes based upon the fixation of mono or divalent ions upon crystals of dinucleotides such as proposed by KIM et al. (1973) are unlikely in solution where ions are not site-bound. Yet we know that a double-stranded stretch of 20 to 50 base pairs of the lac operator is chosen without error in an *E. coli* chromosome of 3 million nucleotide pairs (RIGGS, SUZUKI and BOURGEOIS, 1970).

Nearly all the details of DNA structure have come from X-ray fiber diffraction data on oriented fibers which were obtained ten or twenty years ago. Then and now, structures were defined by their diffraction patterns and not conversely. The fiber diagrams for B-DNA, for example, have remained essentially the same for twenty years but the details of the structure attributed to them have appreciably changed four or five times since the days of WATSON and CRICK (1953). FRANKLIN and GOSLING (1953) with Na DNA, found the so-called A fiber pattern at lower relative humidity and the B pattern at higher relative humidity (rh). Lithium DNA did not yield an A diagram but instead showed either crystalline B or another type or diagram – type C (MARVIN et al., 1961). In 1959 a study was published of the low humidity A diagrams of a variety of DNAs (HAMILTON et al., 1959) and since all of the DNAs gave indistinguishable A patterns, it was concluded that all DNA structures were independent of the base composition. This dogma was challenged only when it was found that the X-ray patterns of various DNAs in solution were indeed dependent upon base content (BRAM, 1971b). It was suggested in this study that the transition to the A form might also depend upon base content and this was then found to be the case (PILET and BRAHMS, 1972). The diffraction experiments in solution lead to fiber diffraction studies (BRAM, 1972a; BRAM and TOUGARD, 1972) and the discovery of four new and distinct X-ray diffraction patterns. Furthermore, it was found that the transitions between the various DNA conformations are far more interesting and complex than had been thought (AZOULAY and BRAM, 1973) (BRAM and BAUDY, 1974).

Other physical chemical techniques, such as infra-red and Raman spectroscopy (PILET and BRAHMS, 1972) (ERFURTH, KISER and PETICOLAS, 1972) are now beginning to play important roles in DNA structural determinations and will certainly reduce the many man-years now required for obtaining good structural coordinates. However, this review will be restricted mainly to recent X-ray studies of DNA. It is worth mentioning that infra-red studies have only recently been interpreted to show that the B-conformations depend on the base composition (PILET, BLICHARSKI and BRAHMS, 1975).

II. A Summary of the Principles of DNA X-Ray Diffraction

Oriented fibers of DNA are obtained by exploiting the high viscosity of DNA gels. When tension is applied to a wet gel the long molecules align parallel to each other in the direction of the tension. This tension can then either be maintained or increased while drying. The rate of drying, which we shall see is a very important parameter, can be controlled by adjusting the temperature and the relative humidity during drying. For example, a gel can be dried to a fiber in half an hour over 44 % rh, at 37°C, while a day may be required to dry it at 4°C over 79 % rh. The fiber is then mounted in a collimated beam of X-rays and the intensity diffracted is recorded on a film behind the fiber. The direction on the film parallel to the long axis of the fiber is called the meridian, and the perpendicular to this direction, the equator. One usually fixes the water content of the fibers by carrying out the experiments in helium which has been equilibrated with standard saturated salt solutions. Specimens for solution or gel X-ray scattering can be obtained by mixing solvent with DNA fibers; they are then irradiated in sealed thin-walled glass capillary tubes (sometimes after dialysis *vs.* the solvent).

The diffraction from a discontinuous helix such as DNA has the following properties: (a) it is confined to layer lines separated by a distance proportional to 1/pitch; (b) on the lower layers the diffraction is described by a Bessel function of the same order as the layer number; (c) the pattern has a cross-shaped form in its center; (d) there is a strong meridional intensity spot at a distance corresponding to 1/inter-unit separation; (e) for the double helix the intensity diffracted is modified by a fringe function corresponding to the relative position of the two strands. The reader can obtain an estimate of the pitch and interbase pair separation by taking the ratio between the distance to the first real meridional reflection on the photographs (which corresponds to 3.4Å in the B and 3.3 Å in the C diffraction) and the distance separating the layer lines. This ratio is 11.3 for a helix of 10 base pairs per turn and is about 9 for a helix of eight bases per repeat. The two B diagrams of Fig. 2a and b show most of the helical diffraction features, but the classical B distribution of weak intensity on the first and third layers and strong on the second is observed only with DNA relatively rich in GC. This alteration of weak and strong intensity on the inner layer lines was one of the major reasons why WATSON and CRICK (1953) concluded that DNA must have a double helical structure whose chains were separated by a rotation of about 180°.

In the A-DNA fibers, the molecules are always packed into a crystalline lattice; consequently the intensity on the various layer lines is discontinuously sampled, making the helical nature of the diffraction less evident.

III. Generalities of DNA Transitions

In vivo, many other molecules are present about DNA, especially large amounts of water, proteins and mono and divalent ions. DNA is in a complex of either nucleoprotein or polyamines. These various proteins, macro-ions and polyamines will exclude water from the DNA surface to a degree dependent on their interaction with DNA. Consequently, it follows that much of the DNA will be more or less dehydrated due to steric exclusion by proteins. Those regions which remain relatively hydrated will exist in a B-like conformation, while those regions of reduced water content will undergo structural transitions in response to the local environment.

Upon drying pure sodium or potassium DNA, a B to A transition occurs at less than 6 - 10 H_2O per nucleotide. In the more than 20 years since the discovery of the A form (RILEY and OSTER, 1951), all X-ray experiments have found it co-existent with a crystalline lattice. Consequently it was suggested that intramolecular lattice interactions are responsible for this configuration of DNA (BRAM and BAUDY, 1974). To test this suggestion, further experiments were carried out on DNA in 80 % ethanol and isopropanol. Here also, only crystalline A patterns were discerned (see Fig. 1). Furthermore, spontanous orientations was sometimes observed (BRAM and BAUDY, 1974) indicating that lattice formation has a preferred direction.

Previously, it seemed possible that DNA could adopt the A form in local regions of a chromosome when provided with the correct environment by bound proteins. However, a requirement for crystallization would eliminate this possibility. Furthermore, it has been known for some time that high concentrations of histones (WILKINS, ZUBAY and WILSON, 1959) and polyaminoacids (ZUBAY, WILKINS and BLOUT, 1961) block a transition to the A form. Probably the strongest evidence against the existence of the presence of A-DNA *in vivo* comes from the recent finding in our laboratory that certain histone proteins will block the A transition in fibers at concentrations of one to two percent, which represents about one protein per thousand base pairs. Such small amounts of protein probably act as an impurity, which prevents crystallization. It is therefore important to consider what is known about the structure and transitions of DNA when the formation of the A state is inhibited.

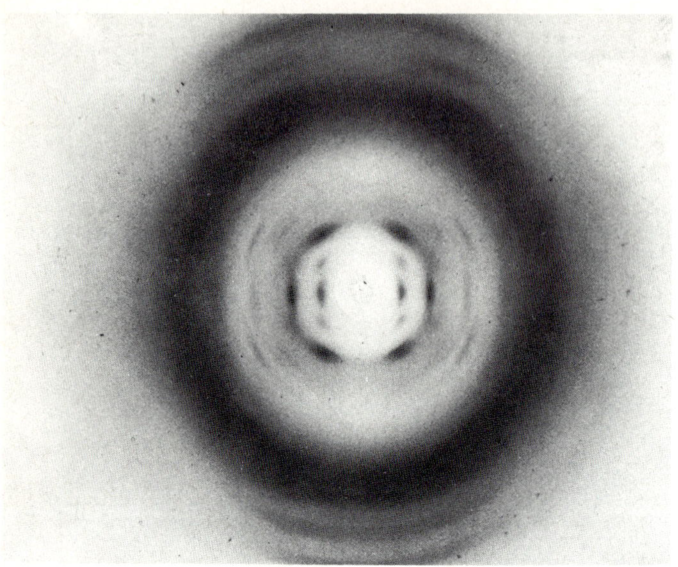

Fig. 1. An X-ray fiber diagram of a sodium calf thymus DNA fiber in 90 % isopropanol by volume. The fiber was photographed with a torridal mirror diffractometer after equilibration with 98 % relative humidity and subsequent immersion in a large excess of 90 % isopropanol containing 5 mM NaCl. The fiber diagram is identical to those of A fibers in air except for the ring at 4 Å from isopropanol (compare to the A diagram of Fig. 3). The diffraction patterns of DNA in ethanol, isopropanol or in air equilibrated with water have always been crystalline

IV. B-like Structures

The B configuration is probably the most prevalent structure *in vivo* . Yet there exist for DNAs of various base content a large family of distinct B-like structures which are related, in that they all have about ten base pairs per turn of 34 A.

Although it had been inferred for some time, the first conclusive evidence that DNA in solution had a B-like configuration came from small- and wide-angle X-ray scattering experiments (BRAM and BEEMAN, 1971). The mass per unit length obtained was exactly that expected from B-DNA, and the wide-angle scattering pattern was much more similar to that calculated for a B structure than an A. Nevertheless, the wide-angle scattering patterns were not the same as those calculated with the atomic coordinates based upon LiDNA fiber diffraction diagrams at 66 % rh (BRAM, 1971a). These results have more recently been repeated and confirmed by MATHAIS et al. (1973). The spacing of the first scattering maxima was observed to be at 13.6 A instead of at 12.6 A as calculated with the B coordinates. Therefore, it had to be concluded

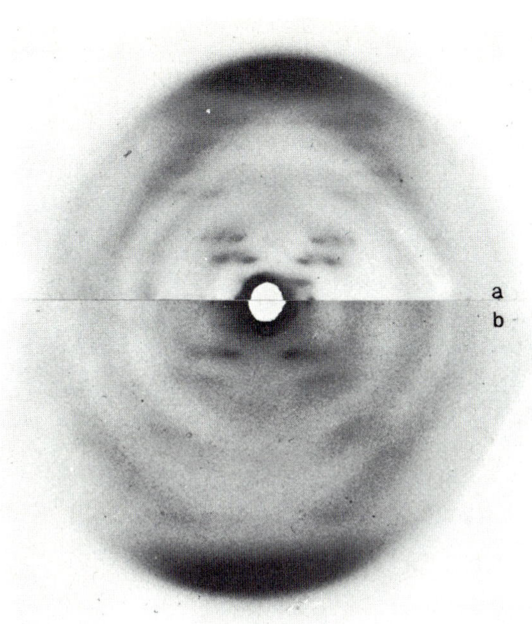

Fig. 2 a and b. Two sodium B diffraction diagrams at 44 % rh, after drying under tension at 37°C, 44 % rh. (a) *C. perfringens* DNA (66 % AT). Note the strong intensity on the first three layer lines. (b) *S. lutea* DNA (29 % AT). Note the very weak intensity on the first and third layer lines. The NaCl content in the fiber was 3 % of the weight of the anhydrous DNA

that either the pitch or the radius of the DNA was larger than that given by the atomic coordinates. Since I did not want to dispute the crystallography at that time, I accepted the published coordinates for a base pair, and suggested that the best explanation involved a variation in the pitch. But now after examining many very high-humidity B fiber patterns and finding that all have the same pitch of 34 A (BRAM, 1973a and b), it seems much more likely that the coordinates were incorrect. Recently rather different coordinates have been published (ARNOTT and HUKINS, 1974) and the scattering calculated with the new values agrees better with experiment.

In 2 M Li or NaCl, where the hydration about the DNA is lowered, the pitch or the effective radius of the DNA is about 5 % smaller than in dilute salt solution (BRAM, 1971 b). Likewise, in chromatin, the wide-angle scattering patterns showed that the structure is more compact than in pure DNA solutions, but here a decrease in both the effective radius and the pitch would best agree with the data. In any case, it is unwise to attribute the changes in an experimental measurement, whether it be X-ray scattering or especially circular dichroism, to a detailed structural transition (for instance to a C form) expecially when the structural parameters may not be directly related to the measured values.

It has been shown that solutions of DNAs with different base composition have different wide-angle X-ray scattering patterns and as a direct consequence must have different structures (BRAM, 1971b). The ratio of the X-ray intensity maximum near 10 A to that observed near 13 A increased almost two-fold over the range of 31 to 100 % AT. Since this ratio of intensity (10/13 A) is directly related to the ratio of intensity between the third layer line and the second layer, it would be expected that the relative intensity of the third layer lines in the B-fiber diagrams would increase strongly with AT. Indeed, it was found that the intensity on the third and also on the first layer line of the very AT-rich DNA B patterns was two to three times greater than in the GC-rich DNA diffraction (BRAM, 1973a and b). The intensity on the odd layer lines at very high rh seems to increase semicontinuously with AT. This same tendancy also appears in the B diagrams observed at low rh (see Fig. 2a and b); here the crystalline nature of the diffraction patterns might aid in structural determinations. Since the intensity in solution parallels that of the fiber diagrams, this rules out the possibility that lattice effects are involved and requires that these variations in layer-line intensity reflect differences in the molecular structure with base composition. On the other hand, all of the fiber diagrams have the same meridional spacing of 3.4 A and layer-line separation of 34 A. This tells us that the base-pair separation and pitch are not functions of AT. The only metamorphoses consistent with these constraints are changes in either the base-pair tilt or the angle between bases in a base-pair. Detailed model calculations and comparison with the X-ray data (or perhaps the utilization of infrared or Raman techniques) will be necessary to decide between these two possibilities. In any case, the differences in the X-ray patterns result from either (1) a structural variation over large regions or clusters of base-pairs, or (2) a difference in the structure of each individual GC or AT pair or grouping of a few nucleotides. The apparently progressive rise of the odd layer-line intensity with AT, which is probably related to the monotonic decrease in density of DNA with AT on both CsCl (SCHILDKRAUT, MARMUR, and DOTY, 1962) and NaI gradients (ANET and STRAGER, 1969), would be more consistent with (2). The second mechanism also finds support in recent gel electrophoresis studies on various DNAs which show that the mobility increases continuously with AT; these studies were also interpreted to imply that the cross-section morphology of DNA is a function of the base content (ZEIGER et al., 1972). This latter mechanism provides a direct answer to the question of how DNA is recognized by proteins; each base-pair has a distinct and recognizable configuration. This would imply that DNA is almost infinitely polymorphic in the B state.

B- X-ray diffraction diagrams are also observed in fibers at lower water content (see Fig. 2). With lithium DNA it is a stable state (LANGRIDGE et al., 1956) but with sodium DNA, where it has only recently been shown to exist (BRAM, 1972a), it is a metastable state (AZOULAY and BRAM, 1973) (BRAM and BAUDY, 1974). Low-humidity sodium B states can be obtained by quickly dehydrating a gel containing 3 % or more NaCl to 66 % or less relative humidity. Depending on the fiber-pulling conditions, the fibers contain variable amounts of A-DNA. It is still an open question as to whether or not the low humidity conformations are equal to those at higher water contents.

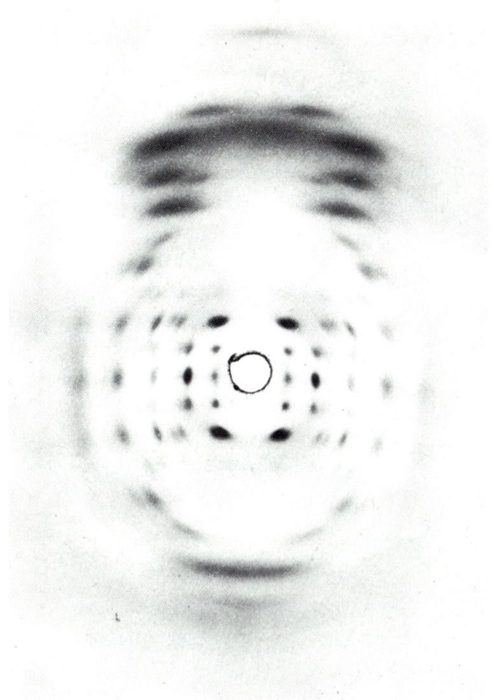

Fig. 3. An A-diffraction pattern from calf thymus DNA at 79 % rh. This fiber was first quickly dried from a wet gel under tension and exhibited a C-diffraction pattern at 44 % rh. The relative humidity was then raised to 79 % and photographed after two hours. A C→A transformation has occurred. The fiber was tilted by about 20° from the vertical

V. The A Form

If one considers form to mean a particular unique structure, then the A configuration is the only real form of DNA. The A-diffraction diagram is the only one that has been found to be independent of base composition (HAMILTON et al., 1959) and salt content (COOPER and HAMILTON, 1966). It is the equilibrium form of all pure sodium DNA and also probably of potassium DNA at lower water contents (FRANKLIN and GOSLING, 1953). However, some synthetic (LANGRIDGE, 1969) and some very AT-rich DNAs (BRAM, 1972) have not adopted the A form under the limited range of conditions so far studied. The B to A transition is favored by higher GC content (PILET and BRAHMS, 1972) and lower excess salt contents (COOPER and HAMILTON, 1966). Although it is generally held that low water content is the most important factor, fibers of *M. lysodeikticus* DNA which is very rich in GC have remained in the A form at relative humidities above 98 %, and very high NaCl contents (above about 8 % NaCl) seem to block the A transition of calf-thymus DNA at all humidities. Both high AT and salt contents and also high water contents probably inhibit the A-transition by making crystallization less favorable. As previously mentioned, the A form of DNA has only been found in a crystalline lattice. The A-fiber diagram is characterized by a layer-line separation of 28 A and an intensity distribution of weak intensity on the inner part of the first layer line, a very strong second layer

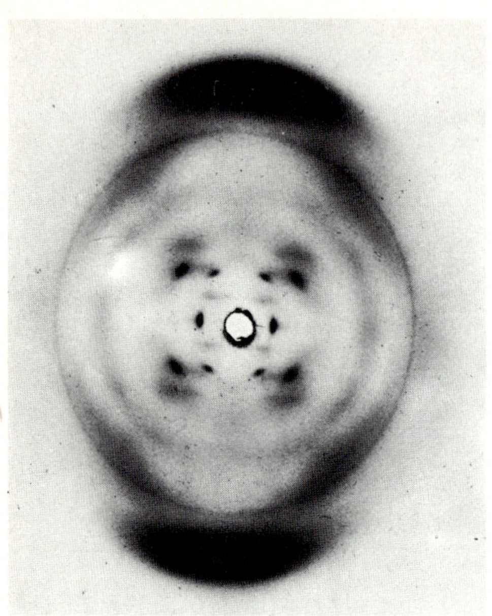

Fig. 4. A sodium-C diagram from calf thymus DNA containing 1,5 % NaCl at 44 % rh. The diffraction is indistinguishable from the lithium-C diagrams

line, and almost zero intensity on the central portion of the third layer (Fig. 3 is an example of an A-X-ray diffraction diagram). The A conformation features 11 base-pairs per turn, each of which is tilted 20° from the perpendicular to the helix axis and is separated from its nearest neighbors by 2.6 A. The base-pairs are also displaced off the helix axis by 4 A and the sugar ring is puckered in the so-called (C2-endo) conformation (ARNOTT et al., 1968).

Crystallographers have shown that the structure of the hybrid of RNA and DNA is very similar to that of A-DNA (MILMAN, CHAMBERLIN, and LANGRIDGE, 1967) and have suggested that this is consequently the form of DNA active in transcription (ARNOTT et al., 1968) where hybridization of RNA to DNA is the most important step. Also, the lack of polymorphism in the A form would make it a more readily used substrate for polymerase enzymes. Since the transition to the A state may be blocked in the nucleus by divalent ions or histone-like proteins, a catalyst may be required to drive the DNA into the A state. A likely candidate for this catalyst is RNA itself, which forms only an A-like state with DNA.

VI. C-like Structures

Of the three classical states, the C state is the most polymorphic and least well-defined. Calf thymus lithium DNA adopts structures whose pitches vary between 29.2 and 32.2 A (MARVIN et al., 1961, Table 1). The C state, like the other classes of DNA structures, is characterized by its X-ray diffraction pattern, which features an intensity distribution of weak, very strong, weak on the first three layer lines which are separated by about 31 A. A strong intensity on the layer line just below the 3.3 A meridional reflection is also particular to the C diffraction (see Fig. 4). The X-ray diffraction patterns are consistent with a slight base-tilting of 5 - 10° and a pitch which is about 10 % less than B-DNA (MARVIN et al., 1961).

It had been accepted that C-type structures existed only at low water contents and with lithium DNA. Although the first condition is probably correct if one considers the local DNA environment, the latter is false. Magnesium calf thymus DNA exhibits C-like diffraction patterns at lower relative humidity (SKURATOVSKI and MOKULSKY, 1971) and it has been shown that sodium calf thymus salmon sperm and *E.coli* DNA adopt metastable C conformation (AZOULAY and BRAM, 1973 and BRAM and BAUDY, 1974) (see Fig.4).

The lithium and magnesium ions, like the applied tension or quick-drying required for observing a sodium C pattern, probably block a transition to the A form at lower relative humidities. It is not unreasonable to suggest that the C state may be an intermediate between the B and A states. This suggestion is supported by the fact that several sodium C to A transitions have been observed at low relative humidities where the B state is not stable. The C to A transition can be provoked by releasing the tension applied to the fiber which then quickly contracts by about 30 % in lenght, and X-ray photographs show A patterns (see Fig. 3).

VII. Polynucleotide Polymorphism

Synthetic DNAs exhibit an even greater variety of structures than natural DNA. The well-known polynucleotide, poly [d(A-T) · d(A-T)] was found to adopt at least three new conformations at lower water contents (DAVIES and BALDWIN, 1963). The most interesting of these – the D form of DNA – may be another low-humidity equilibrium state. Poly [d(G-C)] · poly [d(G-C)] , the other synthetic alternating co-polymer, which can exist in natural DNA, may also adopt another distinct form (POHL and JOVIN, 1972). Homopolymers such as poly dA · poly dT and poly dI · poly dC have not been found to adopt the A form in fibers but instead yield new types of X-ray diffraction diagrams (LANGRIDGE, 1969). In solution, poly dG · poly dC gave a pattern very different from the B type X-ray scattering; the cross-section radius of gyration was 20 % larger and the intensity distribution was radically altered (BRAM, 1971c). Thus, stretches of repeating DNA sequences may exist in natural DNA and might play a role in DNA recognition.

Table 1. The various states of DNA studied by X-ray fiber diffraction

State	Pitch (Å)	Intensity distribution	Favorable conditions
A	28	W, VS, VW	Lower rh, lower salt Na, K, Rb salts; 80 % alcohol
B	34	Strong function of base composition	High rh
	34	Strong function of base composition	Li salt with low rh
	34	Strong function of base composition	Na salt with low rh, fast drying, applied tension
C	29-32	W, S, W	Li or Mg salt, low rh
	29-32	W, S, W	Na salt, low rh applied tension
D	24	S, W, W	poly [d(A-T)] · poly [d(A-T)] 44 - 92 % rh
T	28	S, S, M	Less than 69 % rh C.perfringens DNA, fast drying, tension and more than 1.5 % NaCl
P	30	W, S, S	Less than 79 % rh C.perfringens DNA, fast drying tension, less than 1.5 % NaCl
J_1	33	S, S, M	Less than 2 % NaCl Less than 76 % rh, C.johnsonii DNA
J_2	31	S, S, VW	More than 2 % NaCL, less than 76 % rh C.johnsonii DNA

VIII. New Conformations of AT-rich DNA

Fiber-diffraction studies, until very recently, have been confined to DNAs containing less than 65 % AT and have used essentially the same equilibrium conditions for drawing fibers. (In fact, nearly all fiber diffraction had been restricted to calf thymus DNA.) However, when very AT-rich DNA is drawn into thicker fibers (250μ compared to the 100μ previously employed) using non-equilibrium conditions, several new and distinct X-ray diagrams are obtained (BRAM, 1972a; BRAM and TOUGARD, 1972). The four new diffraction patterns, P, T, J_1 and J_2, and their corresponding structures were arbitrarily named in order to break with the ABC dogma which had dominated DNA structure studies. (P was chosen to denote the fact that it was first found at the Pasteur Institute, T was chosen to denote the high AT content of the DNA and the very turned nature of this structure, and the J_1 and J_2 patterns were observed with *C. johnsonii* DNA). Since there has not been enough DNA or time, the *C. johnsonii* patterns have not been further studied and this discussion will be restricted to

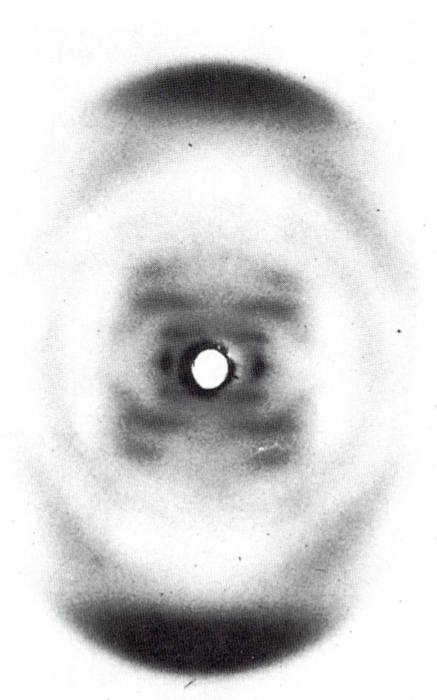

Fig. 5. A T-diffraction pattern from *C. perfringens* DNA containing 3 % NaCl at 66 % rh. Note the strong intensity on the first and second layer lines and the weaker intensity on the third layer

the T and P states. Even with the latter states only a preliminary effort has been made to determine the details of their structure because it seems more reasonable first to characterize the transitions to and from these new conformations.

The T and P conformations are more highly wound and compact than the classical states; there are eight to nine base-pairs – each separated by 3.4 A – per turn. They are distinguished by their X-ray diffraction intensity distributions and layer-line separations (see Table 1 and Fig. 5). P diffraction diagrams are obtained with *C. perfringens* DNA containing less than about 1 % NaCl and T diagrams at higher salt contents (BRAM and TOUGARD, 1972). One of their most interesting characteristics is that they are metastable states with respect to the A form. If a wet gel of *C. perfringens* DNA is slowly dried, the A form is almost always obtained; the T and P states appear only when not enough time has been allowed for equilibrium or crystallization to be attained (AZOULAY and BRAM, 1973). Some fibers have undergone a reversible P to B transition upon raising the rh when maintained under tension, but the hysteresis loop T or P to B and then to A is usually obtained upon first raising and then lowering to low rh.

C. perfringens DNA, when converted to the magnesium salt, exists in yet another state which gives an X-ray pattern very similar to, but not equal to, that of C-DNA (BRAM and BAUDY, 1974). This makes a total of at least five distinct conformations (P-, A-, B-, T-, and C-like) for this one AT-rich DNA at the same water content. Because of the very polymorphic nature of AT-rich DNA, I suggested that DNA rich in AT base-pairs served as a class of recognition sites, and several of the properties of DNA-protein interactions support such a model (BRAM, 1972b).

I would like to call attention to some contempory fiber-diffraction studies of T2 DNA (63 % AT). It has been known for some time that the glucosylated bases T2 DNA block the transformation to the A form in fibers (HAMILTON et al., 1959). MOKULSKY, KAPITONOVA, and MOKULSKAYA (1972) found that a lower rh, T2 DNA adopts another configuration with 8 - 9 base-pairs/turn which gives an X-ray fiber pattern at 76 % rh very similar to that found for the *C. perfringens* DNA P state. (Their structure was named the T form and it may be necessary to interchange the names P and T for *C. perfringens* DNA to avoid confusion.) The independent observation of a T2 DNA pattern similar to those found at the Pasteur Institute provides a mutual corroboration and also implies that these new forms may also exist for other DNAs under different conditions. Although the existence of these new forms for T2 DNA was attributed to the glucosylated residues, MOKULSKY et al., 1972), this cannot be true for *C. perfringens* DNA.

IX. Biological Relevance

DNA polymorphism would provide an effective way of assigning classes of DNA base compositions or sequences for particular roles (BRAM, 1971b, 1972b). In such a scheme coding DNA would have only a limited base composition range, which should be between about 35 and 60 % AT. Another consequence is that certain DNA function-associated proteins would bind only regions within a very narrow base composition range. Since AT-rich DNA is much more polymorphic than typical DNA, it was suggested that all recognition sites would be in regions richer in AT than the bulk chromosome DNA (BRAM, 1971b).

This type of scheme has received direct and indirect support. The base compositions of all known transcribed RNAs are between 35 and 55 % AT (BRAM, 1972b). The binding sites of RNA polymerase in phages λ, T5 and T7 are all about 65 % in AT (LE TALAER and JEANTEUR, 1971; LE TALAER, KERMICI, and JEANTEAU, 1973). That local inhomogeneities in base composition serve for general punctuation and recognition DNA is directly inferred from high-resolution melting studies carried out by C. REISS (REISS and MICHEL, 1974; BRAM et al., 1974). It was possible to assign each of the genetic regions of phage λ to specific melting modes and to map their locations. REISS has found that each gene of λ corresponds to a melting mode and that each gene is separated by a short DNA region of inhomogenous base content (REISS, 1975).

The very specific recognition of DNA may well be due to sequence-specific structural variations. One generally considers DNA recognition and regulation to involve a protein binding to an already present and fixed structural feature on the DNA. This is a lock-and-key-type scheme. *In vivo*, naked DNA exists in a B state and here the various regions of a DNA molecule which are not at all uniform in base content will exhibit structures dependent on the *local* base content and sequence. In this context, the suggestion that the angle between base-pairs is different for A-T and G-C means that the local groove geometry of the DNA is a function of sequence. The various structures along a given DNA will be sampled by proteins or other recognizer molecules and some will interact to form weak-binding complexes. For the complex to become a strong one, water must be excluded from the DNA surface, and, as another consequence of the protein-binding, a new local ionic environment will result. The DNA may then quickly change its conformation in response to its new surroundings. Such a transition would be very rapid and would not allow enough time for equilibrium to be attained. (In the case of the lac repressor-operator interaction, the forward rate constant is about 10^{10} M^{-1} sec.$^{-1}$ (RIGGS, BOURGEOIS, and COHEN, 1970)). This scheme, which is a type of induced-fit mechanism, has previously been discussed in the framework of a structural chromosomal model (BRAM, 1972b).

CRICK (1974) has recently suggested that the poly-morphism of DNA is either not very important, or has profound biological implications. Although I feel that the evidence supports the latter, this is a question which remains to be resolved.

References

ANET, R., STRAYER, D.R.: Sodium iodide density gradients for the preparative buoyant separation of DNA mixtures. Biochem. Biophys. Res. Commun. 37, 52-58 (1969).

ARNOTT, S., FULLER, W., HODGSON, A., PRUTTON, I.: Molecular conformation and structural transition complementary helices and their possible biological significance. Nature 220, 561-564 (1968).

ARNOTT, S., HUKINS, D.W.L.: Optimised parameters for A-DNA and B-DNA. Biochem. Biophys. Res. Comm. 47, 1504-1509 (1974).

AZOULAY, R., BRAM, S.: Etude par diffraction de rayons-X des etats métastables et des cycles d'hysteresis de l'ADN. C. R. Acad. Sci. 276, 2977 (1973).

BRAM, S.: The secondary structure of DNA in solution and in nucleohistone. J. Mol. Biol. 58, 277-288 (1971a).

BRAM, S.: Secondary structure of DNA depends on base composition. Nature New Biology 232, 174-176 (1971b).

BRAM, S.: Polynucleotide polymorphism in solution. Nature New Biology 233, 161-164 (1971c).

BRAM, S.: II Polymorphism of natural DNA. Biochem. Biophys. Res. Comm. 48, 1088-1092 (1972a).

BRAM, S.: The function of the structure of DNA in chromosomes. Biochimie 8, 1005-1011 (1972b).

BRAM, S.: Etude préliminaire de diagrammes de diffraction de rayons-X des fibres de différents ADN à forte humidité. C. R. Acad. Sci. 276(d), 657-659 (1973a).

BRAM, S.: The variation of the B-kind DNA X-ray fiber diagrams with base composition. Proc. Nat. Acad. Sci. (Wash.) 70, 2167-2170 (1973b).

BRAM, S., BAUDY, P.: X-ray diffraction studies of DNA at reduced water contents. Nature 250, 414-416 (1974).

BRAM, S., BEEMAN, W.W.: On the cross-section structure of deoxyribonuclei acid in solution. J. Mol. Biol. 55, 311-324 (1971).

BRAM, S., BUTLER-BROWNE, G., BRADBURY, E.M., BALDWIN, J., REISS, C, IBEL, K.: Chromatin neutron and X-ray diffraction studies and high-resolution melting studies of DNA-histone complexes. Biochemie 56, 987-994 (1974).
BRAM, S., TOUGARD, P.: Polymorphism of natural DNA. Nature New Biology 239, 128-131 (1972).
COOPER, P.J., HAMILTON, L.D.: The A-B conformational change in the sodium salt of DNA. J. Mol. Biol. 16, 562-563 (1966).
CRICK, F.H.C.: General model for the chromosomes of higher organisms. Nature 234, 25-27 (1971).
CRICK, F.H.C.: The double helix: A personal view. Nature 248, 766-769 (1974).
DAVIES, D.R., BALDWIN, R.L.: X-ray studies of two synthetic DNA copolymers. J. Mol. Biol. 6, 251-255 (1963).
ERFURTH, S.C., KISER, E.J., PETICOLAS, W.L.: Determination of the back bone structure of nucleic acids and nucleic acid olygomers by laser raman scattering. Proc. Nat. Acad. Sci. (Wash.) 69, 938-941 (1972).
FRANKLIN, R.G., GOSLING, R.G.: Molecular configuration in sodium thymonucleate. Nature 171, 740-741 (1953).
HAMILTON, L.D., BARCLAY, R.K., WILKINS, M.H.F., BROWN, G.L., WILSON, H.R., MARVIN, D.A., EPHRUSSI-TAYLOR, H., SIMMONS, N.S.: Similarity of the structure of DNA from a variety of sources. J. Biophys. Biochem. Cytol. 5, 397 (1959).
KIM, S.H., BERMAN, H.M., SEEMAN, N.C., MARSHALL, N.D.: Seven basic conformations of nucleic acid structures. Acta Cryst. B 29, 703-710 (1973).
LANGRIDGE, R.: Nucleic acids and polynucleotides. J. Cell. Physiol. 74 Suppl., 1-20 (1969).
LANGRIDGE, R., SEEDS, W.E., WILSON, H.R., HOOPER, C.W., WILKINS, M.H.F., HAMILTON, L.D.: The molecular structure of desoxyribonucleic acid. J. Biophys. Biochem. Cytol. 3, 767-778 (1956).
LE TALAER, J.Y., JEANTEUR, P.: Purification and base composition analysis of phage lambda early promotors. Proc. Nat. Acad. Sci. (Wash.) 68, 3211-3215 (1971).
LE TALAER, J.Y., KERMICI, M., JEANTEUR, P.: Isolation of *E. coli* RNA polymerase-binding sites on T5 and T7 DNA: Further evidence for a sigma-dependent recognition of A-T-rich sequences. Proc. Nat. Acad. Sci. (Wash.) 70, 2911-2915 (1973).
MANIATIS, T., VENABLE, J.H., LERMAN, L.S.: The structure of ψ DNA. J. Mol. Biol. 84, 37-64 (1973).
MARVIN, D.A., SPENCER, M., WILKINS, M.H.F., HAMILTON, L.: The molecular configuration of DNA. III X-ray diffraction study of the C. form of the lithium salt. J. Mol. Biol. 3, 547-565 (1961).
MILMAN, G., CHAMBERLIN, M.J., LANDRIDGE, R.: The structure of a DNA-RNA hybrid. Proc. Nat. Acad. Sci. (Wash.) 57, 1804-1810 (1967).
MOKULSKY, M.A., KAPITONOVA, K.A., MOLKULSKAY, T.D.: Secondary structure of T2 phage DNA. Mol. Biophys. 6, 883-892 (1972).
NEVILLE, D.M., DAVIES, D.R.: The interaction of acridine dyes with DNA: An X-ray diffraction and optical investigation. J. Mol. Biol. 17, 57 (1966).
PILET, J., BLICHARSKI, BRAHMS, J.: Conformations and structural transitions in polynucleotides. Biochemistry 14, 1869-1876 (1975).

PILET, J., BRAHMS, J.: The dependence of B-A conformational change in DNA on base composition. Nature New Biology 236, 99-100 (1972).

POHL, F.M., JOVIN, T.M.: Salt-induced co-operative conformational change of a synthetic DNA: Equilibrium and kinetic studies with poly(dG-dC). J. Mol. Biol. 67, 375-396 (1972).

REISS, C., MICHELL, F.: An apparatus for studying the thermal transition of nucleic acids at high resolution. Analyt. Biochem. 62, 499-508 (1974).

RILEY, D.P., OSTER, G.: An X-ray diffraction investigation of aqueous systems of desoxyribonucleic acid. Biochem. Biophys. Acta 7, 526-546 (1951).

RIGGS, R.D., SUZUKI, H., BOURGEOIS, S.: Lac-repressor-operator interactions. I. Equilibrium Studies. J. Mo. Biol. 48, 47 (1970).

SCHILDKRAUT, C.L., MARMUR, J., DOTY, P.: Determination of the base composition of deoxynucleic acid from its buoyant density in CsCl. J. Mol. Biol. 4, 430-443 (1962).

SKURATOVSKI, I.I., MOKULSKY, M.A.: An X-ray structural study of the magnesium salt of DNA. Dokladi Acad. Sci. USSR 200, 637-640 (1971).

WATSON, J.D., CRICK, F.H.C.: A structure for deoxyribose nucleic acid. Nature 17, 737-738 (1953).

WILKINS, M.H.Z., ZUBAY, G., WILSON, H.R.: X-ray diffraction studies of the molecular structure of nucleohistone and chromosomes. J. Mol. Biol. 1, 179-185 (1959).

ZEIGER, R.S., SALOMON, R., DINGMAN, C.W., PEACOCK, A.C.: Role of base composition in the electrophoresis of microbial and crab DNA in polyacrylamide gels. Nature New Biology 238, 65-69 (1972).

ZUBAY, G., WILKINS, M.H.F., BLOUT, E.R.: An X-ray diffraction study of DNA and a synthetic polypeptide. J. Mo. Biol. 4, 69-72 (1962).

Regulation of Bacteriophage T4 Gene Expression

Heman J. Witmer

I. Introduction

The lytic process of coliphage T4 begins with adsorption of the virion to the bacterial cell wall and subsequent injection of the viral DNA into the host (COHEN, 1963, 1968; MATHEWS, 1971). Synthesis of bacterial proteins stops within seconds (BENZER, 1953; BILEZIKIAN, KAEMPFER, and MAGASANIK, 1967; HAYWARD and GREEN, 1965; KAEMPFER and MAGASANIK, 1967; KENNELL, 1970; LEVIN and BURTON, 1961; ROUVIERE et al., 1968; SHER and MALLETTE, 1954). Bacterial DNA and RNA continue to be synthesized for several minutes, but at much-reduced rates (ADESNIK and LEVINTHAL, 1970; COHEN, 1963, 1968; KENNELL, 1968; LANDY and SPIEGELMAN, 1968; MATHEWS, 1971; NOMURA, HALL, and SPIEGELMAN, 1960; VOLKIN and ASTRACHAN, 1956). Transcription of the viral DNA begins immediately (OLESON, PISPA, and BUCHANAN, 1969) and the first viral proteins are completed less than one minute after infection (HOSODA and LEVINTHAL, 1968). Replication of the viral chromosome starts 5 min after infection but viral DNA replication does not reach its maximum rate until 10 min (COHEN, 1963, 1968; MATHEWS, 1971). Intracellular progeny viruses first appear 20 or so min after infection and their number increases steadily until 30 to 35 min, at which time the infected cell lyses (COHEN, 1963, 1968; MATHEWS, 1971).

The entire viral genome is not expressed simultaneously. Rather, the viral genes are turned-on in a well-defined and well-regulated temporal sequence, as evidenced by the fact that different species of viral-specific RNAs and proteins first appear at characteristic times in the latent period (ADESNIK and LEVINTHAL, 1970; HOSODA and LEVINTHAL, 1968).

The relative intracellular levels of specific RNAs and proteins can be used to gauge the degree of gene expression. Depending upon when in the infection process they become fully expressed, the genes of coliphage T4 can be split into two large groups. Early genes are turned on during the first 5 min of the lytic process (i.e. prior to the onset of viral DNA replication) and most are fully expressed by the time DNA replication assumes its maximum rate (ADESNIK and LEVINTHAL, 1970; COHEN, 1963, 1968; HOSODA and LEVINTHAL, 1968; MATHEWS, 1971). There is little synthesis of early protein later than 10 min after infection, although most species of early RNA persist beyond that time (ADESNIK and LEVINTHAL, 1970; HOSODA and LEVINTHAL, 1968). Expression of the late genes begins 5 min after infection but these genes are not fully expressed until after DNA synthesis reaches its maximum rate (ADESNIK and LEVINTHAL, 1970; COHEN, 1963, 1968; HOSODA and LEVINTHAL, 1968; MATHEWS, 1971).

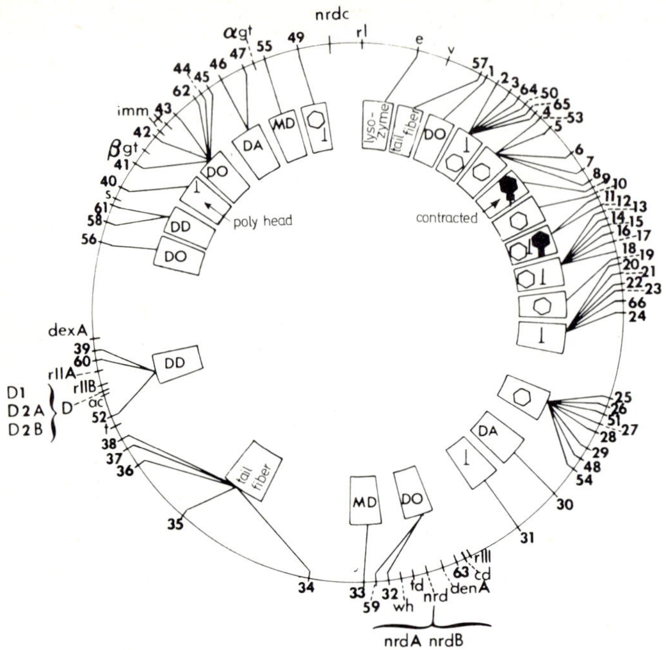

Fig. 1. The genetic map of coliphage T4. The map is taken mainly from EDGAR and WOOD (1966) and from MATHEWS (1971). The sequence of genes from frd through cd as well as the map position of gene nrdC is taken from TESSMAN and GREENBERG (1972). The map positions of genes denA and dexA are according to RAY et al. (1972) and WARNER et al. (1972), respectively. The separation of D2 into D2A and D2B is according to SEDEROFF, BELLE, and EPSTEIN (1971). The inner circle designates the gross phenotype of the mutant. Abbreviations: DA means DNA arrest (DNA synthesis starts at its normal time but stops after a relatively short interval); DD means DNA-delay (DNA synthesis starts somewhat later than usual but proceeds at normal rate); DO means no DNA synthesis. In the case of structural genes, the inner circle depicts phage structural components observed upon electron microscopic examination of lysates

The early genes of T4 can be subdivided further. Immediate early genes are the first to be turned on and, during the initial minute or so of the infection process, these are the only viral genes that are functional (BRODY et al., 1970; GRASSO and BUCHANAN, 1969; SALSER, BOLLE, and EPSTEIN, 1970; WITMER, 1971b). In contrast, delayed early genes are not turned on until 1.5 min after infection (BRODY et al., 1970; GRASSO and BUCHANAN, 1969; SALSER, BOLLE, and EPSTEIN, 1970; WITMER, 1971b). In like fashion, late genes can be divided into two groups. Quasi-late genes begin to be expressed prior to the onset of viral DNA replication but, like all late genes, DNA replication is required for their optimal expression (SALSER, BOLLE, and EPSTEIN, 1970). Expression of true-late genes is co-incident with and obligately dependent upon DNA replication (BOLLE et al., 1968b).

While the above discussion is an over-simplified rendition of the T4 lytic process, it nevertheless, serves to illustrate that regulation of viral gene expression is complex and seemingly involves controls at the level of transcription as well as translation. In this paper, the nature of these controls will be explored.

II. The Genetic Map of Coliphage T4

Before discussing regulation of viral gene expression, something should be said concerning the arrangement of genes on the T4 chromosome. The genetic map of coliphage T4 is topographically equivalent to a circle (Fig. 1). To a large degree, early and late genes are clustered into different sections of the chromosome. Clustering is also evident on a more microscopic scale, in that functionally related early and late genes tend to be closely linked within their respective major cluster. The evolutionary basis for clustering is obscure but this phenomenon probably reflects, among other things, the different requirements of various genes for optimal expression.

Most of the known early genes seem to be involved with DNA metabolism within the phage-infected cell (Table 1). However, much of the early region of the T4 chromosome is genetically blank, so it is impossible to estimate the degree to which early genes are involved with other processes.

Late genes code principally for structural proteins (Fig. 1) although some code for enzymes (Table 2). It is interesting to note that late genes that code for enzymes or serve a regulatory function map among the early genes (Table 2).

III. Resolution of the Early and Late Classes of T4-specific RNA

It has long been appreciated that viral-specific enzymes involved with nucleotide metabolism and DNA synthesis (early proteins) appear much sooner during the latent period than the structural proteins and enzymes responsible for lysis of the infected cell (late proteins) (COHEN, 1963, 1968). At the transcriptional level, such observations could mean one of two things. (a) All RNAs are transcribed simultaneously but are translated sequentially. (b) The viral messengers that code for different classes of protein appear at different times in the lytic process.

One of the earliest attempts to clarify this situation was made by KANO-SUEOKA and SPIEGELMAN (1962). In their study, parallel cultures of T2 infected bacteria were pulse-labeled with either $[^{14}C]$ uridine during the early period (from 3 to 5 min) or $[^{3}H]$ uridine during the late period (from 13 to 15 min. The cultures were mixed and the RNA isolated. The purified RNA was passed

Table 1. Some well-studied early genes

Gene	Enzyme or Function	Immediate or delayed early
1	deoxynucleoside monophosphate kinase[a]	delayed early[1]
30	polynucleotide ligase[a]	immediate early[2]
42	dCMP hydroxymethylase[a]	immediate early[3]
43	DNA polymerase[a]	delayed early[4]
46	exonuclease (?)[a]	unknown
47	exonuclease (?)[a]	unknown
56	dCTPase, dUTPase[a]	delayed early[5]
ac	unknown	immediate early[6]
D1	unknown	delayed early[7]
D2 (A+B)	unknown	immediate early[8]
dexA	exonuclease A[b]	unknown
imm	resistance to superinfection[c]	immediate early[9]
wh (=frd)	dihydrofolate reductase[a]	immediate early[10]
td	dTMP synthetase[a]	delayed early[11]
nrd A	ribonucleotide reductase (subunit A)[a]	unknown
nrd B	ribonucleotide reductase (subunit B)[b]	unknown
nrd C	thioredoxin[d]	unknown
den A(-nd)	endonuclease II[e]	delayed early[12]
den B	endonuclease IV	delayed early[13]
cd	dCMP deaminase[a]	delayed early[14]
rII (A+B)	unknown	delayed early[15]
αgt	dHMP-α-glycosyl transferase[a]	delayed early[16]
βgt	dHMP-β-glycosyl transferase[a]	delayed early[17]
IP1	internal protein I	immediate early[18]
IP2	internal protein II	immediate early[18]
IP3	internal protein III	immediate early[19]

[a]See MATHEWS (1971) for references, [b]WARNER et al. (1972), [c]VALLEE and CORNETT (1972), [d]TESSMAN and GREENBERG (1972), [e] HERCULES et al. (1970); JORGENSEN et al. (1970); RAY et al. (1972).

[1]SAKIYAMA and BUCHANAN (1972); TRIMBLE, GALIVAN, and MALEY (1972), [2]JAYARAMAN (1972), [3]JAYARAMAN (1972); TRIMBLE, GALIVAN, and MALEY (1972), [4]JAYARAMAN (1972), [5]PETERSON, COHEN, and ENNIS (1972), [6]WITMER, unpublished data, [7]SEDEROFF, BOLLE, GOODMAN, and EPSTEIN (1971), [8]SCHMIDT et al. (1970); SEDEROFF, BOLLE, GOODMAN, and EPSTEIN (1971), [9]PETERSON, COHEN, and ENNIS (1972), [10]TRIMBLE, GALIVAN, and MALEY (1972), [11]TRIMBLE, GALIVAN, and MALEY (1972), [12]WITMER, unpublished data, [13]YOUNG (1970a), [14]TRIMBLE, GALIVAN, and MALEY (1972), [15]JAYARAMAN (1972); SCHMIDT et al. (1970); WITMER (1971b), [16]YOUNG (1970a); YOUNG and VANHOWE (1970); SAKIYAMA and BUCHANAN (1972), [17] BLACK and GOLD (1971), [18]BLACK and GOLD (1971).

Table 2. Unclustered late genes

Gene	Remarks
e	T4-specific lysozyme[a]
s	mutants lyse, even in absence of functional lysozyme[b]
t	mutants do not lyse, even in presence of functional lysozyme[c]
31	correct aggregation of major head protein[d]
40	mutants show "polyhead" phenotype[e]
49	endonuclease required for packaging of DNA into heads[f]
57	regulation of tail fiber assembly[g]
63	enzyme required for attachment of tail fibers to base plate[h]

[a]STREISINGER et al. (1971), [b]EMRICH (1968), [c]JOSSLIN (1970), [d]LAEMMLI, BEGUIN, and GUJER-KELLENBERGER (1970), [e]EPSTEIN et al. (1963), [f]FRANKEL, BATCHELER, and CLARK (1971), [g]EDGAR and LIELAUSIS (1968), [h]WOOD and HENNINGER (1969).

through a methylated albumin-kieselgur chromatography column which fractionates nucleic acids according to size as well as nucleotide sequence and composition. The two radioactively labeled RNAs displayed markedly different elution profiles from such columns. While some of this difference was understandable in terms of a size-differential between $[^3H]$ RNA and $[^{14}C]$ RNA, the magnitude of the difference was too great to be explained solely in this fashion. KANO-SUEOKA and SPIEGELMAN (1962), therefore, concluded that fundamentally different classes of T2-specific RNA are transcribed during the early and late periods.

A more quantitative approach to this problem was made possible by the development of DNA-RNA hybridization competition techniques (HALL and SPIEGELMAN, 1961; NYGAARD and HALL, 1964). Initial attempts to apply this procedure to problems of bacteriophage transcription were made by HALL, NYGAARD, and GREEN (1964), HALL et al. (1963), KHESIN and SHEMYAKIN (1962), and KHESIN et al. (1963).

In the study by HALL, NYGAARD, and GREEN (1964), T2-infected cells were pulse-labeled with $[^{32}P]$ inorganic phosphate either from 0 to 6.5 min after infection or from 15 to 19 min after infection. Unlabeled RNA extracted 6.5 min after infection competed essentially 100 % against $[^{32}P]$ RNA labeled during the initial 6.5 min of the lytic process. However, only 50 % to 60 % of the $[^{32}P]$ RNA labeled between 15 and 19 min after infection was competable by unlabeled 6.5-min RNA. In contrast, unlabeled 19-min RNA competed completely against both $[^{32}P]$ RNAs tested. Hence, it appears that 19-min RNA contains a component (the "late" RNA) which is not present in significant amounts 6.5 min after infection. Nevertheless, most of the T2 RNA present 19 min after infection is equivalent to those sequences which constitute all the 6.5-min RNA (i.e. the "early" RNA).

A somewhat more detailed analysis of this problem was subsequently made with T4-infected cells by BOLLE et al. (1968a). In one set of experiments, T4-infected bacteria were pulse-labeled with [^{3}H] uridine from 0 to 5 min after infection. Unlabeled RNAs isolated 5 and 20 min after infection competed completely against the [^{3}H] RNA. However, the initial slope of the competition curve obtained with 20-min RNA was 1/3 to 1/2 that observed with 5-min RNA. Since the amount of T4 RNA increases 40 % between 5 and 20 min, these results suggest that early RNA is 3.2 to 5.3-fold less abundant in 20-min RNA than in 5-min RNA.

BOLLE et al. (1968a) also demonstrated that identical competition curves are observed with unlabeled 5-min RNA and (i) [^{3}H] RNA continuously labeled from 1 to 20 min after infection and (ii) [^{3}H] RNA pulse-labeled from 17 to 20 min after infection. In both cases, 40 to 50 % of the [^{3}H] RNA was not competable by 5-min RNA. Several conclusions may be drawn from these data. First, much of the T4-RNA present 20 min after infection represents a class of RNA not present in significant quantities 5 min after infection. Second, both classes of T4-RNA are metabolically unstable to about the same degree, otherwise their relative proportion in continuously labeled [^{3}H] RNA would have differed from those pulse-labeled [^{3}H] RNA.

Using the [^{3}H] RNA continuously labeled from 1 to 20 min after infection and "mixed competitor" experiments, BOLLE et al. (1968a) were able to estimate the relative increase in concentration of late RNA between 5 and 20 min after infection. First, [^{3}H] RNA was competed by unlabeled 5-min RNA. With the concentration of [^{3}H] RNA employed, 0.6 mg/ml of 5-min RNA reduces the amount of [^{3}H] RNA hybridized to about 50 %. In a parallel experiment, [^{3}H] RNA was simultaneously competed with by 0.6 mg/ml of 5-min RNA and increasing concentrations of 20-min RNA. From the initial slope of the additional competition observed with 20-min RNA *versus* the initial slope of the additional competition observed with concentrations of 5-min RNA between 0.6 mg/ml and 1.6 mg/ml, it was calculated that the late class of T4-RNA increased at least 80-fold between 5 and 20 min after infection. More refined experiments suggested that a several hundred- to several thousand-fold increase may actually occur in the case of most components of late RNA (BOLLE et al., 1968a).

Further evidence for the existence of a "new" class of RNA at late infection times comes from studies on the anealing of [^{3}H] RNA isolated various times after infection to the individual strands of T4 DNA. When phage-infected cells are continuously labeled during the initial 2.5 and 10 min of the lytic process, 95 to 100 % of the [^{3}H] RNA anneals to the *l*-strand of T4 DNA (GUHA and SZYBALSKI, 1968; GUHA et al., 1971). On the other hand, [^{3}H] RNA isolated 12, 15, 20, and 30 min after infection contains increasing amounts of an RNA species that anneal to the *r*-strand. Since the amounts of *l*-strand and *r*-strand hybridizable RNAs present 5 and 20 min after infection agree well with the relative amounts of early and late RNA present at the same time-intervals as estimated by hybridization competition studies, it is evident that early RNA originates predominately from the *l*-strand while late RNA is transcribed principally off the *r*-strand.

It should be mentioned that the opposite polarity of transcription of early and late genes was first demonstrated genetically by observing the direction of polarity exerted by amber mutations in certain viral genes. These polarity studies showed that early genes are transcribed counter-clock-wise (as the map is usually drawn) while late genes are transcribed in the clockwise direction (see STAHL et al., 1970 for references).

IV. Resolution of the Quasi-late RNA

While the experiments related above clearly demonstrate the existence of two classes of viral-specific RNA within T4 infected cells, subsequent studies revealed that 5-min RNA is actually composed of two discrete populations. This was shown by a careful analysis of the competition curves obtained with [^{3}H] RNA labeled 1 to 20 min after infection and unlabeled 5-min RNA (SALSER, BOLLE, and EPSTEIN, 1970). The amount of 5-min RNA required to give a particular degree of competition is 3- to 5-fold higher than that expected if the relative frequencies of all constituents of 5-min competable RNA is the same 5 and 20 min after infection. In other words, most 5-min competable [^{3}H] RNA present at 20 min is a minor component at 5 min but a major component at 20 min. That component in 5-min competable RNA which decreases many-fold in concentration between 5 and 20 min is called true-early. Since true-early RNA is not more than 1/4 to 1/3 of the 5-min competable RNA present at 20 min, it may well be that the concentrations of true-early sequences undergo a 10- to 25-fold reduction in relative abundance by 20 min.

On the basis of hybridization competition studies (BOLLE et al., 1968a; SALSER, BOLLE, and EPSTEIN, 1970), quasi-late RNA can be distinguished from true-late RNA in two ways. First, quasi-late RNA is present by the 5th min whereas true-late RNA is not. Second, the concentration of quasi-late sequences increase only 15- to 20-fold between the 5th and 20th min while true-late RNA sequences increase at least several hundred fold within the same interval.

Which of the DNA strands code for quasi-late RNA? As mentioned above, the amount of *l*-strand RNA present at 20 min is the same as the amount of 5-min competable RNA present at the same time (GUHA et al., 1971; SALSER, BOLLE, and EPSTEIN, 1970). Consequently, most quasi-late RNA probably originates from the *l*-strand. However, some of the T4 phage-specific RNA present at 5 min does originate from the *r*-strand (GUHA et al., 1971; NOTANI, 1973).

V. Immediate-early and Delayed-early RNAs

When *E. coli* is infected with bacteriophage T4, in the presence of chloramphenicol, RNA isolated 5 min after infection competes against 10 % of the material in [^{3}H] RNA labeled during the first 1.5 to 2.0 min of normal lytic process but against only 40 to 50 % of the material present in [^{3}H] RNA labeled during the first 5 min of normal lytic event (GRASSO and BUCHANAN, 1969; SALSER, BOLLE, and EPSTEIN, 1970). Consequently, chloramphenicol prevents the accumulation of those RNA sequences normally transcribed between the, say, 2nd and 5th min of a normal lytic process. That subset of T4 phage-specific prereplicative (early) RNA that fails to accumulate during treatment with chloramphenicol is called delayed-early due to its absence form 2-min RNA. That subset which accumulates in the presence of chloramphenicol is termed immediate-early.

VI. Transcription of the Prereplicative Genes of Bacteriophage T4: The Read-through Mechanism

T4 phage-specific [^{3}H] RNA labeled during the first minute or so of a normal lytic process or during the initial 5 min of an infection carried out in the presence of chloramphenicol sediments through sucrose density gradients at 8 to 10S (BRODY et al., 1970; MILANESI et al., 1970; WITMER, 1971b). All of the T4 phage-specific RNA transcribed under these conditions is immediate-early. On the other hand, most of the T4 RNA labeled during the initial 5 min of a normal lytic event migrates through sucrose at 14 to 16S and a sizeable component sediments as rapidly as 30S (BRODY et al., 1970; MILANESI et al., 1970; WITMER, 1971b). In addition, these longer-chained RNA molecules contain immediate- and delayed-early sequences (BRODY et al., 1970).

If T4 RNA is pulse-labeled with [^{3}H] uridine from 1.0 to 2.0 min after infection, pulse-labeled delayed-early [^{3}H] RNA sediments at 12 to 14S whereas pulse-labeled total early [^{3}H] RNA sediments at 8S (BRODY et al., 1970). Therefore, the earliest appearing delayed-early transcripts are preferentially associated with the longest-chained RNAs present at 2 min. The apparent length of pulse-labeled delayed-early RNA is 4 times longer than that expected for chains initiated during the pulse. Similar results were obtained for RNAs pulse-labeled from 1.0 to 2.5 min and from 1.0 to 3.0 min. The simplest interpretation for such data is that delayed-early sequences present by the 3rd min arise principally by the extension of RNA chains that are initiated at immediate-early promoters. Immediate-early promoters have also been called Class I (BAUTZ and BAUTZ, 1970a) or early promoters (O'FARRELL and GOLD, 1973a).

From the above results, a general model for transcription of, at least, the first delayed-early sequences emerges (Fig. 2). According to this model, immediate-early and delayed-early genes are interdigitated on the T4 chromosome. The host RNA polymerase initiates transcription solely at immediate-early promoters. Upon completion of immediate-early transcripts, most RNA polymerase molecules continue directly into adjacent delayed-early

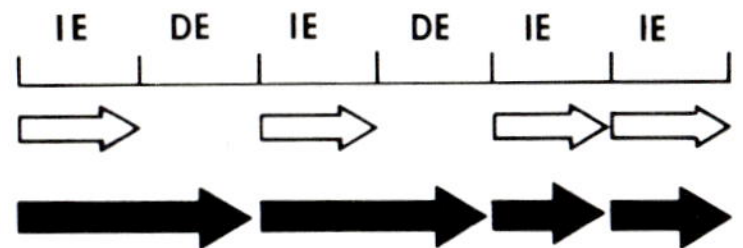

Fig. 2. The read-through mechanism. See text for details. White arrows depict transcription *in vitro* in the presence of rho and *in vivo* in the presence of chloramphenicol. Black arrows represent transcription *in vitro* in the absence of rho and *in vivo* in the absence of chloramphenicol. The fact that adjacent immediate-early (IE) genes are not co-transcribed in polycistronic fashion is inferred from SEDEROFF, BOLLE, GOODMAN, and EPSTEIN (1971). DE means delayed-early

regions. In other words, most immediate-early genes are co-transcribed with delayed-early genes as portions of polycistronic messengers.

In this connection, it has been demonstrated that messengers for internal proteins II and III (BLACK and GOLD, 1971), β-glucosyl-transferase (BLACK and GOLD, 1971), and deoxynucleoside mono-phosphate kinase (SAKIYAMA and BUCHANAN, 1973) are associated with RNA chains that are too long to be monocistronic. Using deletion mutations, it has been demonstrated that the adjacent rIIA and rIIB genes are frequently co-transcribed (SCHMIDT et al., 1970) and that many transcripts from the rII region are too long to come only from that region of the chromosome (SEDEROFF, BOLLE, and EPSTEIN, 1971). While these data suggest that transcription from immediate-early promoters is generally polycistronic, the D2 (immediate-early) region is unique in that it is not co-transcribed with any delayed-early genes (SEDEROFF, BOLLE, and EPSTEIN, 1971).

Additional support for the read-through mechanism comes from studies on the *in vitro* transcription of the mature T4 DNA by the host RNA polymerase.

DNA-dependent RNA polymerase, isolated from uninfected *E. coli*, is a complex enzyme that contains at least 5 distinct subunits (Table 3). The holoenzyme has a molecular weight of 500,000 and the subunit structure, $\alpha_2\beta'\beta\sigma\omega$. Chromatography of the holoenzyme through cellulose phosphate columns separates σ from the so-called core enzyme $\alpha_2\beta'\beta\omega$ (BURGESS et al., 1969).

Sigma posesses no known catalytic properties but this protein seems to be essential for binding of the holoenzyme to specific promoter sites on template DNA (BAUTZ and BAUTZ, 1970a; BAUTZ, BAUTZ, and DUNN, 1969; BURGESS et al., 1969; SUGIURA, OKAMOTO, and TAKANAMI, 1970). Although sigma is commonly referred to as an "initiation factor", this subunit actually dissociates from the enzyme-DNA complex prior to formation of the first phosphodiester bond (DUNN and BAUTZ, 1969).

All "polymerase" activity resides in the core enzyme. The role of the α and ω subunits is currently unknown. The β' and β subunits are involved with binding of the enzyme to DNA and initiation of RNA chains, respectively (ZILLIG et al., 1970a,b).

Table 3. Subunits of DNA-dependent RNA polymerase from *E. coli*

Subunit	Molecular weight	Function
σ	100,000[1]	Binding of holoenzyme to promoter[4]
β'	175,000[2]	Binding of core enzyme to template[5]
β	165,000[2]	Initiation[6]
α	40,000[2]	Unknown
ω	10,000[2]	Unknown
ρ	200,000[3]	Termination[7]

[1]BURGESS et al. (1969), [2]BURGESS (1969), [3]ROBERTS (1969). Rhos appears to be composed of 4 identical subunits each with a molecular weight of 50,000, [4]BAUTZ and BAUTZ (1970a,b); BAUTZ, BAUTZ, and DUNN (1969); BURGESS et al. (1969); SUGIURA, OKAMOTO, and TAKANAMI (1970), [5]ZILLIG et al. (1970b), [6]ZILLIG et al. (1970b), [7]ROBERTS (1969).

Another transcription protein, rho (ROBERTS, 1969), can be isolated from uninfected *E. coli* (Table 3). Like sigma, rho has no known catalytic properties. Rho causes RNA chains to be terminated at apparently discrete sites on T4 phage DNA (see below). However, there is still some question as to whether or not rho behaves *in vivo* in the same manner that it functions *in vitro* (O' FARRELL and GOLD, 1973b).

When mature T4 DNA is transcribed *in vitro* by the host RNA polymerase, only the early genes are transcribed (GEIDUSCHEK et al., 1966). RNA chains are initiated exclusively at immediate-early promoters (MILANESI et al., 1970; MILANESI, BRODY, and GEIDUSCHEK, 1969; WITMER, 1971b). When rho is present, transcription is confined to immediate-early genes and the *in vitro* product is 1,500 to 2,000 nucleotides long (RICHARDSON, 1970a; WITMER, 1971b). RNA made in the presence of rho competes against at least 80 % of the sequences present in labeled "chloramphenicol RNA" (WITMER, 1971b, and unpublished data).

In the absence of the termination protein, RNA molecules initiated at immediate-early promoters reach an average length of 4,500 to 5,000 nucleotides but some achieve lenghts of 7,500 to 8,000 nucleotides (MILANESI et al., 1970; MILLETTE and TROTTER, 1970; MILLETTE et al., 1970; RICHARDSON, 1970a,b,c; WITMER, 1971a, b). Unlabeled RNA made in the absence of rho competes against 95 % of the material in [^{3}H] RNA transcribed during the initial 5 min of a normal lytic process (MILANESI, BRODY, and GEIDUSCHEK, 1969; MILANESI et al., 1970; WITMER, 1971b). T4 RNA chains made *in vitro* do not acquire significant levels of delayed-early sequences until an average chain length of 1,500 to 2,000 nucleotides is achieved.

Studies on the transcription of sheared T4 DNA templates also support the notion that delayed-early genes are promoter distal (BRODY, GOLD, and BLACK, 1971; MILANESI et al., 1970; TRIMBLE,

GALIVAN, and MALEY, 1972). In these experiments, DNA molecules are sheared to certain sizes prior to transcription. Such studies have shown that delayed-early genes are much more shear-sensitive than transcription of immediate-early genes.

Besides supporting the basic theme of the read-through mechanism, the *in vitro* experiments presented above strongly suggest that almost all of the T4 phage early genes can be transcribed from promoters on immediate-early genes. This, in turn, would seem to suggest that the read-through mechanism is of paramount importance during, at least, the initial 5 min of a normal lytic event.

VII. Delayed-early Promoters

Even though most delayed-early sequences present 5 min after infection aries from extension of immediate-early (promoter proximal) messenger segments, it is now evident that a number of immediate- and delayed-early genes are additionally transcribed from special promoters that do not function until 1.5 to 2 min. The first pertinent observation along this line was made by SCHMIDT et al. (1970). By measuring the amount of deletion-specific T4 RNA, these authors were able to show that transcription of the delayed-early rIIB region began well before the first transcripts of the adjacent rIIA region were completed. These authors concluded that gene rIIB is transcribed partially from a remote immediate-early promoter and partially from a promoter located at or near the 3'-end of the rIIB gene. Subsequent *in vitro* experiments reported by WITMER (1971b) and JAYARAMAN (1972) suggested that the latter promoter did not function in highly purified *in vitro* systems.

In a more extensive analysis, O'FARRELL and GOLD (1973a) showed that immediate-early gene IPIII (internal protein III) and delayed-early genes 43 (DNA polymerase) and 45 (function unknown) are similar to gene rIIB in that they are transcribed from immediate- and delayed-early promoters. More significantly, O'FARRELL and GOLD (1973a) were able to demonstrate that gene 32 was transcribed exclusively from a delayed-early promoter. Subsequent studies by COHEN, NATALE, and BUCHANAN (1974) and BAROS and WITMER (unpublished data) have shown that gene 1 (deoxynucleoside monophosphate kinase) is also transcribed exclusively from a delayed-early promoter.

Delayed-early promoters differ from immediate-early promoters in that the former require the prior systhesis of a phage-specific protein (MATTSON, cited in O'FARRELL and GOLD, 1973a) and they do not, as a rule, function in the conventional *in vitro* systems where mature T4DNA is the template (JAYARAMAN, 1972; O'FARRELL and GOLD, 1973a; WITMER, 1971b). A temperature-sensitive mutation in the gene that controls expression of delayed-early promoters is known but the mechanism of action remains unknown.

Gene 1 is unusual in that it can be transcribed *in vitro* from mature T4 DNA, even in the highly purified systems (COHEN, NATALE, and BUCHANAN, 1974; NATALE and BUCHANAN, 1972; TRIMBLE and MALEY, 1973; TRIMBLE, GALIVAN, and MALEY, 1972). *In vitro* transcription of gene 1 is not inhibitable by rho (TRIMBLE and MALEY, 1973) and transcription of this gene is relatively shear-resistant (TRIMBLE, GALIVAN, and MALEY, 1972). Thus, while the *in vitro* studies just related are consistent with the interpretation that gene 1 is proximal to an immediate-early promoter a number of *in vitro* experiments suggest that this gene is transcribed exclusively from a delayed-early promoter (see above) and that its *in vivo* expression requires the prior synthesis on one or more functional phage-specific proteins (LEMBACH and BUCHANAN, 1970; SAKIYAMA and BUCHANAN, 1973; BAROS and WITMER, unpublished data). The apparent discrepancy has yet to be resolved but COHEN, NATALE, and BUCHANAN (1974) have recently described an *in vitro* system that seems to conserve the constraints initially placed upon the *in vivo* transcription of gene 1.

Several years ago, TRAVERS (1969; 1970a,b) reported that crude extracts from T4 phage-infected cells contained a factor that would direct the bacterial core enzyme to initiate transcription at sites other than immediate-early promoters. RNA made in the presence of the T4 factor competed against only 10 % of the material transcribed *in vitro* by the bacterial holoenzyme; material made *in vitro* by the bacterial holoenzyme competed against only 30 % of the RNA made *in vitro* in the presence of the T4 factor and host core enzyme. On the other hand, RNA isolated 5 min after a normal lytic process contained all sequences that were transcribed *in vitro* by the host core enzyme in the presence of the T4 factor. Presumably, the factor described by TRAVERS was directing initiations *in vitro* at delayed-early promoters but there has been no experimental verification of these results (BAUTZ et al., 1970).

It should be mentioned that at least two other early genes are probably transcribed from delayed-early promoters. These are genes 41 and 57. The exact intracellular role of gene 41 is unknown but it is required for DNA replication (EPSTEIN et al., 1963; OISHI, 1968). In a highly purified *in vitro* system, gene 41 is not transcribed (JAYARAMAN, 1972). Although gene 57 codes for a factor that regulates tail fiber morphogenesis (EDGAR and LIELAUSIS, 1968), there is good reason to believe that it is early. First, gene 57 mRNA is transcribed and translated at early times in the T4 lytic process (CASCINO et al., 1970; HOSODA and LEVINTHAL, 1968; WILHELM and HASELKORN, 1969, 1971). Mature T4 DNA cannot serve as a template for transcription of gene 57 *in vitro* (WILHELM and HASELKORN, 1971).

VIII. The Interrelationship between Translation and Elongation of Chains from Immediate-early Promoters

Chloramphenicol, and other antibiotic inhibitors of protein synthesis, selectively inhibit transcription of delayed-early genes, provided the antibiotics are added before or with the phage (GRASSO and BUCHANAN, 1969; PETERSON, COHEN, and ENNIS, 1972; TRIMBLE, GALIVAN, and MALEY, 1972; SALSER, BOLLE, and EPSTEIN, 1970; BLACK and GOLD, 1972; BRODY et al., 1970). In the case of genes transcribed exclusively from delayed-early promoters, it is relatively easy to understand the chloramphenicol effect since expression of these promoters appears to require the prior synthesis of a functional phage-specific element.

However, most delayed-early sequences arise by RNA chains that are initiated at immediate-early promoters (see Section VI). The basis for the chloramphenicol effect in this case has been the subject of some controversery. Originally, it was proposed that the prior synthesis of a functional phage protein was required to override normal cell termination processes (SCHMIDT et al., 1970; WITMER, 1971b). However, this position is untenable because delayed-early genes, of the promoter distal variety, can be transcribed in the presence of amino acid analogs (GRASSO and BUCHANAN, 1969; SAUERBIER and HERCULES, 1973). More recently, it has been proposed that promoter distal transcription is eliminated by chloramphenicol by virtue of its ability to induce polarity (BLACK and GOLD, 1971).

In uninfected bacteria, polarity can arise from one of two mechanisms. In the case of the tryptophan operon of *E. coli*, messenger distal to a blockade on translation are still transcribed but they are rapidly hydrolyzed by an endoribonuclease that is encoded or controlled by the suA gene (MORSE, 1970; MORSE and PRIMAKOFF, 1970). In the lactose operon of *E. coli* , transcription seems to be coupled to translation and the degree of messenger elongation is, within certain limits, determined by the amount of concurrent ribosomal movement along the growing polyribonucleotide chain (KENNELL and SIMMONS, 1972).

In the author's laboratory, Anna BAROS and Janet FORBES have been studying transcription of the early genes of bacteriophage T4 in an attempt to clarify the mechanism of polarity. They find that chloramphenicol inhibits transcription of promoter distal delayed-early genes in both suA^+ and suA^- bacteria, implying that rapid hydrolysis of nontranslated messenger sequences is not a satisfactory explanation for drug-induced polarity (BLACK and GOLD, 1971). When cells starved for an essential amino acid are infected with T4 phages, promoter distal delayed-early genes are transcribed (BAROS and FORBES, unpublished data). Since starvation for an essential amino acid reduces the level of functional phage protein synthesis to about the same level as does treatment with chloramphenicol, these latter data must be interpreted to mean that a phage-specific protein is probably not involved with the extension of promoter proximal messenger segments into distal regions.

As mentioned above, transcription of the lactose operon of *E. coli* is coupled to ribosomal translocation and messengers of that operon are prematurely terminated when ribosomal movement is inhibited (KENNELL and SIMMONS, 1972; PASTUSHOK and KENNELL, 1974). Thus, one could evision that a similar situation prevails insofar as transcription of promoter distal delayed-early genes of T4 is concerned. It could be reasoned, therefore, that chloramphenicol places such a severe constraint on ribosomal movement that transcription from immediate-early promoters is limited to proximal regions; on the other hand, sufficient translocation occurs during starvation for an essential amino acid that messengers elongate to their full extent.

To estimate the degree of ribosomal movement, WITMER, BAROS, and FORBES have performed the following experiments. Parallel cultures of bacteria were infected with T4. One culture was infected under conditions that permit normal protein synthesis and this culture was labeled with a mixture of [^{3}H] amino acids. The second culture was infected under conditions that retard protein synthesis (either treatment with chloramphenicol or starvation for an essential amino acid, in this case, L-leucine) and it was labeled with a mixture of [^{14}C] amino acids. After an appropriate labeling time, the two cultures were mixed and crude extracts were prepared. Extract proteins were fully reduced and alkylated in the presence of 6M guanidine hydrochloride and chromatographed through columns of G200 Sephadex.

Polypeptides synthesized during a normal lytic process have an average molecular weight of 23,000 to 26,000. Those made during amino-acid starvation had an average molecular weight of 8,000 to 12,000. The small amount of material synthesized in chloramphenicol-treated cells had a molecular weight of only 1,000. These do, indeed, suggest that ribosomal movement is more extensive during starvation for an essential amino acid than during treatment with chloramphenicol.

Chloramphenicol, added 2 or 3 min post-infection, no longer inhibits promoter distal delayed-early transcription (BRODY et al., 1970; GRASSO and BUCHANAN, 1969; YOUNG and VANHOWE, 1970; YOUNG, 1970a; SAKIYAMA and BUCHANAN, 1973). These data imply that promoter-distal transcription becomes uncoupled from translation soon after infection. The basis for this phenomenon remains moot. However, it should be mentioned that promoter distal delayed-early transcription remains chloramphenicol-sensitive in cells starved for an essential amino acid, implying that phage-specific modifications of the transcription-translation complex may be required for uncoupling (WITMER, BAROS and FORBES, unpublished data).

In the conventional *in vitro* systems, promoter distal delayed-early transcription can take place even in the absence of concurrent translation (MILANESI et al., 1970; O'FARRELL and GOLD, 1973b; TRIMBLE and MALEY, 1973; TRIMBLE, GALIVAN, and MALEY, 1972; JAYARAMAN, 1972; WITMER, 1971b), a situation that does not exist *in vivo* earlier than 2 or 3 min post infection. Therefore, the usual *in vitro* systems are free of certain constraints initially placed upon promoter-distal delayed-early transcription. It is interesting to speculate concerning the basis for this discrepancy. First, it is possible that T4 phage DNA, as injected into the host, contains transcriptional barriers and that these barriers can be transiently overcome by concurrent transcription-translation. In a normal infection, some modification is introduced into the parental DNA that permanently removes the barriers in question. In the *in vitro* situation, it is conceivable that those modifications required to inactivate certain transcriptional barriers are mimicked, say, during the DNA isolation procedure. Second, it is still possible that chloramphenicol-induced premature termination *in vivo* is mediated by a termination factor such as rho and that these factors are either absent from or nonfunctional in the *in vitro* systems thus far developed. In the latter connection, it is relevant to note that a substantial body of evidence suggests that rho-dependent termination *in vitro* and chloramphenicol-induced termination *in vivo* occur at similar sites on the T4 phage chromosome (JAYARAMAN, 1972; TRIMBLE and MALEY, 1973; RICHARDSON, 1970; TRAVERS, 1970a; WITMER, 1971b).

Nonsense mutations and some early genes and many late genes show polarity effects (STAHL et al., 1970). Presumably this means that the polarity effect observed with certain ambers arise from a mechanism that is distinct from the premature-termination mechanism that is responsible for chloramphenicol-induced polarity. We are currently studying the mechanism of polarity in T4 late genes.

IX. Requirements for True-late Transcription

It has long been known that synthesis of the late proteins does not take place when cells are infected with DO or MD mutants of coliphage T4 (EPSTEIN et al., 1963; HOSODA and LEVINTHAL, 1968). An extensive series of investigations have shown that the blockade on late gene expression produced by these mutations is due to control at the level of transcription.

When cells are infected with phage defective in genes 1, 32, 41, 42, 43, or 56 (all DO), mixed competitor experiments show that the concentration of true-late RNA 20 min after infection is no higher than the concentration of true-late RNA 5 min after infection by wild-type phage (BOLLE et al., 1968b). In agreement with these experiments, transcription from the *r*-strand of T4 DNA is markedly reduced by DO and MD mutations (GUHA et al., 1971).

Continued DNA synthesis is not required for continued transcription of true-late genes. In DA mutants, for example, substantial amounts of true-late RNA are made although viral DNA synthesis is confined to a brief time interval (BOLLE et al., 1968b; GUHA et al., 1971). Nevertheless, maximum rates of true-late transcription do require maximum rates of DNA replication as evidenced by the fact that the rate of true-late RNA transcription declines several-fold immediately after DNA replication is stopped (CASCINO, RIVA, and GEIDUSCHEK, 1970; CASCINO et al., 1971; LEMBACH, KUNINAKA and BUCHANAN, 1969; RIVA, CASCINO, and GEIDUSCHEK, 1970a,b).

On the basis of the above results, it becomes evident that some modification of the viral chromosome, unique to actively repli-

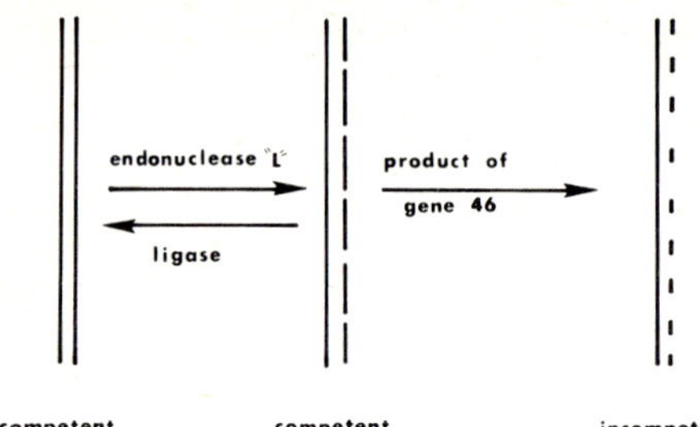

Fig. 3. Uncoupled true-late transcription. See text for details. For purposes of presentation, the single-strand scissions that are believed to render DNA competent for late transcription are considered to be asymmetrically distributed on one strand of the DNA molecule, but there is no evidence that this is so

cating DNA, is necessary for optimal transcription to true-late genes. The nature of this modification is suggested by studies in systems where true-late transcription and DNA replication can be uncoupled to a limited degree.

Phage simultaniously defective in genes 43 and 46 are able to express all of their true-late function, to a limited extent, despite the fact that no measurable DNA synthesis occurs in these systems (CASCINO, RIVA, and GEIDUSCHEK, 1970; RIVA, CASCINO, and GEIDUSCHEK, 1970b). This uncoupled true-late transcription is markedly enhanced by a further mutation in gene 30 (CASCINO, RIVA, and GEIDUSCHEK, 1970).

Under conditions of uncoupled transcription, single-strand breaks accumulate in the non-replicating T4 DNA through the action of some unknown endonuclease (CASCINO, RIVA, and GEIDUSCHEK, 1970; RIVA, CASCINO, and GEIDUSCHEK, 1970b). It has been proposed that these single-strand lesions render DNA "competent" for true-late transcription (Fig. 3). Under normal conditions, the introduction of these lesions is coincident with DNA replication and may, in fact, serve as starting points for DNA replication (KORNBERG, 1969).

Assuming that single-strand lesions at specific points on the T4 chromosome serve as starting points for late transcription, one is able to understand how mutations in genes 30 and 46 enable late transcription to become uncoupled from DNA replication (Fig. 3). First, gene 30 codes for polynucleotide ligase (Table 1) and this enzyme would render DNA incompetent by sealing the single-strand breaks. Second, gene 46 codes for or controls the action of an exonuclease required for degradation of the host chromosome (WIBERG, 1966) as well as continued replication of the viral DNA (see MATHEWS, 1971 for references). However, the product of gene 46 also leads to rapid degradation of non-replicating viral DNA (CASCINO, RIVA, and GEIDUSCHEK, 1970); hence, insofar as uncoupled late transcription is concerned, mutations in gene 46 lead to "conservation" of the DNA.

Although DNA replication is obligately required for abundant true-late transcription, it is not the sole requirement. The true-late genes of T4 cannot be expressed in the cases of phage defective in either gene 33 or 55, although DNA replication proceeds normally in these viruses (BOLLE et al., 1968b; GUHA et al., 1971). Uncoupled transcription of the late genes also requires the products of genes 33 and 55 (CASCINO, RIVA, and GEIDUSCHEK, 1970; RIVA, CASCINO, and GEIDUSCHEK, 1971b). As will be demonstrated later, genes 33 and 55 code for small polypeptides that become associated with RNA polymerase and enable it to transcribe the true-late genes.

Like the early genes, many of the late genes belong to transcriptional units. The known late transcriptional units are 53 → 5, 9 → 11, 13 → 15, 51 → 29, 48 → 54, and 34 → 36 (STAHL et al., 1970). All have been identified on the basis of polarity effects of amber mutations in late genes.

Little can be said concerning the *in vitro* transcription of late genes because this phenomenon has been observed only with crude DNA-RNA polymerase complexes obtained from phage-infected cells (SNYDER and GEIDUSCHEK, 1968). The demonstration that late transcription in these crude systems obligately requires the product of gene 55 was one of the first indications that 55 controlled expression of the late genes at the level of transcription.

X. Transcription of the Quasi-late Genes

As mentioned previously, quasi-late RNA is that component of T4 5-min RNA that increases many-fold in concentration by 20 min after infection. Hybridization competition studies conducted with unlabeled RNA isolated 20 min after infection by DO-defective T4 and [^{3}H] RNA labeled from 1 to 20 min after infection by T4$^+$ phage show that the quasi-late species are not more prevalent 20 min after infection by DO-defective phage than they are 5 min after infection by T4$^+$ phages (SALSER, BOLLE, and EPSTEIN, 1970). These data, coupled with the more recent results of NOTANI (1973), indicate that transcription of quasi-late sequences can begin in the absence of DNA replication as well as the maturation defective proteins but that full expression of quasi-late genes has the same requirements as does true-late transcription.

XI. Turn-off of Early Gene Expression

The early enzymes of bacteriophage T4 reach their maximum specific activities 10 to 12 min after infection (COHEN, 1963, 1968; HOSODA and LEVINTHAL, 1968; WIBERG et al., 1962, 1973). When nonpermissive cells are infected with amber-defective phages that display either the DO or MD phenotypes, synthesis of most early enzymes continues until the 20th min (HOSODA and LEVINTHAL, 1968; WIBERG et al., 1962). This has prompted the suggestion that shut-off of early gene expression occurs in two discrete steps (WIBERG et al., 1973). The first step (S1) occurs 10 to 12 min after infection and is somehow obviated by the absence of late-gene expression. The second step (S2) occurs at 20 min and is presumably due to an early function because it occurs independently of late-gene expression.

When MD or DO phage acquire an additional mutation in the regA gene, the S2 event does not take place (KARAM and BOWLES, 1974; WIBERG et al., 1973). Mutations in the regA gene extend the functional life-time of many, but not all, early messengers (KARAM and BOWLES, 1974). A report by SAUERBIER and HERCULES (1973) also indicates that the absence of a regA product extends the chemical life-time of many early messengers. However, KARAM and BOWLES (1974) report that mutations in the regA gene do not affect the chemical half-life of phage-specific RNA.

During a normal lytic event, functional prereplicative messengers accumulate until the 8th or 10th min and, thereafter, they undergo a many-fold reduction in relative abundance (JAYARAMAN and GOLDBERG, 1969, 1970; SAKIYAMA and BUCHANAN, 1972; TRIMBLE, GALIVAN, and MALEY, 1972; YOUNG, 1970a,b; YOUNG and VANHOWE, 1970). Using pulse-labeling techniques, BOLUND and SKÖLD (1973) and BOLUND (1973) have demonstrated that *de novo* transcription of most prereplicative genes probably starts to decline around the 6th min post infection. Thus, in a normal lytic event, prereplicative transcription either ceases or declines many-fold several minutes before the onset of late-gene expression. When cells are infected with wild-type viruses, and chloramphenicol is added 3 min later, transcription of most prereplicative genes seems to proceed indefinitely (BRODY et al., 1970; YOUNG, 1970a; YOUNG and VANHOWE, 1970; SAKIYAMA and BUCHANAN, 1972), implying that "shut-off" of prereplicative transcription is itself controlled by a prereplicative function.

In a general study, WILHELM and HASELKORN (1971) infected cells with DO or MD phages, RNA was isolated at 19 min and used to direct the *in vitro* synthesis of T4 phage-specific proteins. The proteins synthesized were resolved by electrophoresis through polyacrylamide gels. These authors found that most prereplicative messengers were still relatively abundant, at late times, in both DO and MD infected cells. A few prereplicative messengers were present, at late times, in DO-infected cells but not in MD-infected cells; messengers for alpha-glycosyltransferase have this behavior (YOUNG, 1970a; YOUNG and VANHOWE, 1970). Yet a third subclass of prereplicative messengers was absent, at late times, in both MD and DO-infected cells; specific examples within this subclass are probably messengers for dihydrofolate reductase (MATHEWS, 1962), the D2 region (KASAI and BAUTZ, 1967), and prereplicative transcripts of the endolysin gene (see Section XIV).

Originally, the data just presented have been interpretated to mean that MD and DO conditions resulted in a prolongation of prereplicative transcription. In other words, those parameters required for late expression were also required for the "shut-off" of prereplicative transcription.

However, several lines of evidence now indicate that MD and DO conditions do not result in extended transcription times for most prereplicative genes. The first line of evidence involves the rescue of ambers by 5-flurouridine (FU). In the amber codon (UAG), FU can be incorporated in place of uridine to give FUAG; occasionally, FUAG can be misread as CAG (glutamine). If glutamine can replace the normal acimo acid, the mutated function can be restored, at least partially. It is obvious that FU rescue of ambers can occur only if the gene in question is being transcribed. Experiments have been performed in which parallel cultures of nonpermissive bacteria are infected with FU-rescuable DO ambers. Rescue is monitored either by measuring viable phage production or DNA synthesis. DO ambers can be effectively rescued only if FU is added during the initial 5 min of the lytic process; when the analog is added at later times, the amount of rescue rapidly drops to nil (EDLIN, 1965; BOLUND and SKÖLD, 1973). Comparable experiments have been performed using amber DO phage that also carry a non-FU rescuable temperature sensitive lesion in gene 55 (BOLUND, 1973); the results indicate that the absence of a functional gene 55 product does not prolong transcription times of those genes carrying FU-revertible ambers.

A second line of evidence involves pulse-labeling procedures (BOLUND, 1973). When nonpermissive bacteria are infected with MD or DO phages, RNA can be effectively pulse-labeled only during the initial 5 min of the lytic process. Between the 5th and 12th min post infection, the rate at which RNA can be pulse-labeled decreases ca. 10-fold.

In the author's laboratory, experiments have been performed by William MARICONDIA and Mitchel WEINER in which cultures of *E. coli* B (a nonpermissive host) are infected with T4Dam^{+} T4DamDO, or T4DamMD phages. $[^3H]$ adenine is added 1 min after infection and its incorporation into RNA is minitored. In cells infected with wild-type viruses, $[^3H]$ RNA accumulates linearly for the first 8 min; a 4-min hiatus follows during which little additional accumulation occurs. Between the 12th and 20th min, an additional burst of $[^3H]$ RNA accumulation is noted in wild-type infected cells which presumably corresponds to late transcription. In amber infected cells, $[^3H]$ RNA accumulates with wild-type kinetics until the 8th min. Thereafter, no additional accumulation is evident. Beginning at the 12th min, an exponential decay of $[^3H]$ RNA ensues. By the 40th min 70 % of the material originally present at 12 min has decayed ($t_{1/2}$ = 12 to 14 min); the remainder of the material is apparently stable. DNA-RNA hybridization competition experiments reveal that the "decaying" and "stable" fractions contain different prereplicative sequences. This selective decay of prereplicative sequences may be due to the product of the regA gene (SAUERBIER and HERCULES, 1973).

As mentioned above, shut-off of early protein synthesis appears to occur in two discrete steps. S2 is mediated by the regA gene of bacteriophage T4 and seems to involve the selective decay of prereplicative messengers. Presumably, S2 marks the latest time at which functional transcripts of certain prereplicative genes are still present. The nature of the S1 event is somewhat unsettled. The absence of late-gene expression disrupts S1 but does not disrupt the shut-off of prereplicative transcription. Therefore, S1 cannot involve some derangement of transcriptional level controls. Transposed into other terms, the absence of late-gene expression permits prereplicative messengers to be translated for a longer-than-normal time interval. The mechanism by which this is mediated remains moot. However, it is possible that S1 is realy a rapid replacement of prereplicative messengers in polyribosomes by late messengers; schemes whereby the competition of messengers for ribosomes could "shut-off" synthesis of certain proteins has been invoked by others (O'FARRELL and GOLD, 1973a).

XII. T4-specific Modifications of DNA-dependent RNA Polymerase

Certain bacterial viruses (T3 and T7) code for an entirely new RNA polymerase that replaces the host enzyme (CHAMBERLIN and McGARTH, 1970). This viral-coded polymerase contains only one subunit and it is resistant to the antibiotic, rifampicin. T4 does not seem to code for an entirely new enzyme. Rather, it introduces a number of modifications into the host enzyme (BAUTZ and DUNN, 1969; CROUCH et al., 1969; KHESIN, 1970; SCHACHNER and ZILLIG, 1971; SCHACHNER, SEIFERT, and ZILLIG, 1971; SEIFERT, 1970; SEIFERT et al., 1969; TRAVERS, 1970a; WALTER, SEIFERT, and ZILLIG, 1968; ZILLIG et al., 1970a,b). DNA-dependent RNA polymerase isolated from T4-infected cells, although modified, is immunologically identical to the host enzyme (WALTER, SEIFERT, and ZILLIG, 1968) and, throughout the latent period, transcription of the viral chromosome shows the same sensitivity to rifampicin as does transcription in the uninfected host cell (HASELKORN, VOGEL, and BROWN, 1969). The functional basis for most of the known T4-specific modifications of host RNA polymerase, as well as the relevant T4-specific enzymes, is unknown.

One of the most striking differences between DNA-dependent RNA polymerase isolated from uninfected and T4-infected cells is that the latter enzyme is devoid of sigma (BAUTZ and DUNN, 1969; SCHACHNER, SEIFERT, and ZILLIG, 1971; SEIFERT et al., 1969). DNA-dependent RNA polymerase isolated as early as 1 min after infection contains only 5 to 8 % of the uninfected level of sigma and this subunit is totally absent from RNA polymerase isolated 2 or more min after infection. In conjunction with loss of sigma, RNA polymerase purified from infected cells displays virtually no activity *in vitro* with native T4 DNA, as the template.

The loss of sigma appears to be triggered by adsorption because RNA polymerase isolated from cells to which T4 ghosts have been absorbed shows significantly reduced levels of sigma and a much lower activity *in vitro* with T4 DNA (SEIFERT et al., 1969). In addition, loss of sigma still occurs in the presence of chloramphenicol (SEIFERT et al., 1969), indicating that *de novo* synthesis of T4 proteins is not requires.

Absence of sigma from DNA-dependent RNA polymerase isolated from infected cells does not mean that sigma has been destroyed. Sigma can be recovered from extracts of phage-infected cells by addition of a large molar excess of highly purified bacterial core enzyme (STEVENS, 1972). Nevertheless, sigma present in T4-infected cells seems to be modified in some fashion (SEIFERT, 1970).

The apparent loss of sigma from RNA polymerase so early in the latent period is difficult to understand in view of the fact that sigma is apparently required for initiation at immediate-early genes (BAUTZ, BAUTZ, and DUNN, 1969; BAUTZ et al., 1970). Addition of rifampicin at any time during the early period causes an immediate cessation in transcription of immediate-early genes 30 and 42 (WITMER, unpublished data), implying that RNA chains are continuously initiated at these immediate-early genes throughout the early period. It may be that sigma-mediated initiation is confined to a relatively brief time interval at the beginning of the latent period and that T4 does institute its own system for initiation of RNA chains at immediate-early genes.

On the other hand, it may well be that the absence of sigma from "T4 RNA polymerase" is merely an artifact. It could be that changes do occur in the core enzyme and/or sigma which is significantly reduce the affinity of sigma for the core enzyme but which do not prevent sigma from directing initiation *in vivo*. According to this alternative, the generally employed extraction procedures detach sigma from the core enzyme. It is relevant to note in this connection that a protein electrophoretically similar to sigma is reported to be present in RNA polymerase extracted from T4-infected cells by mild procedures (TRAVERS, 1970a).

Beginning 2 to 3 min after infection, amino acids are added to all subunits of the core enzyme (SCHACHNER and ZILLIG, 1971; SCHACHNER, SEIFERT, and ZILLIG, 1971; ZILLIG et al., 1971a,b). These modifications markedly alter the finger-print maps of trypic digests of each subunit but they have essentially no effect on the molecular weights of the subunits or the immunological properties of the core enzyme. These modifications do require *de novo* synthesis of T4-specific proteins since none occur in the presence of antibiotic inhibitors of protein synthesis.

When core enzyme isolated from uninfected and T4-infected cells are subjected to electrophoresis through polyanylamide gels that contain sodium dodecyl sulfate, the α, β, and β' subunits of both core enzymes have nearly identical mobilities (BAUTZ and DUNN, 1969; GOFF and WEBER, 1970; SEIFERT et al., 1969; TRAVERS, 1970b; ZILLIG et al., 1970a,b). This indicates, as was mentioned above, that little change in molecular weight occurs upon addition of amino acids to the subunits of the core enzyme. However, when the core enzymes are subjected to electrophoresis through polyacylamide gels that contain 6M urea (where separation is due principally to net electrical change), the α subunit of T4-modified core enzyme migrates more rapidly than the corresponding bacterial (BAUTZ and DUNN, 1969; GOFF and WEBER, 1970; SEIFERT et al., 1969) subunit. This enhanced mobility on 6M urea gels is due to the addition of AMP to the bacterial α subunit (GOFF and WEBER, 1970). The adenyl for this interesting reaction is unknown. Adenylation is complete by 5 min after infection (GOFF and WEBER, 1970).

Between 5 and 7 min after infection, the electrophoretic mobility of the β' subunit on 6M urea gels also increases (TRAVERS, 1970a; WITMER, unpublished data). The chemical basis for this altered electrophoretic mobility is currently unknown. Since the initiation of true-late RNA synthesis seems to involve preferential binding of modified RNA polymerase to specifically nicked areas

Table 4. T4-coded subunits of core RNA polymerase[1]

Polypeptide	No. per core enzyme	Time of Max. synthesis	Function	Gene
1	0.2 to 0.5	11 to 12 min.	late transcription	55
2	1.0	11 to 12 min.	?	?
3	0.9	11 to 12 min.	?	?
4	?	11 to 12 min.	late transcription	33

[1] STEVENS (1972).

on competent DNA molecules (CASCINO, RIVA, and GEIDUSCHEK, 1970; RIVA, CASCINO, and GEIDUSCHEK, 1970a), the apparent alteration of the β' subunit at a time when true-late transcription is just beginning may be significant. It was mentioned before that β' subunit is important for proper binding of the core enzyme to DNA (ZILLIG et al., 1970b) so it is possible that the apparent modification of the β' subunit is required for proper binding of RNA polymerase to competent DNA.

Apart from the modification of already existing bacterial subunits, T4 codes for four small polypeptides (Table 4) that seem to be subunits of the core enzyme (STEVENS, 1972). Two of these polypeptides (Nos. 2 and 3) serve no known function but polypeptides 4 and 1 represent the products of genes 33 and 55, respectively. All are synthesized maximally 11 to 12 min after infection so all may be involved with late transcription. It would be stressed that polypeptides 1 and 4 are the only known changes in RNA polymerase that can be correlated with altered transcriptional behavior of the enzyme.

XIII. T4-specific Modification of the Host Translational Machinery

Ribosomal proteins can be divided into two groups. Structural proteins are tightly bound to ribosomal RNA and cannot be removed with 2 M NH_4Cl. The so-called "factor fraction" is those ribosomal proteins that are removable by washing the ribosomes in 2 m NH_4Cl; the factor fraction contains, among other things, the polypeptide chain-initiation proteins.

SMITH and HASELKORN (1969) first noticed that T4-specific proteins are added to host ribosomes minutes after infection. In a more detailed analysis, DUBE and RUDLAND (1970) showed that several T4-specific proteins appear in the factor fraction sometime between 2 and 7 min after infection and another phage-specific protein appears in the structural proteins 10 to 15 min after infection.

Ribosomes isolated from T4-infected cells are reported to translate *E. coli* mRNA and RNA-phage RNA less efficiently *in vitro* than ribosomes prepared from uninfected cells (DUBE and RUDLAND, 1970; HSU and WEISS, 1969; KLEM, HSU, and WEISS, 1970; SHEDL, SINGER, and CONWAY, 1970). On the other hand, T4 ribosomes show no loss of efficiency when tested *in vitro* with T4 RNA (DUBE and RUDLAND, 1970; HSU and WEISS, 1969; KLEM, HSU, and WEISS, 1970). This change in template specificity is due to a chloramphenicol-inhibitable (HSU and WEISS, 1969) alteration in the factor fraction (DUBE and RUDLAND, 1970; HSU and WEISS, 1969; KLEM, HSU, and WEISS, 1970).

LEE-HUANG and OCHOA (1971) and POLLACK et al. (1970) have shown that an initiation protein, F3, is either replaced or modified upon infection by T4. Reconstitution experiments conducted with purified phage-specific F3 and *E. coli* ribosomes washed with 2M NH_4CL showed that much of the altered template specificity observed with intact T4 ribosomes is understandable in terms of a change in the F3 factor.

At first glance, it would seem reasonable to propose that the viral specific changes in the factor fraction are responsible for cessation of host-protein synthesis (BENZER, 1953; BILEZIKAN, KAEMPFER, and MAGASANIK, 1967; LEVIN and BURTON, 1961; SHER and MALLETT, 1954). However, this may not be the case. KENNELL (1970) reports that host-specific mRNA is quantitatively excluded from polyribosomes seconds after infection by T4. The mechanism by which this exclusion occurs is unknown. Host-protein synthesis also terminates upon absorption of DNA-less phage ghosts to bacterial cells (FABRICANT and KENNELL, quoted in KENNELL, 1970) which suggests that arrest of host-protein synthesis is caused by the attachment process and not by expression of a viral gene.

A recent paper by GOLDMAN and LODISH (1972) supports the conclusion that T4-specific changes in the factor fraction have little to do with arrest of host-protein synthesis. These workers find that T4 ribosomes display a 27 % to 50 % lower efficiency *in vitro* with all natural messenger RNA's tested. This slight and non-specific reduction in efficiency was shown to be caused by some change in the factor fraction.

What, then, is the functional basis for the observed changes in the factor fraction that take place in the T4 infection process? Although expression of T4 genes is controlled principally at the level of transcription (BOLLE et al., 1968a,b, 1970; GUHA et al., 1971; SALSER, BOLLE, and EPSTEIN, 1970), there is some reason to believe that expression of certain early genes is, at least, partially regulated at the level of translation. For example, COHEN, ZETTER, and WALSH (1972) have demonstrated that the efficiency of translation of gene 1 mRNA varies during the early period and that much more gene 1 mRNA is made than is ever translated. SALSER, GESTELAND, and RICARD (1969) likewise indicate that post-transcriptional control may well play a part in expression of gene 1 during the late period. It is possible that the T4-specific components of the factor fraction are required to affect translational level controls in the case of certain genes.

Perhaps the best evidence for translational level control of T4 gene expression comes from studies on the T4 lytic process in *E. coli* B207, a bacterial mutant that cannot concentrate K^+ from the growth medium (LUBIN and KESSEL, 1960). Under conditions of K^+ depletion, RNA synthesis proceeds normally but protein synthesis is inhibited (ENNIS and LUBIN, 1961). In medium that contains 1mM K^+, the rate of global protein synthesis in T4-infected B207 is only 50 % of the normal level, i.e. that seen in medium containing 33 mM K^+ (COHEN, 1968, 1970). The reduced amount of protein synthesis seen with 1mM K^+ is caused by a reduction in the numbers of functional ribosomes and not to a reduction in the rate of polypeptide chain elongation (ENNIS, 1971). B207 infected with T4 produce roughly 30 progeny viruses per infected cell (COHEN and ENNIS, 1965), indicating that the lytic process goes to completion at low K^+ concentrations.

Under normal conditions, T4-specific dCMP hydroxymethylase, deoxynucleoside triphosphatase, dTMP synthetase, and deoxynucleoside monophosphate kinase show comparable induction kinetics. However, upon infection of B207 in 1mM K^+, radically different induction kinetics are observed (COHEN, 1970). (i) In the case of dCMP hydroxymethylase and deoxynucleoside triphosphatase, enzyme synthesis begins after a slightly prolonged lag. The rate of enzyme synthesis is 50 % of normal and these enzymes accumulate to only 40 to 50 % of their normal level. (ii) Synthesis of dTMP synthetase also starts only after a prolonged lag. However, in this case, the final amount of enzyme made is 100 % of normal although the rate of enzyme synthesis is only 50 % of normal. (iii) Deoxynucleoside monophosphate kinase synthesis shows the most unusual response. After a prolonged lag, enzyme synthesis proceeds at 50 % of the normal rate until 55 % of the normal activity is attained. No further synthesis of kinase is observed for the next 20 to 30 min. Thereafter, synthesis of this enzyme resumes and continues until normal enzyme levels are reached. This second wave of kinase synthesis is inhibitable by chloramphenicol (COHEN, 1970). Measurements on the amount of kinase mRNA synthesized in 1mM K^+ (COHEN, unpublished data) and the effect of rifampicin, added 10 min prior to the second wave of kinase synthesis (WITMER, unpublished data), indicate that the second wave of kinase synthesis at 1mM K^+ takes place in the absence of significant *de novo* synthesis of kinase messenger RNA.

Further evidence for involvement of translational level control in dCMP hydroxmethylase and deoxynucleoside monophosphate kinase synthesis has been provided by COHEN (1972). If the potassium concentration is raised to 33 mM after synthesis of dCMP hydroxymethylase has stopped at 1mM K^+, synthesis of the enzyme resumes and continues until normal levels are reached. Renewed hydroxymethylase synthesis in 33 mM K^+ is preventable by chloramphenicol but not by rifampicin. Therefore, the cessation of hydroxymethylase synthesis observed in 1mM K^+ is not due to a depletion of functional hydroxymethylase messenger but to some defect at the level of translation. By the same token, the initial cessation of kinase synthesis observed in 1mM K^+ appears not to be due to depletion of functional messengers.

While a precise interpretation of the above data is currently impossible, it, nevertheless, seems that translational level control does play a role in synthesis of dCMP hydroxymethylase and deoxynucleoside monophosphate kinase. Whether or not translational level controls are important for synthesis of other genes remains to be seen. In this connection, it is worth noting that considerable levels of functional early mRNA persist in phage-infected cells (BLACK and GOLD, 1971; TRIMBLE, GALIVAN, and MALEY, 1972) well after measurable synthesis of most early proteins stops (HOSODA and LEVINTHAL, 1968). Here again a problem is raised by the fact that ribosome isolated from T4-infected cells do not appear to have any special preference for T4 late RNA (GOLDMAN and LODISH, 1972).

Aside from the alteration of ribosomal proteins, T4 phage also codes for several new species of transfer RNA and amino acid activating enzymes. This area has been reviewed recently (MATHEWS, 1971). Suffice it to say here, these changes seem to be involved with altered frequencies of several codons in T4 phage messenger RNA's (SHERBERG and WEISS, 1972).

XIV. Expression of the Endolysin Gene of Phage T4

The e gene of bacteriophage T4 codes for the phage-specific endolysin (STREISINGER et al., 1961). Even though this is a late function, gene *e* is transcribed at both early and late times in a normal lytic event (BAUTZ et al., 1966; KASAI and BAUTZ, 1967, 1969; JAYARAMAN and GOLDBERG, 1970; WITMER, PADNOS, MARICONDIA, and WEINER, manuscript in preparation). Typically, the early wave of gene *e* transcription occurs between the 3rd and 6th min whereas the late wave occurs between the 12th and 20th min. However, endolysin production is coincident with the late wave of gene *e* transcription (see below). Thus, the endolysin gene of phage T4 seems to have a particularly unusual mode of expression.

The endolysin gene is transcribed *in vitro* by the highly purified *E. coli* RNA polymerase (JAYARAMAN, 1972; WITMER, 1971b). This implies that early transcription of the endolysin gene occurs from an immediate-early promoter. The time course of early *e* transcription is normal in cells infected with MD and DO infected cells whereas the late wave is inapparent (WITMER, PADNOS, MARICONDIA, and WEINER, unpublished data). More importantly, chloramphenicol, added 2 or 3 min after the phage, has no effect on the time course of early *e* transcription whereas transcription of genes rIIA, rIIB, 42 and 43 is prolonged indefinitely (WITMER et al., unpublished data). Thus, we currently believe that the early wave of gene *e* transcription is controlled by a mechanism distinct from that controlling transcription of most other early genes (see Section XII).

Despite the fact early *e* RNA is present in polyribosomes (WITMER et al., manuscript in preparation) translation of the early *e* transcripts seemingly does not occur *in vivo* because all lysozyme synthesis is coincident with the late wave of *e* transcription (SALSER, GESTELAND, and RICARD, 1969). Thus, it appears that some translational barrier exists *in vivo* that prevents synthesis of lysozyme from early *e* RNA.

T4 RNA isolated during the first 10 min of a normal lytic process cannot direct the *in vitro* synthesis of endolysin using crude extracts prepared from uninfected cells (BLACK and GOLD, 1971; GESTELAND, SALSER, and BOLLE, 1967; GOLD and SCHWEIGER, 1970; SALSER, GESTELAND, and BOLLE, 1967; SALSER, GESTELAND, and RICARD, 1969; YOUNG, 1970a; YOUNG and VANHOWE, 1970) or T4 phage-infected cells (WITMER, unpublished data). On the other hand, RNA extracted at later times will direct the synthesis of high levels of endolysin. Consequently, it would seen that a blockade on early *e* RNA translation is a function of the messenger that contains the early *e* transcripts and is not a function of the translational machinery *per se.*

It has been proposed that early *e* transcription is accomplished by extension of RNA chains that are initiated at a remote point from *e* (JAYARAMAN, 1972; SCHMIDT et al., 1970; WITMER, 1971b). KASAI and BAUTZ (1969) have demonstrated that those regions of the T4 chromosome immediately to the left and right of *e* are not transcribed during the late period, implying that late *e* RNA is monocistronic. It is tempting to speculate that *e*mRNA is translatable only in its late (monocistronic) form. However, SALSER, GESTELAND, and RICARD (1969) report that the species of late T4 phage RNA responsible for the *in vitro* synthesis of endolysin is 1,500 nucleotides long. Since T4 phage endolysin contains only 169 amino acids (TSUGITA and INOUYE, 1969), the results of SALSER, GESTELAND, and RICARD (1968) are inconsistent with the notion that late *e* transcripts are monocistronic.

References

ADESNIK, M., LEVINTHAL, C: RNA metabolism in T4-infected *Escherichia coli*. J. Mol. Biol. 48, 187 (1970).

BAUTZ, E.K.F., BAUTZ, F.A.: Studies on the function of RNA polymerase σ factor in promoter selection. Cold Spring Symp. Quant. Biol. 35, 227 (1970a).

BAUTZ, E.K.F., BAUTZ, F.A.: Initiation of RNA synthesis: The function of σ in the binding of RNA polymerase to promoter sites. Nature 226, 1219 (1970b).

BAUTZ, E.K.F., BAUTZ, F.A., DUNN, J.J.: *E. coli* σ factor: A positive control element in phage T4 development. Nature 223, 1022 (1969).

BAUTZ, E.K.F., DUNN, J.J.: DNA-dependent RNA polymerase from T4-infected *E. coli*: An enzyme missing a factor required for transcription of T4 DNA. Biochem. Biophys. Res. Comm. 34, 230 (1969).

BAUTZ, E.K.F., DUNN, J.J., BAUTZ, F.A., SCHMIDT, D.A., BAZAITIS, A.F.: Initiation and regulation of transcription by RNA polymerase. In: Lepetit Colloquium on RNA polymerase and Transcription (Ed. L. SILVESTRI), p. 90. Amsterdam: North-Holland Publ. Co. 1970.

BAUTZ, E.K.F., KASAI, T., REILLY, E., BAUTZ, F.A.: Gene-specific mRNA II. Regulation of mRNA synthesis in *E. coli* after infection with bacteriophage T4. Proc. Nat. Acad. Sci. (Wash.) 55, 1081 (1966).

BAUTZ, E.K.F., REILLY, E.: Gene-specific messenger RNA: Isolation by the deletion method. Science 151, 328 (1966).

BENZER, S.: Induced synthesis of enzymes in bacteria analyzed at the cellular level. Biochem. Biophys. Acta 11, 383 (1953).

BILEZIKIAN, J.P., KAEMPFER, R.O.R., MAGASANIK, B.: Mechanism of tryptophanase induction in *E. coli*. J. Mol. Biol. 27, 495 (1967).

BLACK, L.W., AMAD-ZADEH, C.: Internal proteins of bacteriophage T4D: Their characterization and relation to head structure and assembly. J. Mol. Biol. 57, 71 (1969).

BLACK, L.W., GOLD, L.M.: Pre-replicative development of the bacteriophage T4: RNA and protein synthesis *in vivo* and *in vitro*. J. Mol. Biol. 60, 365 (1971).

BOLLE, A., EPSTEIN, R.H., SALSER, W., GEIDUSCHEK, E.P.: Transcription during bacteriophage T4 development: Synthesis and relative stability of early and late RNA. J. Mol. Biol. 31, 325 (1968a).

BOLLE, A., EPSTEIN, R.H., SALSER, W., GEIDUSCHEK, E.P.: Transcription during bacteriophage T4 development: Requirements for late messenger synthesis. J. Mol. Biol. 33, 339 (1968b).

BOLUND, C.: Influence of gene 55 on the regulation of synthesis of some early enzymes in bacteriophage T4-infected *E. coli*. J. Virol. 12, 49 (1973).

BOLUND, C., SKÖLD, O.: Regulation of early mRNA synthesis in bacteriophage T4-infected bacteria: Dependence on bacteriophage-specific protein synthesis. J. Virol. 12, 39 (1973).

BRODY, E.N., GOLD, L.M., BLACK, L.W.: Transcription and translation of sheared bacteriophage T4 DNA *in vitro*. J. Mol. Biol. 60, 389 (1971).

BRODY, E.N., SEDEROFF, R., BOLLE, A., EPSTEIN, R.H.: Early transcription in T4-infected cells. Cold Spring Harbor Symp. Quant. Biol. 35, 203 (1970).

BRUNER, R., CAPE, R.E.: The expression of two classes of late genes of bacteriophage T4. J. Mol. Biol. 53, 69 (1970).

BURGESS, R.R.: Separation and characterization of the subunits of ribonucleic acid polymerase. J. Biol. Chem. 244, 6168 (1969).

BURGESS, R.R., TRAVERS, A.A., DUNN, J.J., BAUTZ, E.K.F.: Factor stimulating transcription of RNA polymerase. Nature 221, 43 (1969).

CASCINO, A., GEIDUSHEK, E.P., CAFFERATA, R.L., HASELKORN, R.: T4 DNA replication and viral gene expression. J. Mol. Biol. 61, 357 (1971).

CASCINO, A., RIVA, S., GEIDUSCHEK, E.P.: DNA ligation and the coupling of T4 late transcription to replication. Cold Spring Harb. Symp. Quant. Biol. 35, 213 (1970).

CHAMBERLIN, M., McGARTH, J.: Characterization of a T7-specific RNA polymerase isolated from *E. coli* infected with T7 phage. Cold Spring Harbor Symp. Quant. Biol. 35, 259 (1970).

COHEN, S.S.: The biochemistry of viruses. Ann. Rev. Biochem. 32, 83 (1963).

COHEN, S.S.: Virus-induced enzymes. New York: Columbia University Press. 1968.

COHEN, P.S.: Regulation of cessation of early enzyme synthesis in bacteriophage T4-infected cells *E. coli* : Separate mechanism for different enzymes. Virology 41, 453 (1970).
COHEN, P.S.: Translational regulation deoxycytidylate hydroxymethylase and deoxynucleotide kinase synthesis in T4 infected cells. Virology 47, 780 (1972).
COHEN, P.S., ENNIS, H.L.: The requirement for potassium for bacteriophage protein and deoxyribonucleic acid synthesis. Virology 27, 282 (1965).
COHEN, P.S., NATALE, P.J., BUCHANAN, J.M.: Transcriptional regulation of T4 bacteriophage-specific enzymes synthesized *in vitro*. J. Virol. 14, 292 (1974).
COHEN, P.S., ZETTER, B.R., WALSH, M.L.: Evidence that more kinase mRNA is transcribed than translated during T4 infection of *E. coli*. Virology 49, 808 (1972).
CROUCH, R.J., HALL, B.D., HAGER, G.: Control of gene transcription in T-even bacteriophages: Alterations in RNA polymerase accompanying phage infection. Nature 223, 476 (1969).
DUBE, S.K., RUDLAND, P.S.: Control of translation by T4 phage: Altered binding of disfavoured messengers. Nature 223, 820 (1970).
DUNN, J.J., BAUTZ, E.K.F.: DNA-dependent RNA polymerase from *E. coli*: Studies on the role of σ in chain initiation. Biochem. Biophys. Res. Commun. 36, 925 (1969).
EDGAR, R.S., LIELAUSIS, I.: Some steps in the assembly of bacteriophage T4. J. Mol. Biol. 32, 263 (1968).
EDGAR, R.S., WOOD, W.B.: Morphogenesis of bacteriophage T4 in extracts of mutant-infected cells. Proc. Nat. Acad. Sci. (Wash.) 55, 498 (1966).
EDLIN, T.: Gene regulation during bacteriophage T4 development I. Phenotypic reversion of T4 amber mutations by 5-fluorouracil. J. Mol. Biol. 12, 363 (1965).
EMRICH, J.: Lysis of T4-infected bacteria in the absence of lysozyme. Virology 35, 158 (1968).
ENNIS, H.L.: Role of potassium in the regulation of polysome content and protein synthesis in *E. coli*. Arch. Biochem. Biophys. 143, 190 (1971).
ENNIS, H.L., LUBIN, M.: Dissociation of ribonucleic acid and protein synthesis in bacteria deprived of potassium. Biochim. Biophys. Acta 50, 399 (1961).
EPSTEIN, R.H., BOLLE, A., STEINBERG, C.M., KELLENBERGER, E., BOY de la TOUR, E., CHEVALLEY, R., EDGAR, R.S., SUSMAN, M., DENHARDT, E.H., LIELAUSIS, A.: Physiological studies of conditional lethal mutants of bacteriophage T4D. Cold Spring Harbor Symp. Quant. Biol. 28, 375 (1963).
FRANKEL, F.R., BATCHELER, F.R., CLARK, C.K.: The role of gene 49 in DNA replication and head morphogenesis in bacteriophage T4. J. Mol. Biol. 62, 439 (1971).
GEIDUSCHEK, E.P., SNYDER, L., COLVILL, A.J.E., SARNAT, M.: Selective synthesis of T-even bacteriophage early messenger RNA *in vitro*. J. Mol. Biol. 19, 541 (1966).
GESTELAND, R.F., SALSER, W., BOLLE, A.: *In vitro* synthesis of T4 lysozyme by suppression of amber mutations. Proc. Nat. Acad. Sci. (Wash.) 58, 2o36 (1967).
GOFF, C., WEBER, K.: A T4-induced RNA polymerase α-subunit modification. Cold Spring Harbor Symp. Quant. Biol. 35, 101 (1970).

GOLD, L.M., SCHWEIGER, M.: The initiation of T4 deoxyribonucleic acid dependent β-glucosyl transferase *in vitro*. J. Biol. Chem. 244, 5100 (1969a).

GOLD, L.M., SCHWEIGER, M: Synthesis of phage-specific and β-glucosyl transferase directed by T-even DNA *in vitro*. Proc. Nat. Acad. Sci. (Wash.) 62, 892 (1969b).

GOLD, L.M., SCHWEIGER, M: Control of β-glucosyl transferase and lysozyme synthesis of T4 deoxyribonucleic acid-dependent ribonucleic acid and protein synthesis *in vitro*. J. Biol. Chem. 245, 2255 (1970).

GOLDMAN, E., LODISH, H.F.: Specificity of protein synthesis by bacterial ribosomes and initiation factors. Absence of change after T4 infection. J. Mol. Biol. 67, 35 (1972).

GRASSO, R.J., BUCHANAN, J.M.: Synthesis of early RNA in bacteriophage T4 infected *E. coli*. Nature 224, 882 (1969).

GUHA, A., SZYBALSKI, W.: Fractionation of the complementary strands of coliphage T4 DNA based on the asymmetric distribution of Poly U and Poly U, G binding sites. Virology 34, 608 (1968).

GUHA, A., SZYBALSKI, W., SALSER, W., BOLLE, A., GEIDUSCHEK, E.P., PULITZER, J.: Controls and polarity of transcription during bacteriophage T4 development. J. Mol. Biol. 59, 329 (1971).

HALL, B.D., GREEN, M., NYGAARD, A.P., BOEZI, J.: The copying of DNA in T2-infected *E. coli*. Cold Spring Harbor Symp. Quant. Biol. 28, 201 (1963).

HALL, B.D., NYGAARD, A.P., GREEN, M.H.: Control of T2-specific RNA synthesis. J. Mol. Biol. 9, 143 (1964).

HALL, B.D., SPIEGELMAN, S.: Sequence complementarity of T2 DNA and T2-specific RNA. Proc. Nat. Acad. Sci. (Wash.) 47, 137 (1961).

HASELKORN, R., VOGEL, M., BROWN, R.D.: Conservation of rifampicin sensitivity of transcription during T4 development. Nature 221, 836 (1969).

HAYWARD, W.S., GREEN, M.H.: Inhibition of *E. coli* and bacteriophage lambda messenger synthesis by T4. Proc. Nat. Acad. Sci. (Wash.) 54, 1675 (1965).

HERCULES, K., MUNRO, J.L., MENDELSOHN, S., WIBERG, S.S.: Non-lethal mutations in bacteriophage T4 affecting degradation of host DNA. Fed. Proc. 29, 465 (1970).

HOSODA, J., LEVINTHAL, D.: Protein synthesis by *E. coli* infected with bacteriophage T4D. Virology 34, 709 (1968).

HSU,W.-T., WEISS, S.B.: Selective translation of T4 template RNA by ribosomes form T4-infected *E. coli* . Proc. Nat. Acad. Sci. (Wash.) 64, 345 (1969).

JAYARAMAN, R.: Transcription on bacteriophage T4 DNA by *E. coli* RNA polymerase *in vitro*: identification of some immediate early and delayed early genes. J. Mol. Biol. 70, 253 (1972).

JAYARAMAN, R., GOLDBERG, E.B.: A genetic assay for mRNA's for phage T4. Proc. Nat. Acad. Sci. (Wash.) 64, 198 (1969).

JAYARAMAN, R., GOLDBERG, E.B.: Transcription of bacteriophage T4 genome *in vivo*. Cold Spring Harbor Symp. Quant. Biol. 35, 197 (1970).

JORGENSEN, S.E., KOERNER, J.F., SHUSTAD, D.P., WARNER, H.R.: Isolation of phage T4 defective in host DNA degradation. Fed. Proc. 29, 465 (1970).

JOSSLIN, R.: Lysis mechanism of phage T4: Mutants affecting lysis. Virology 40, 719 (1970).

KAEMPFER, R.O.R., MAGASANIK, B.: Effect of T-even phage on the inducible synthesis of β-galactosidase in *E. coli*. J. Mol. Biol. 27, 453 (1967).
KANO-SUEOKA, T., SPIEGELMAN, S.: Evidence for a non-random reading of the genome. Proc. Nat. Acad. Sci. (Wash.) 48, 1942 (1962).
KARAM, J.D., BOWLES, M.G.: Mutation to overproduction of bacteriophage T4 gene products. J. Virol. 13, 428 (1974).
KASAI, T., BAUTZ, E.K.F.: Interdependence of translation and transcription in T4-infected *E. coli*. In: Organizational Macromolecules (Eds. H.J. VOGEL, J.O. LAMPEN, E. BRYSON), p. 111. New York: Academic Press 1967.
KASAI, T., BAUTZ, E.K.F.: Regulation of gene specific RNA synthesis in bacteriophage T4. J. Mol. Biol. 41, 401 (1969).
KASAI, R., BAUTZ, E.K.F., GUHA, A., SZYBALSKI, W.: Identification of the transcribing DNA strand for the rII and endolysin genes of coliphage T4. J. Mol. Biol. 34, 709 (1968).
KENNELL, D.: Inhibition of host protein synthesis during infection of *E. coli* by bacteriophage T4 I. Continued synthesis of host ribonucleic acid. J. Virol. 2, 1262 (1968).
KENNELL, D.: Inhibition of host protein synthesis during infection of *E. coli* by bacteriophage T4 II. Induction of host messenger ribonucleic acid and its exclusion from polysomes. J. Virol. 6, 208 (1970).
KENNELL, D., SIMMONS, C.: Synthesis and decay of mRNA from the *lac* operon of *E. coli* during amino acid starvation. J. Mol. Biol. 70, 451 (1972).
KHESIN, R.B.: Studies on RNA synthesis and RNA polymerase in normal and phage-infected *E. coli* cells. In: Lepefit Colloquium on RNA polymerase and transcription (Ed. L. SILVESTRI), p. 167. Amsterdam: North-Holland Publ. Co. 1970.
KHESIN, R.B., GORLENKO, Zh.M., SHEMYAKIN, M.B., BASS, I.A., PROSOROV, A.A.: Connection between protein synthesis and regulation of messenger RNA formation *E. coli* upon development of T2 phage. Biokhimiya. 28, 1070 (1963).
KHESIN, R.B., SHEMYAKIN, M.F.: Some properties of informational ribonucleic acids and their complexes with deoxyribonucleic acids. Biokhimiya. 27, 761 (1962).
KLEM, E.B., HSU, E.-T., WEISS, S.B.: The selective inhibition of protein initiation by T4 phage-induced factors. Proc. Nat. Acad. Sci. (Wash.) 67, 696 (1970).
KORNBERG, A.: Active center of DNA polymerase. Science 163, 1410 (1969).
LAEMMLI, E.K., BEGUIN, F., GUJER-KELLENBERGER, G.: A factor preventing the major head protein of bacteriophage T4 from random aggregation. J. Mol. Biol. 47, 69 (1970).
LANDY, A., SPIEGELMAN, S.: Exhaustive hybridization and its application to an analysis of ribonucleic acid synthesized in T4-infected cells. Biochemistry 7, 585 (1968).
LEE-HUANG, S., OCHOA, S.: Messenger discriminating species of factor F3. Nature 234, 236 (1971).
LEMBACH, K.J., BUCHANAN, J.M.: The relationship of protein synthesis to early transcriptive vents in bacteriophage T4-infected *E. coli* B. J. Biol. Chem. 245, 1575 (1970).
LEMBACH, K.J., KUNINAHA, A., BUCHANAN, J.M.: The relationship of DNA replication to the control of protein synthesis in protoplasts of T4-infected *E. coli* B. Proc. Nat. Acad. Sci. (Wash.) 62, 446 (1969).

LEVIN, A.P., BURTON, K.: Inhibition of enzyme formation following infection of *E. coli* with phage T2r$^+$. J. Gen. Microbiol. 25, 307 (1961).
LUBIN, M., ENNIS, H.L.: On the role of intracellular potassium in protein synthesis. Biochim. Biophys. Acta 80, 614 (1964).
LUBIN, M., KESSEL, D.: Preliminary mapping of the genetic locus for potassium transport in *E. coli*. Biochem. Biophys. Res. Comm. 2, 249 (1960).
MATHEWS, C.K.: Deoxyribonucleic acid metabolism and virus-induced enzyme synthesis in a thymine-requiring bacterium infected by a thymine-requiring bacteriophage. Biochemistry 5, 2092 (1962).
MATHEWS, C.K.: Bacteriophage Biochemistry. New York: Van Norstrand-Reinhold Co. 1971.
MILANESI, G., BRODY, E.N., GEIDUSCHEK, E.P.: Sequence of the *in vitro* transcription of T4 DNA. Nature 221, 1041 (1969).
MILANESI, G., BRODY, E.N., GRAU, O., GEIDUSCHEK, E.P.: Transcription of the bacteriophage T4 template *in vitro*: separation of "delayed-early" from "immediate-early" transcription. Proc. Nat. Acad. Sci. (Wash.) 66, 181 (1970).
MILLETTE, R.L., TROTTER, C.D.: The initiation and release of RNA by DNA-dependent RNA polymerase. Proc. Nat. Acad. Sci. (Wash.) 66, 701 (1970).
MILLETTE, R.L., TROTTER, C.D., HERRLICH, P., SCHWEIGER, M.: *In vitro* synthesis, termination, and release of active messenger RNA. Cold Spring Harbor Symp. Quant. Biol. 35, 135 (1970).
MORSE, D.E.: "Delayed-early" mRNA for the tryptophan operon? An effect of chloramphenicol. Cold Spring Harbor Symp. Quant. Biol. 35, 495 (1970).
MORSE, D.E., PRIMAKOFF, P.: Relief of polarity in *E. coli* by "suA". Nature 226, 28 (1970).
NATALE, P.J., BUCHANAN, J.M.: DNA-directed synthesis *in vitro* of T4 phage-specific enzymes. Proc. Nat. Acad. Sci. (Wash.) 69, 2513 (1972).
NOMURA, M., HALL, B.D., SPIEGELMAN, S.: Characterization of RNA synthesized in *E. coli* after bacteriophage T2 infection. J. Mol. Biol. 2, 306 (1960).
NOTANI, G.W.: Regulation of bacteriophage T4 gene expression. J. Mol. Biol. 73, 231 (1973).
NYGAARD, A.P., HALL, B.D.: Formation and properties of RNA-DNA complexes. J. Mol. Biol. 9, 125 (1964).
O'FARRELL, P.Z., GOLD, L.M.: Bacteriophage T4 gene expression: Evidence for two classes of prereplicative cistrons. J. Biol. Chem. 248, 5502 (1973a).
O'FARREL, P.Z., GOLD, L.M.: Transcription and translation of prereplicative bacteriophage T4 genes *in vitro*. J. Biol. Chem. 248, 5512 (1973b).
OISHI, M.: Studies of DNA replication *in vivo* III. Accumulation of a single-strand isolation product of DNA replication by conditional lethal mutants of T4. Proc. Nat. Acad. Sci. (Wash.) 60, 1000 (1968).
OLESON, A.E., PISPA, J.P., BUCHANAN, J.M.: Transient stimulation of RNA polymerase in *E. coli* after infection with bacteriophage T4. Proc. Nat. Acad. Sci. (Wash.) 63, 473 (1969).
PASTUSHOK, C., KENNELL, D.: Residual polarity and transcription/translation coupling during recovery from chloramphenicol or fusidic acid. J. Bacteriol. 117, 631 (1974).

PETERSON, R.F., COHEN, P.S., ENNIS, H.L.: Properties of phage T4 messenger RNA synthesized in the absence of protein synthesis. Virology 48, 201 (1972).
POLLACK, Y., GRONER, Y., AVIV, H., REVEL, M.: Role of initiation factor B (F3) in the preferential translation of T4 late messenger RNA in T4 infected *E. coli*. FEBS Letters 9, 218 (1970).
PULITZER, J.F.: Function of T4 gene 55 I. Characterization of temperature-sensitive mutations in the "maturation" gene 55. J. Mol. Biol. 49, 473 (1970).
RAY, P., SINHA, K., WARNER, H.R., SNUSTAD, D.P.: Genetic location of a mutant of bacteriophage T4 deficient in the ability to induce endonuclease II. J. Virol. 9, 184 (1972).
REVEL, M., AVIV, H., GRONER, Y., POLLACK, Y.: Fractionation of translation initiation factor B (F3) into cistron specific species. FEBS Letters 9, 213 (1970).
RICHARDSON, J.P.: Rho factor function in T4 RNA transcription. Cold Spring Harbor Symp. Quant. Biol. 35, 127 (1970a).
RICHARDSON, J.P.: Reinitiation of RNA synthesis *in vitro*. Nature 225, 1109 (1970c).
RICHARDSON, J.P.: Rates of bacteriophage T4 RNA chains growth *in vitro*. J. Mol. Biol. 49, 235 (1970b).
RIVA, S., CASCINO, A., GEIDUSCHEK, E.P.: Coupling of late transcription to viral DNA replication in bacteriophage T4 development. J. Mol. Biol. 54, 85 (1970a).
RIVA, S., CASCINO, A., GEIDUSCHEK, E.P.: Uncoupling of late transcription from DNA replication in bacteriophage T4 development. J. Mol. Biol. 54, 103 (1970b).
ROBERTS, J.W.: Termination factor for RNA synthesis. Nature 224, 1168 (1969).
ROUVIERE, J., WYNGAARDEN, J., CANTONI, J., GROS, R., KEPES, A.: Effect of T4 infection on messenger RNA synthesis in *E. coli*. Biochim. Biophys. Acta 166, 94 (1968).
SAKIYAMA, S., BUCHANAN, J.M.: *In vitro* synthesis of deoxynucleotide kinase programmed by bacteriophage T4 RNA. Proc. Nat. Acad. Sci. (Wash.) 68, 1376 (1971).
SAKIYAMA, S., BUCHANAN, J.M.: Control of the synthesis of T4 phage deoxynucleotide kinase messenger ribonucleic acid *in vivo*. J. Biol. Chem. 247, 7806 (1972).
SAKIYAMA, S., BUCHANAN, J.M.: Relationship between molecular weight of T4 phage-induced deoxynucleotide kinase and the size of its messenger. J. Biol. Chem. 248, 3150 (1973).
SALSER, W., BOLLE, A., EPSTEIN, R.H.: Transcription during bacteriophage T4 development: A demonstration that distinct subclasses of "early" RNA appear at different times and that some are "turned off" at late times. J. Mol. Biol. 49, 271 (1970).
SALSER, W., GESTELAND, R.F., BOLLE, A.: *In vitro* syntheses of bacteriophage lysozyme. Nature 215, 588 (1967).
SALSER, W., GESTELAND, R.F., RICARD, B.: Characterization of lysozyme messenger and lysozyme synthesized *in vitro*. Cold Spring Harbor Symp. Quant. Biol. 34, 771 (1969).
SAUERBIER, W., HERCULES, K.: Control of gene function in bacteriophage T4 IV. Post-transcriptional shut-off of expression of early genes. J. Virol. 12, 538 (1973).
SCHACHNER, M., SEIFERT, W., ZILLIG, W.: A correlation of changes in host and T4 bacteriophage specific RNA synthesis with changes of DNA-dependent in *E. coli* infected with bacteriophage T4. Eur. J. Biochem. 22, 520 (1971).

SCHACHNER, M., ZILLIG, W.: Fingerprint maps of tryptic peptides from subunits of *E. coli* and T4-modified DNA-dependent RNA polymerase. Eur. J. Biochem. 22, 513 (1971).
SCHERBERG, N.H., WEISS, S.B.: T4 transfer RNA's: Codon regognition and translational propterties. Proc. Nat. Acad. Sci. (Wash.) 69, 1114 (1972).
SCHMIDT, D.A., MAZAITIS, A.J., KASAI, T., BAUTZ, E.K.F.: Involvement of a phage T4 factor and an anti-terminator protein in the transcription of early T4 genes *in vivo*. Nature 25, 1012 (1970).
SCHWEIGER, M., GOLD, L.M.: Bacteriophage T4 DNA-dependent *in vitro* synthesis of lysozyme. Proc. Nat. Acad. Sci. (Wash.) 63, 1351 (1969).
SCHWEIGER, M., GOLD, L.M.: *E. coli* and Bacillus subtilis phage deoxyribonucleic acid-directed deoxycytidylate deaminase synthesis in *E. coli* extracts. J. Biol. Chem. 245, 5022 (1970).
SEDEROFF, R., BOLLE, A., EPSTEIN, R.H.: A method for the detection of specific T4 messenger RNAs hybridization competition. Virology 45, 440 (1971).
SEDEROFF, R., BOLLE, A., GOODMAN, H.M., EPSTEIN, R.H.: Regulation of rII and region D transcription in T4 bacteriophage: A sucrose gradient analysis. Virology 46, 817 (1971).
SEIFERT, W.: Changes in structure and function of RNA polymerase from *E. coli* after T4 infection. In: Lepetit Colloq. on RNA polymerase and transcription (Ed. L. SILVESTRI), p. 158. Amsterdam: North-Holland Publ. Co. 1970.
SEIFERT, W., QUASBA, P., WALTER, G., PALM, P., SCHACHNER, M., ZILLIG, W.: Kinetics of alteration and modification of DNA-dependent RNA polymerase in T4-infected *E. coli* cells. Europ. J. Biochem. 9, 319 (1969).
SEKIGUCHI, M., COHEN, S.S.: The synthesis of messenger RNA without protein synthesis II. Synthesis of phage-induced RNA and sequential enzyme production. J. Mol. Biol. 8, 638 (1964).
SHEDL, P.D., SINGER, R.E., CONWAY, T.W.: A factor required for the translation of F2 RNA in extracts of T4-infected cells. Biochem. Biophys. Res. Commun. 38, 631 (1970).
SHER, I.H., MALLETTE, M.S.: The adaptive nature of the formation of lysine decarboxylase in *E. coli*. Arch. Biochem. Biophys. 52, 331 (1954).
SKÖLD, O.: Regulation of early RNA synthesis in bacteriophage T4-infected *E. coli* cells. J. Mol. Biol. 53, 339 (1970).
SMITH, F.L., HASELKORN, R.: Proteins associated with ribosomes in T4-infected *E. coli*. Cold Spring Harbor Symp. Quant. Biol. 34, 91 (1969).
SNYDER, L., GEIDUSCHEK, E.P.: *In vitro* synthesis of T4 late messenger RNA. Proc. Nat. Acad. Sci. (Wash.) 59, 460 (1968).
STEVENS, Audrey: New small polypeptides associated with DNA-dependent RNA polymerase of *E. coli* after infection with bacteriophage T4. Proc. Nat. Acad. Sci. (Wash.) 69, 603 (1972).
STREISINGER, G., MUKAI, F., DREYER, W.J., MILLER, R., HORIUCHI, S.: Mutations affecting the lysozyme of phage T4. Cold Spring Harbor Symp. Quant. Biol. 26, 25 (1961).
SUGIURA, M., OKAMOTO, T., TAKANAMI, M.: RNA polymerase factor and the selection of initiation site. Nature 225, 598 (1970).
STAHL, F.W., CASEMANN, J.M., YEGIAN, C., STAHL, M.M., NAKATA, A.: Co-transcribed cistrons in bacteriophage T4. Genetics 64, 157 (1970).

TESSMAN, I., GREENBERG, D.B.: Ribonucleotide reductase genes of phage T4: Map location of the thioredoxin gene nrdC. Virology 49, 337 (1972).
TRAVERS, A.A.: Bacteriophage sigma factor for RNA polymerase. Nature 223, 1107 (1969).
TRAVERS, A.A.: RNA polymerase and T4 development. Cold Spring Harbor Symp. Quant. Biol. 35, 241 (1970a)
TRAVERS, A.A.: Positive control of transcription by a bacteriophage sigma factor. Nature 225, 1009 (1970b).
TRAVERS, A.A., BURGESS, R.R.: Cyclic reuse of the RNA polymerase sigma factor. Nature 222, 537 (1969).
TRIMBLE, R.B., GALIVAN, J., MALEY, F.: The temporal expression of $T2r^+$ bacteriophage genes *in vivo* and *in vitro*. Proc. Nat. Acad. Sci. (Wash.) 69, 1659 (1972).
TRIMBLE, R.B., MALEY, F.: *In vitro* synthesis of deoxynucleotide kinase, dihydrofolate reductase, and deoxycytidylate hydroxymethylase from RNA transcripts of T2 phage DNA. Biochem. Biophys. Res. Commun. 52, 1063 (1973).
TSUGITA, A., INOUYE, M.: Complete primary structure of phage lysozyme from *E. coli* phage T4. J. Mol. Biol. 44, 415 (1968).
VALLEE, M., CORNETT, J.B.: A new gene of bacteriophage T4 determining immunity against superinfecting ghosts and phage in T4-infected *E. coli*. Virology 48, 777 (1972).
VOLIN, E., ASTRACHAN, L.: Phosphorus incorporation in *E. coli* ribonucleic acid after infection with bacteriophage T2. Virology 2, 149 (1956).
WAIS, A.C., GOLDBERG, E.B.: Growth and transformation of phage T4 in *E. coli* B/4, Salmonella, Aerobacter, Proteus, and Serratia. Virology 39, 153 (1969).
WALTER, G., SEIFERT, W., ZILLIG, W.: Modified DNA-dependent RNA polymerase from *E. coli* infected with bacteriophage T4. Biochem. Biophys. Res. Commun. 30, 240 (1968).
WARNER, H.R., SNUSTAD, D.P., KOERNER, J.F., CHILDS, J.D.: Identification and genetic characterization of mutants in bacteriophage T4 defective in the ability to induce exonuclease A. J. Virol. 9, 399 (1972).
WATSON, J.D.: Molecular Biology of the Gene, 2nd ed. New York: W.A. Benjamin Inc. 1970.
WIBERG, J.S.: Mutants of bacteriophage T4 unable to cause breakdown of host DNA. Proc. Nat. Acad. Sci. (Wash.) 55, 614 (1966).
WIBERG, J.S., DIRKSEN, M.L., EPSTEIN, R.H., LURIA, S.E., BUCHANAN, J.M.: Early enzyme synthesis and its control in *E. coli* infected with some amber mutants of bacteriophage T4. Proc. Nat. Acad. Sci. (Wash.) 48, 292 (1962).
WIBERG, J.S., MENDELSOHN, S., WARNER, V., HERCULES, K., ALDRICH, C., MUNRO, J.L.: SP62, a viable mutant of bacteriophage T4D defective in regulation of phage early enzyme synthesis. J. Virol. 12, 775 (1973).
WILHELM, J.M., HASELKORN, R.: *In vitro* synthesis of T4 proteins: Lysozyme and the products of genes 22 and 57. Cold Spring Harbor Symp. Quant. Biol. 34, 793 (1969).
WILHELM, J.M., HASELKORN, R.: *In vitro* synthesis of T4 proteins: The products of genes 9, 18, 19, 23, 24, and 38. Virology 43, 198 (1970a).
WILHELM, J.M., HASELKORN, R.: *In vitro* synthesis of T4 proteins: Control of transcription of gene 57. Virology 43, 209 (1970b).

WARNER, H.R., SNUSTAD, D.P., KOERNER, J.F., CHILDS, J.D.: Identification and genetic characterization of mutants in bacteriophage T4 defective in the ability to induce exonuclease A. J. Virol. 9, 399 (1972).

WITMER, H.J.: Effect of ionic strenght and temperature on the *in vitro* transcription of T4 DNA. Biochim. Biophys. Acta 246, 29 (1971a).

WITMER, H.J.: *In vitro* transcription of T4 deoxynucleic acid by *E. coli* ribonucleic acid polymerase: Sequential transcription of immediate early and delayed early cistrons in the absence of the release factor, rho. J. Biol. Chem., 246, 5220 (1971b).

WOOD, W.B., HENNINGER, M.: Attachment of tail fibers in bacteriophage T4 assembly: Some properties of the reaction *in vitro* and its genetic control. J. Mol. Biol. 39, 603 (1969).

YOUNG, E.T.: Control of functional messenger synthesis. Cold Spring Harbor Symp. Quant. Biol. 35, 189 (1970a).

YOUNG, E.T.: Cell free synthesis of bacteriophage glucosyl transferase. J. Mol. Biol. 51, 591 (1970b)

YOUNG, E.T., VANHOWE, G.: Control of synthesis of functional glucosyl transferase and lysozyme messenger after T4 infection. J. Mol. Biol. 51, 605 (1970).

ZILLIG, W., FUCHS, E., PALM, P., RABUSSAY, D., ZECHEL, K.: On the different subunits of DNA-dependent RNA polymerase of *E. coli* and their role in the complex function of the enzyme. In: Lepetit Colloq. on RNA polymerase and transcription (Ed. L. SILVESTRI), p. 151. Amsterdam: North Holland Publ. Co. 1970a.

ZILLIG, W., ZECHEL, K., RABUSSAY, D., SCHACHNER, M., SETHI, V.S., PALM, P., HEIL, A., SEIFERT, W.: On the role of different subunits of DNA-dependent RNA polymerase from *E. coli* in the transcription process. Cold Spring Harbor Symp. Quant. Biol. 35, 47 (1970b).

Defective Bacteriophages: The Phage Tail-like Particles

Wolfgang Lotz

I. Introduction

A bacterium is called lysogenic if it has the latent potential to produce bacteriophage, i.e. if it harbors and reproduces the genome of the phage ("prophage") along with that of the bacterium. Lysogeny is detected by the fact that this potential is occasionally expressed, so that any culture produced by growing the progeny of a single lysogenic cell will contain some phage (CAMPBELL, 1968).

In recent years it has become more and more apparent that not only intact, but also defective lysogeny is a wide-spread phenomenom among the bacteria (BRADLEY, 1967; GARRO and MARMUR, 1970). It is generally inferred that a given bacterial strain is defectively lysogenic and harbors a defective phage genome, if it has the abbility to produce non-viable "phage-like" particles.

A more detailed definition of such particles has been given by GARRO and MARMUR (1970), who suggested that defective bacteriophages are particles which contain either all or some of the normal phage components, but which fail to form plaques on any known host, even at high multiplicities of infection.

As pointed out by CAMPBELL (1968), the presence of a defective prophage can be detected (1) if it confers to its host immunity to superinfecting phages of the homologous type, and/or (2), if it recombines with a genetically marked superinfecting phage. Both operations can only be applied where the intact counterpart to the defective prophage is known. For example, in the case of the phage Lambda/*E. coli* system, sophisticated techniques for the study of defective phages have been developed (CAMPBELL, 1968).

In contrast, the intact counterparts are not known for "naturally" occurring defective prophages. In this case the detection of defective lysogeny relies heavily on electron microscopic observations, which are necessary for the detection and characterization of the phage-like particles. It is conceivable that an extensively defective prophage coding only for submicroscopic particles would be "overlooked" with this method. This may explain why most of the defectively lysogenic strains so far described in the literature are producers of relatively large phage-like structures, such as phage tails with or without heads.

GARRO and MARMUR (1970) divided the presently known defective bacteriophages into the following three groups: (I) particles which contain almost exclusively DNA of the bacterial host; (II) those containing phage-specific DNA; and (III) particles which may not contain DNA and which resemble bacteriophage tails without a head.

Of these groups, the last one will be dealt with in the review. Most of the defective bacteriophages so far described fall into this category. Due to the absence of a phage head containing nucleic acid, the tail-like particles can be clearly and relatively easily classified as "defective". They are furthermore of special interest, because many of them have bactericidal activity and can be regarded as "high molecular weight bacteriocins" (BRADLEY, 1967).

Of the group III defective phages, the "pyocin R" or pyocin R-type particles, which are produced by *Pseudomonas aeruginosa* have been studied most extensively. Constant references is therefore made to these particles. The review covers four main aspects: (I) the structure, (II) the production of the defective phages, (III) their bactericidal activity, and (IV) their possible evolution from intact bacteriophages.

Two classical reviews have been published in recent years covering various topics on defective bacteriophages: a review by BRADLEY (1967) on the "Ultrastructure of bacteriophages and bacteriocins", and a review by GARRO and MARMUR (1970) on "Defective bacteriophages". In addition, TIKHONENKO (1970) covered various structural aspects of defective bacteriophages in her book on the "Ultrastructure of bacterial viruses", and BRANDIS and ŠMARDA (1971) presented a review on the bactericidal pyocin R-type particles in their book on "Bacteriocins and bacteriocin-like substances". The action of phage-like bacteriocins has been discussed in NOMURA's review on "Colicins and related bacteriocins" (1967) and in REEVES' book on "The bacteriocins" (1972). Recently, a comprehensive survey on "Particulate bacteriocins" has been presented by ACKERMANN and BROCHU (1973).

II. Structure

For the purpose of this review, the phage tail-like particles are divided into three groups: (I) particles which resemble sheathed phage tails, (II) particles resembling "long tubes", and (III) particles resembling sheathless phage tails. Bacterial strains know to produce these structures are listed in Table 1.

A. Electron Microscopical Characterization

1. Structure of Particles Resembling Sheathed Phage Tails

A first description of phage-like particles was presented by G. CHAPMAN, J. HILLIER, and F.H. JOHNSON in 1951. Electron microscopy of chromium-shadowed particles from single plaques of

Table 1. Bacterial strains producing bacteriophage tail-like particles

Host	Morphology	References
Aerobacter cloacae	sheathed tails	BRADLEY, 1967
Archangium violaceum	sheathed tails 'long tubes'	REICHENBACH, 1967
Chondrococcus columnaris	'long tubes'	PATE et al., 1967
Chromobacterium violaceum	sheated tails 'long tubes'	ACKERMANN and GAUVREAU, 1972 RUCINSKY et al., 1972
Clostridium botulinum	sheathed tails 'long tubes'	IIDA and INOUE, 1968 INOUE and IIDA, 1968 LAU et al., 1974 UEDA and TAKAGI, 1972
Erwinia carotovora	sheathed tails	CHAPMAN et al., 1951
Escherichia coli	sheathed tails	BRADLEY, 1967
Hydrogenomonas eutropha	'long tubes'	MAYER and AULING, unpublished
Listeria monocytogenes	sheathed tails	BRADLEY and DEWAR, 1966
Photobacterium harveyi	'long tubes'	YAMAMOTO, 1967
Proteus mirabilis	sheathed tails 'long tubes'	ITERSON et al., 1967 TAUBENECK, 1963, 1967 YAMAMOTO, 1967
Proteus vulgaris	sheathed tails 'long tubes'	COETZEE et al., 1968 YAMAMOTO, 1967
Pseudomonas aeruginosa	sheathed tails 'long tubes' sheathless tails	BRADLEY and DEWAR, 1966 HIGERD et al., 1967, 1969 HOMMA et al., 1967 ISHII et al., 1965 KAGEYAMA, 1964 TAKEYA et al., 1967
Pseudomonas fluorescens	sheathed tails 'long tubes' sheathless tails	AMAKO et al., 1970 YAMAMOTO et al., 1967
Rhizobium spec.	sheathed tails 'long tubes' sheathless tails	GISSMANN, 1974 LOTZ and MAYER, 1972 PFISTER, 1974
Rhizobium meliloti	sheathed tails	RAUTENSTEIN and MOSKALENKO, 1970
Saprospira grandis	sheathed tails	DELK and DEKKER, 1972 LEWIN, 1963 LEWIN and KIETHE, 1965
Serratia marcescens	sheathed tails sheathless tails	TRAUB, 1972
Spirillum itersonii	sheathed tails 'long tubes'	CLARK-WALKER, 1969
Spirulina	sheathed tails 'long tubes'	CHANG and ALLEN, 1974

Table 1 (continued)

Host	Morphology	References
Sporocytophaga myxococcoides	sheathed tails	GRÄF, 1965
Streptococcus faecalis	'long tubes' sheathless tails	BRADLEY, 1967
Vibrio colera	sheathed tails 'long tubes'	LANG et al., 1968
Vibrio comma	sheathed tails 'long tubes'	FARKAS-HIMSLEY et al., 1971 JAYAWARDENE and FARKAS-HIMSLEY, 1968
Vibrio eltor	'long tubes'	ADHIKARI and CHATTERJEE, 1972

an *Erwinia carotovora* bacteriophage "revealed an extraordinary diversity of phage and phage-like particles, whereas a single type had been expected".

Since the method of negative staining was not known then, the study of the fine structure of these particles was limited. Nevertheless, it can be concluded from the electron micrographs presented by the authors that, besides the intact bacteriophages, particles resembling contractile phage tails were produced by the bacterial strain.

Since these early studies, other bacterial species were found to produce particles resembling sheathed phage tails. The dimensions of these particles are summarized in Table 2 (p. 58/59). The structures consist of two cylinders fitting into one another: i.e. an inner core which is surrounded by a contractile sheath. The latter generally carries a base plate with six short fibers.

A comparison of values in the columns of Table 2 indicates that the particles obtained from different and unrelated bacterial strains have similar or identical core diameters (6 - 10 nm), sheath diamters (15 - 20 nm; in the extended state) and lengths of fibers (20 - 40 nm). The examples of sheathed phage tail-like particles shown in Fig. 1 demonstrate this similarity in general appearance.

The fibers of particles from *P. aeruginosa* (HIGERD et al., 1969), *Chromobacterium violaceum* (ACKERMANN and GAUVREAU, 1972), and *Rhizobium* strain 16-3 (LOTZ and MAYER, 1972) were often found to carry small bodies spherical in appearance at their free ends (Fig. 1b). The diamter of these bodies has been determined as 1 nm and 3,7 nm for the particles produced by *P. aeruginosa* and *Rhizobium* strain 16-3 respectively.

The particles (so-called "rhapidosomes"; LEWIN, 1963) produced by *Saprospira grandis*, a procaryotic marine micro-organism, are exceptional, because their sheaths apparently exist only in the contracted state and have neither a base plate nor fibers

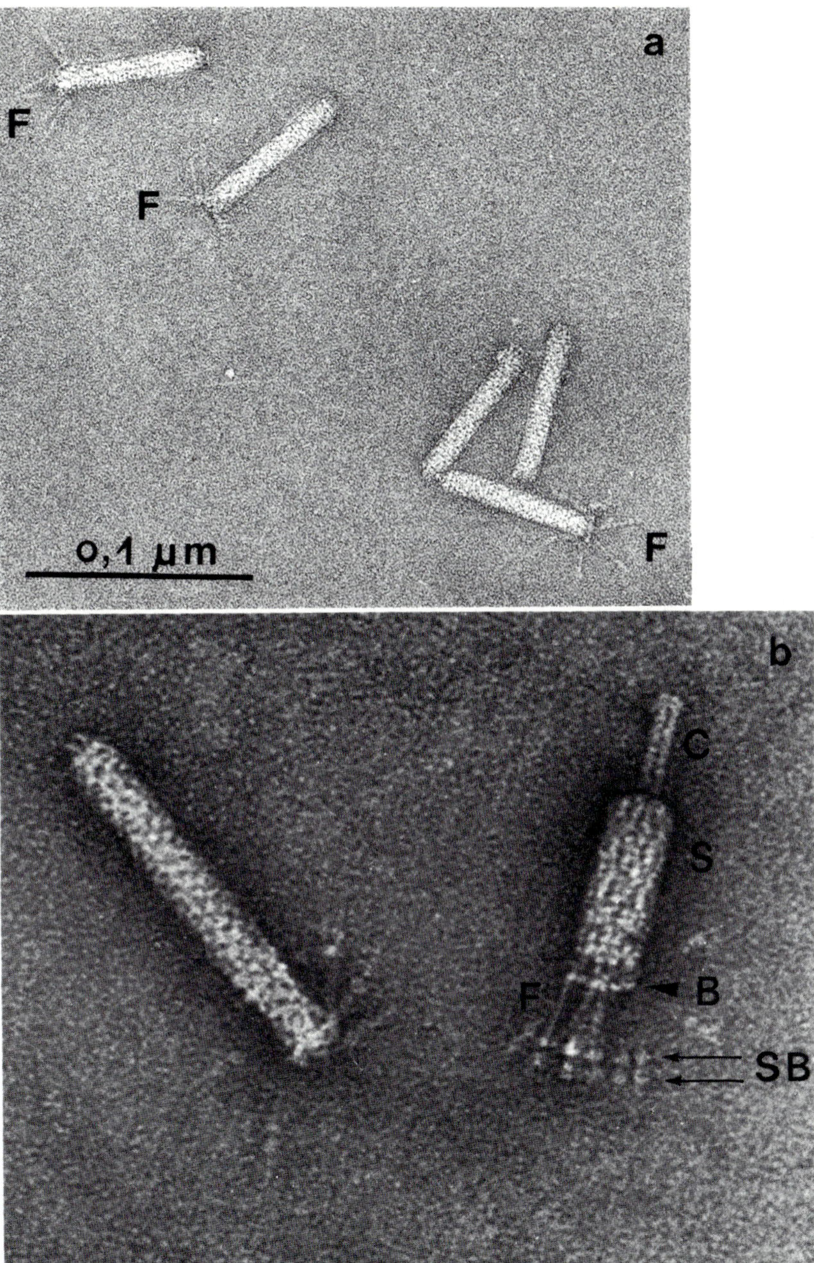

Fig. 1 a and b. Electron micrographs of particles resembling sheathed phage tails. (a) Particles from *Serratia marcescens*, negative staining was with uranyl acetate (URAC) (TRAUB and ACKER, unpublished). (b) Particles from *Rhizobium spec*. with an extended and a contracted sheath, respectively; URAC; magnification: 428.000x (GISSMANN, unpublished). S = sheath, C = core, B = base plate, F = fiber, SB = "spherical bodies" attached to free ends of fibers

Table 2. Particles resembling sheathed phage tails[a]

Host	Extended Particles		Contracted Particles				Baseplate with fibers[b]	References
	Diameter of sheath	Total length	Sheath		Core			
			Diameter	Length	Diameter	Length		
Archangium violaceum	17,5	210	22	100	7,8		+	REICHENBACH, 1967
Chromobacterium violaceum	16,0 16,5	218,5 144	21 19	83,5 64	7,0 6,0	218,5 144	+ +	ACKERMANN and GAUVREAU, 1972
Listeria monocytogenes			20	160	7,5	240	+	BRADLEY and DEWAR, 1966
Proteus mirabilis	20	125-132,5			7,5		+	TAUBENECK, 1967
Proteus vulgaris	18	128	20	56	7			COETZEE et al., 1968
Pseudomonas aeruginosa	15	120	18	46	5,7	120	+	ISHII et al., 1965
Pseudomonas fluorescens	15		18					AMAKO et al., 1970
Rhizobium spec.	16,5	123	24	67	7,0	119	+	LOTZ and MAYER, 1972
Rhizobium meliloti		160	16		5,5			RAUTENSTEIN and MOSKALENKO, 1970
Saprospira grandis			28	200	9,0	430	- !	DELK and DEKKER, 1972
Serratia marcescens	17	124	22	50	8	124	+	TRAUB, 1972
Spirillum itersonii	20	140	24	60			+	CLARK-WALKER, 1969
Spirulina			23-24	(139)[c]	7,5	350	- !	CHANG and ALLEN, 1974
Sporocytophaga myxococcoides	15-20	100-200						GRÄF, 1965
Vibrio cholera			22		10			LANG et al., 1968

Vibrio comma	20	110	24	9-10	JAYAWARDENE and FARKAS-HIMSLEY, 1968, 1969

[a]Measurements in nanometer units. [b]Fibers of most particles are 20 - 40 nm long.
[c]Length of the contracted sheath was calculated by this author from the particles shown in Fig. 5 of CHANG and ALLEN, 1974.

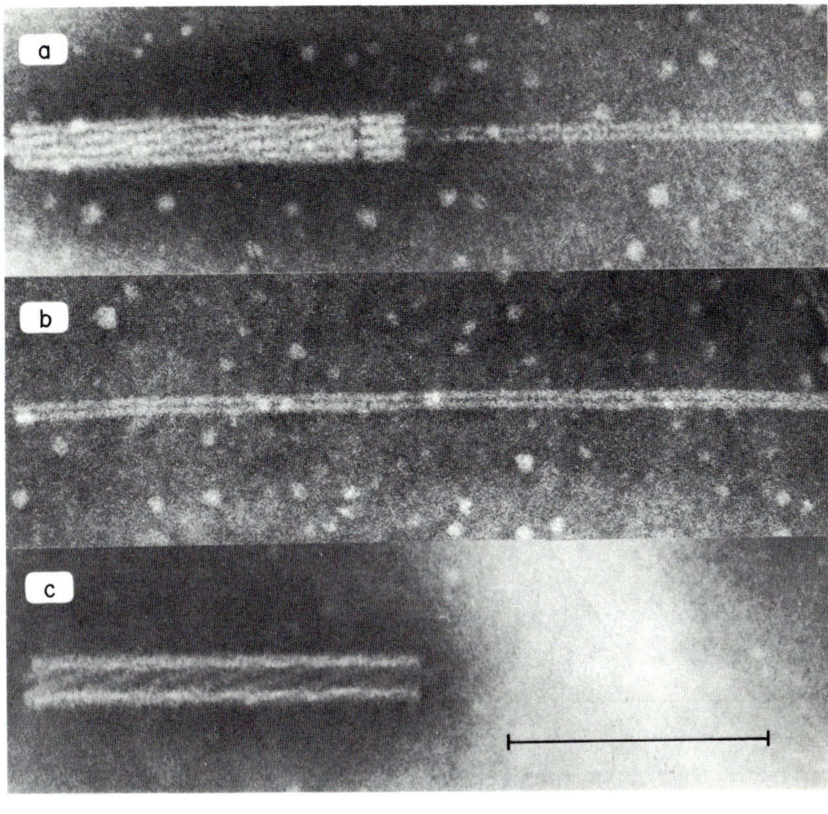

Fig. 2a-c. Electron micrograph of the rhapidosome particles produced by *Saprospira grandis* (DELK and DEKKER, 1972). (a) Intact particle, (b) core, (c) sheath; negative staining was with phosphotungstic acid (PTA); marker = 130 nm

(Fig. 2) (DELK and DEKKER, 1972; REICHLE and LEWIN, 1968). Recently, it has been shown that rhapidosome-like structures are also produced by the marine photoautotrophic blue-green alga *Spirulina* (CHANG and ALLEN, 1974).

After negative staining the contracted sheats of many particles show prominent helical grooves which indicate a helical arrangement of the sheath subunits. Such an arrangement has been confirmed (PFISTER, 1974) for the extended and contracted sheaths of the particles from *Rhizobium* ("INCO" particles; LOTZ and MAYER, 1972) by analyzing corresponding electron micrographs with the optical diffraction technique (KLUG and BERGER, 1964).

The resulting Figs. 3a and 4a show the Fraunhofer diffraction patterns of the high resolution electron micrographs presented in Figs. 3b and 4b. The main

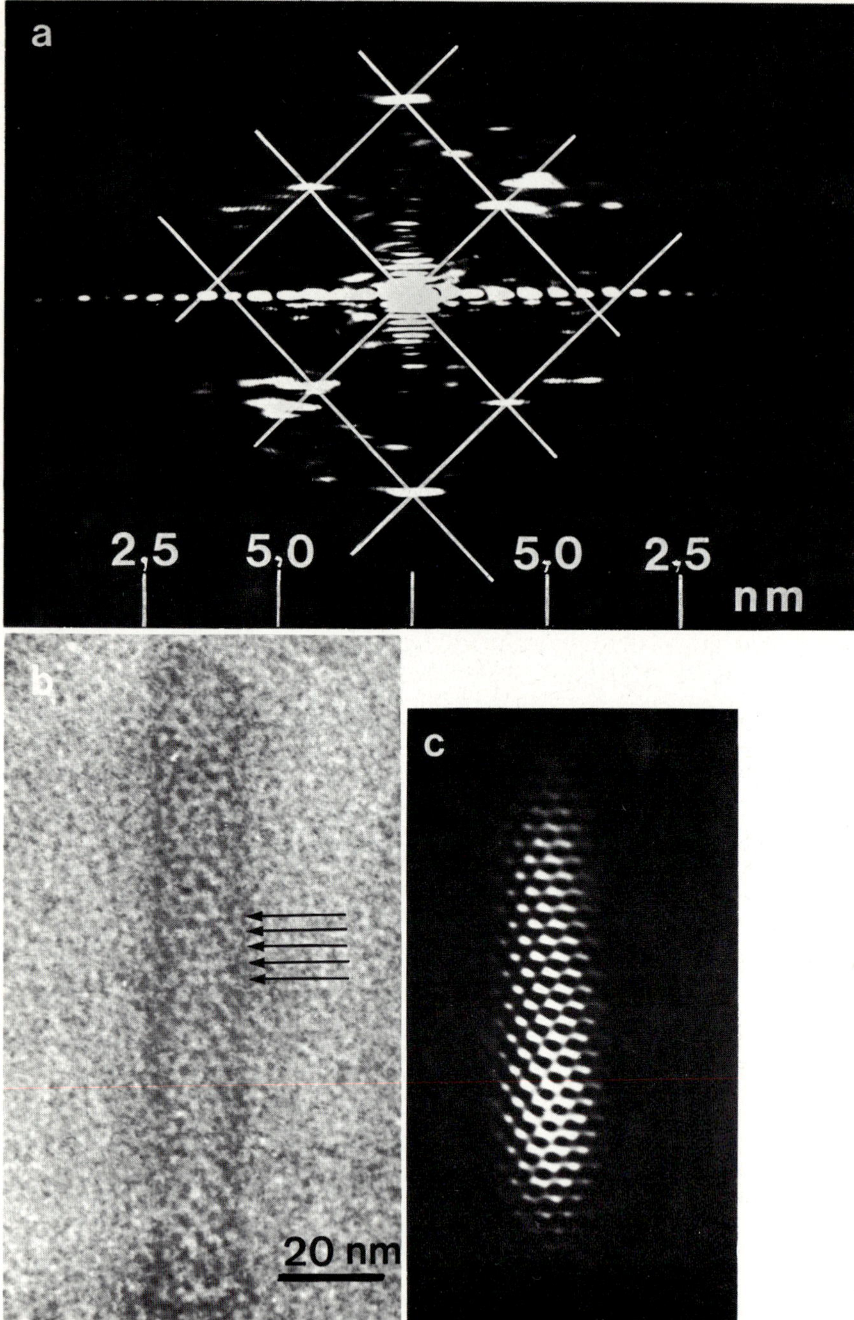

Fig. 3. (a) Optical diffraction pattern of the extended sheath of the INCO particle shown in (b); a reciprocal lattice corresponding to one of the two faces of the sheath is shown. (c) Reconstructed image of the extended sheath. Arrows in (b) point to cross-striations which probably represent the annuli mentioned in Section II.A.1.; URAC, (PFISTER, 1974)

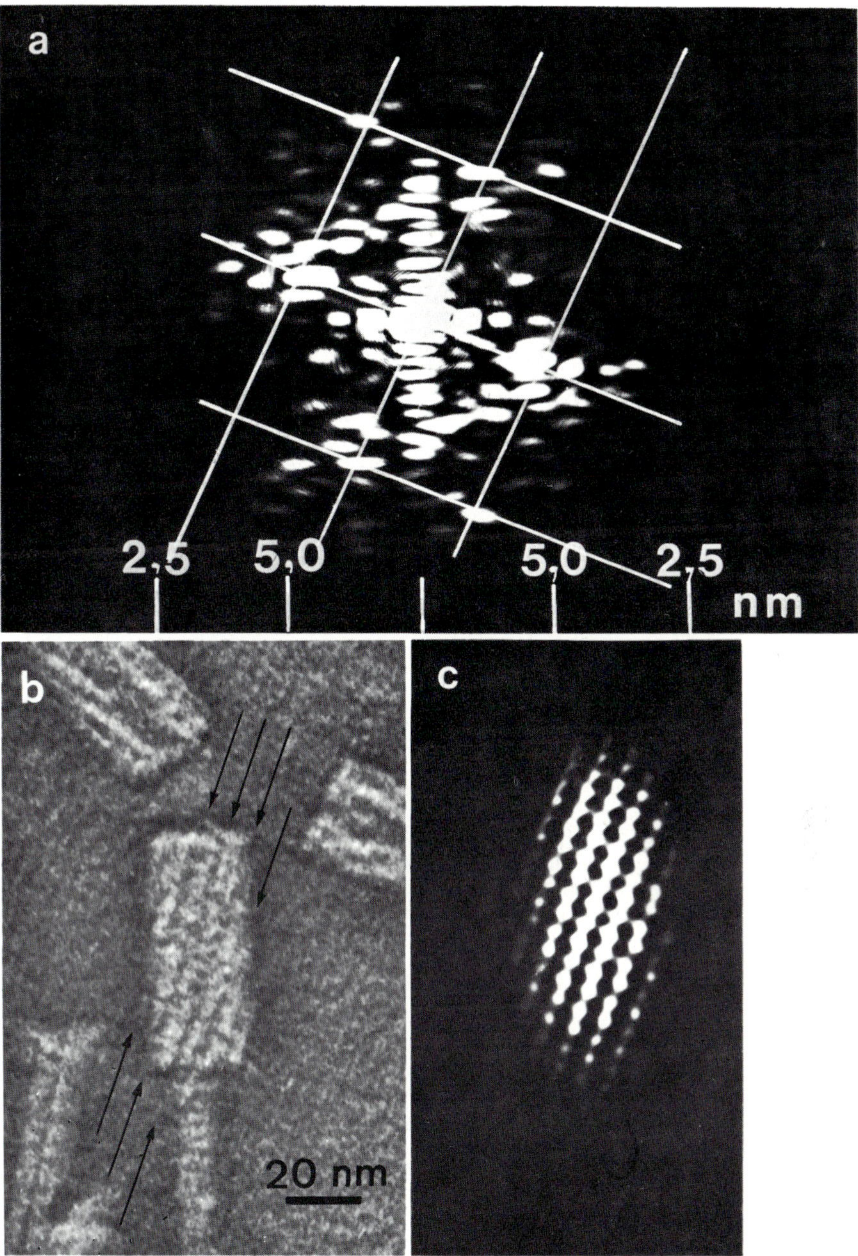

Fig. 4. (a) Optical diffraction pattern of the contracted sheath of the INCO particle shown in (b); a reciprocal lattice corresponding to one of the two faces of the sheath is shown. (c) Reconstructed image of the contracted sheath. Helical grooves can be observed on the sheath shown in (b), see arrows; URAC, (PFISTER, 1974)

Table 3. Particles resembling long (polysheath-like) tubes[a]

Host	Outer Diameter	Inner Diameter[b]	Length	'criss-cross' pattern[c]	References
Archangium violaceum	20,0-23,5	7,3	435	+	REICHENBACH, 1967
Chondrococcus columnaris	28,0-30,0	10,0	50-1500	+	PATE et al., 1967
Chromobacterium violaceum	20,0		210-890		ACKERMANN and GAUVREAU, 1972
Clostridium botulinum	25,0-30,0	10,0			IIDA and INOUE, 1968
Hydrogenomonas eutropha	24	6,0-6,5	2000 (maximum)	+	MAYER and AULING, unpublished
Photobacterium harveyi	22,0		250-320	+	YAMAMOTO, 1967
Proteus mirabilis	20,0	6,4	100-450		ITERSON et al., 1967
Proteus vulgaris	20,0		220-250		YAMAMOTO, 1967
Pseudomonas fluorescens	18,0-22,0	8,0	250-270	+	AMAKO et al., 1970 YAMAMOTO, 1967
Vibrio eltor	27,0-33,0		156-290		ADHIKARI and CHATTERJEE, 1972

[a]Measurements in nanometer units.
[b]Diameter of inner channel.
[c]See Section II.A.2.

maxima of these patterns represent two reciprocal lattices, indicating that the upper- and lower surface of the sheath are pictured in the electron micrographs. For image reconstruction (KLUG and DE ROSIER, 1966) of the sheath surface (Figs. 3c and 4c), the diffraction maxima corresponding to the lattice indicated in Figs. 3a and 4a were used.

The subunits of the extended sheath can be viewed as forming helices (pitch angle: 42°) or as being arranged in transverse annuli (see arrows in "b" of Fig. 3). Due to sheath contraction the INCO sheath shortenes from 112 nm to 60 nm and the pitch angle of the helices decreases from 42° to 16°. In addition to these "small-scale" helices, the subunits of the contracted sheath form "large-scale" helices with a pitch angle of 70°. The corresponding helical "grooves" can be easily discerned on the electron micrograph shown in Fig. 4b (see arrows).

2. *Particles Resembling "Long Tubes"*

Many bacterial strains producing phage tail-like particles, also synthesize structures which resemble "long tubes" (Table 1). The following two types will be described below: (I) tubes, which have an outer diameter similar to that of contracted sheaths (about 20 - 30 nm; Table 3) and resemble the polysheaths of T-even bacteriophages (KELLENBERGER and BOY DE LA TOUR, 1964). (II) tubes, which have an outer diameter similar to the cores of the phage tail-like particles and resemble polycores. In both cases, the outer diamter of the structures is relatively constant, while there is a large variation in their length.

"Polysheaths": The tubes usually appear rigid and can often be seen electron microscopically in disintegrating bacterial cells, which allow entrance of the staining solution (Figs. 5 and 6).

Frequently the tubes have been referred to in the literature as "rhapidosomes". It is proposed here to use the general term "tube" instead, because the rhapidosomes of *S. grandis* consist of two cylinders and appear to have a definite length (DELK and DEKKER, 1972), whereas the long tubes usually consist of only one hollow cylinder and vary in length. It is further proposed to use the term "rhapidosome" only for the characteristic phage tail-like particles (without a base plate) produced by *S. grandis*. If morphologically similar particles are found in other bacterial strains, they could possibly be termed "rhapidosome-like".

After negative staining, the tubes often exhibit a "criss-cross" pattern or cavities which widen and narrow with regular periodicity along their central axis (Fig. 7 and Table 3). Such patterns or periodicities have also been observed with negatively stained contracted sheaths and polysheaths of bacteriophage T4 (MOODY, 1967), isolated sheaths of rhapidosomes from *S. grandis* (DELK and DEKKER, 1972) and isolated sheaths of INCO particles from *Rhizobium* (Fig. 8).

As discussed by MOODY (1967) and DELK and DEKKER (1972), the inference is that helical grooves of both the upper and lower

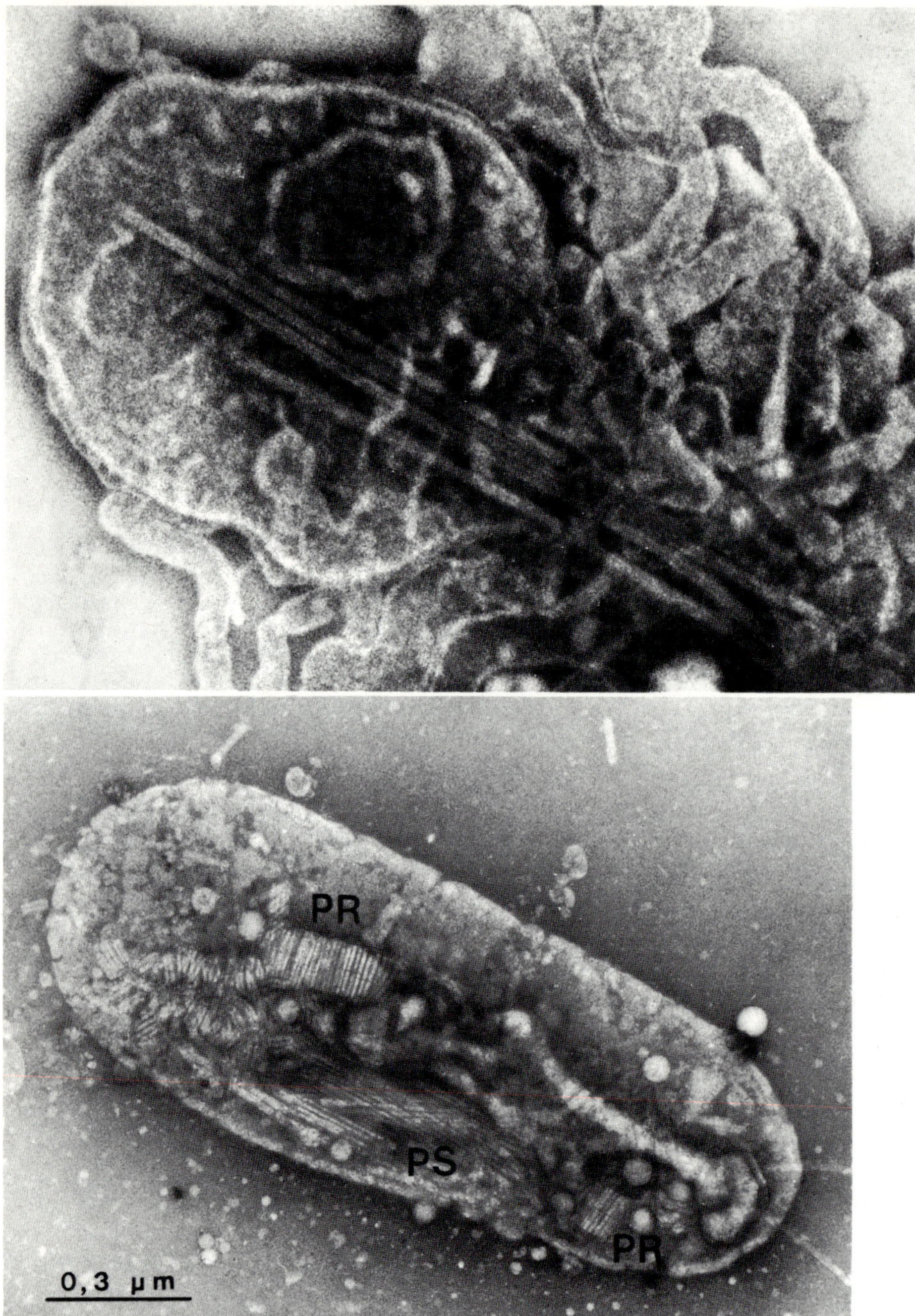

Fig. 5. Electron micrograph of long polysheath-like tubes within a lysed cell of *Hydrogenomonas eutropha*, PTA, 76.000x (MAYER and AULING, unpublished)

Fig. 6. Electron micrograph of a cell from *Rhizobium* strain 16-40 containing polysheath-like tubes (PS) and end-to-end aggregated pairs (PR) of single sheaths packed in a lateral fashion; PTA

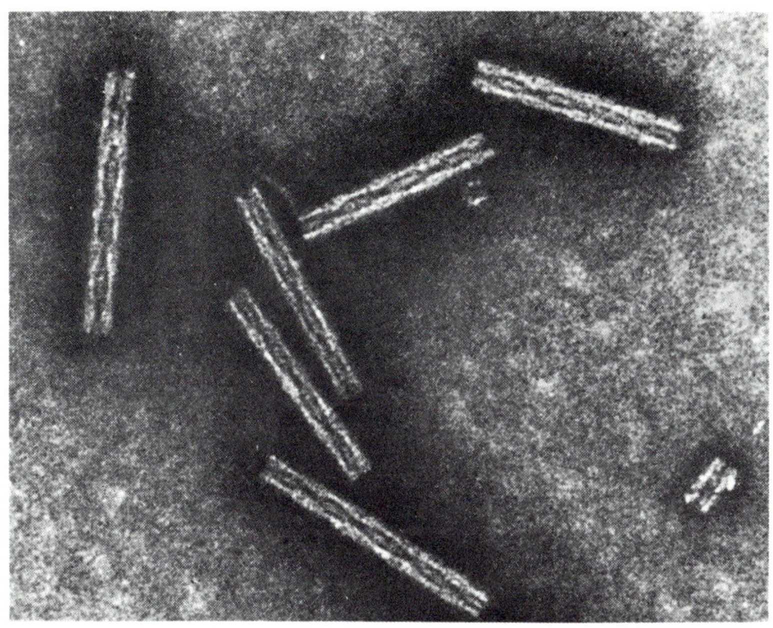

Fig. 7. Electron micrograph of tube-like structures produced by *Chondrococcus columnaris*. The particles show a periodic pattern along their length axis; PTA; 212.000x (PATE et al., 1967)

surface of the sheath deeply embedded in the staining material are visible. The criss-cross pattern observable with many of the long tubes therefore indicates a helical arrangement of their subunits (PATE et al., 1967; REICHENBACH, 1967; TAUBENECK, 1969; YAMAMOTO, 1967).

It has been suggested that the long tubes may be polysheaths arising by "unlimited" self-assembly of excess sheath subunits (AMAKO et al., 1970; CLARK-WALKER, 1969; GUMBERT and TAUBENECK, 1968; ITERSON et al., 1967; REICHENBACH, 1967; TAUBENECK, 1967). On the other hand, it is possible that in many cases such "polysheaths" arise by end-to-end aggregation of single contracted sheath.

For example, polysheaths 1 µm in length and apparently composed of a series of single empty contracted sheaths were frequently observed in purified preparations of pyocin R-type particles (HIGERD et al., 1969). Similar sheath-aggregates were also found after purification of INCO particles by CsCl density gradient centrifugation (Fig. 8). Obviously, purification procedures can lead to sheath contraction and to a dissociation of the particles into single cores and sheaths; the latter would then participate in the formation of the "polysheaths".

The polysheath-like tubes of *P. aeruginosa* were shown by microagglutination experiments (technique of HUMMELER et al., 1962) to have the same antigenicity as the sheaths of pyocin R-type

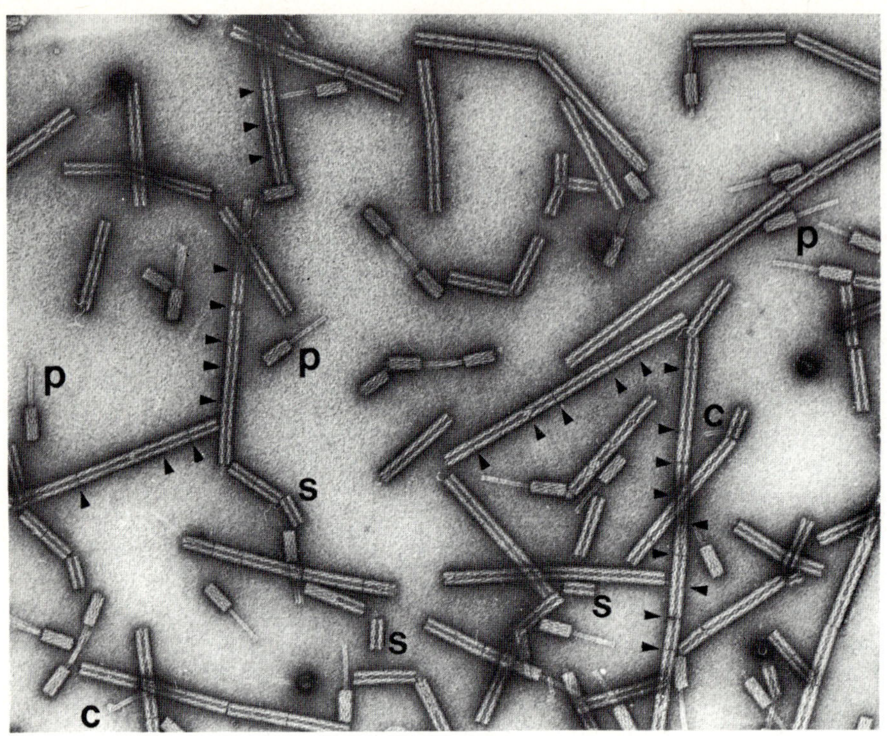

Fig. 8. Electron micrograph of INCO particles purified by CsCL equilibrium density gradient centrifugation. Most of the originally intact and extended particles dissociated into isolated contracted sheats (S) and single cores (C) during some step of the purification procedure. In addition to these substructures contracted particles without a base plate (P) and long tube-like aggregates of single contracted sheaths can be observed; arrows indicate points of contact between single sheaths within an aggregate. The isolated sheaths and the aggregates show a periodic pattern along their length axis. URAC. 66,000x (PFISTER, unpublished)

particles (AMAKO et al., 1970). After mixing a suspension of the pyocin particles and the polysheath-like tubes with antiserum against the pyocin, the authors observed antibody bridges between both structures.

"Polycores": The production of long tubes with a diameter identical to the core of the phage-like particles was reported for *P. aeruginosa* (HIGERD et al., 1969), *S. grandis* (DELK and DEKKER, 1972), *P. mirabilis* (TAUBENECK, 1967), and *Rhizobium* strain 16-40 (Fig. 9; LOTZ and MAYER, unpublished). These structures have been named "polycores", although little is known about them as yet.

3. Particles Resembling "Sheathless Phage Tails"

The dimensions of particles resembling sheathless phage tails are summarized in Table 4. The particles exhibit similar outer diameters (7 - 10 nm), and individual particles from *P. aeruginosa*

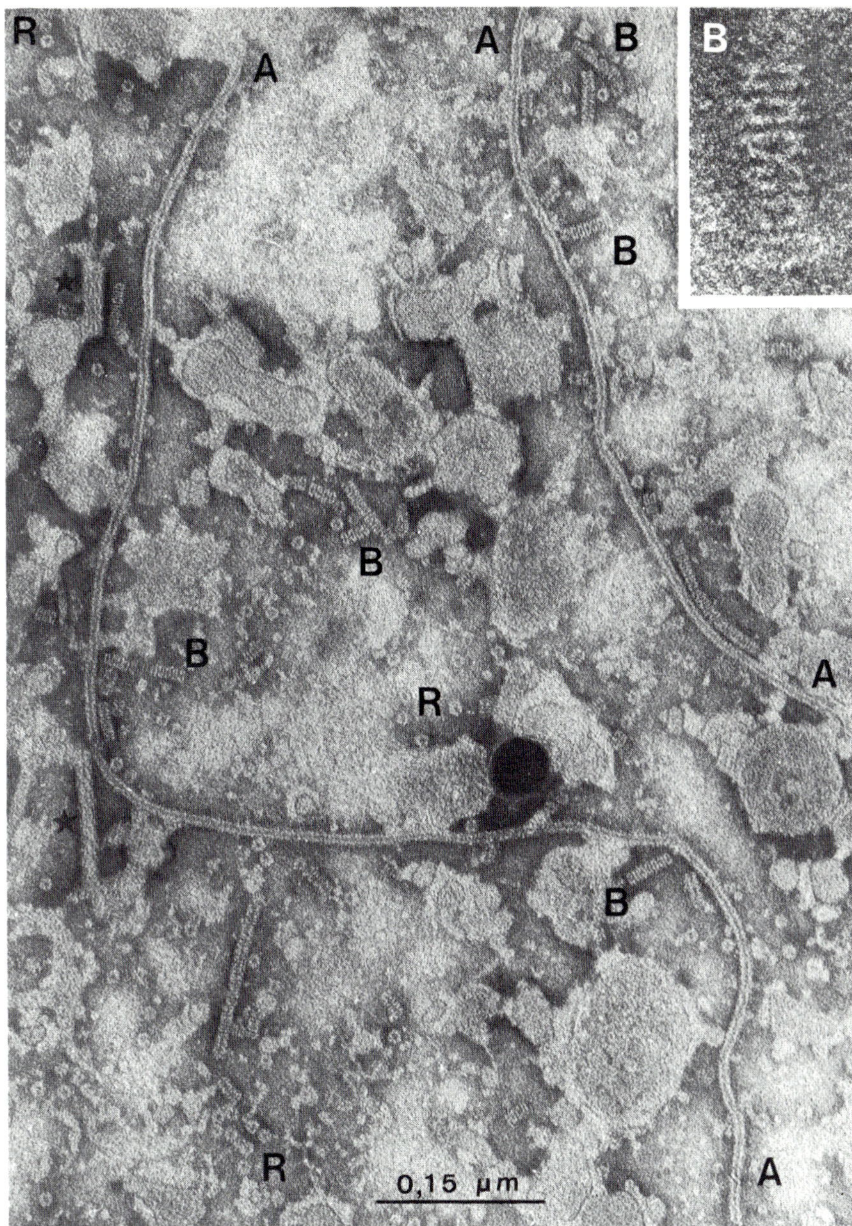

Fig. 9. Electron micrograph of polycore-like tubes produced by strain 16-40 of *Rhizobium*. Two types of rods (A and B) can be discerned. Type A is longer and more flexible than type B. Type B seems to be composed of 'stacked discs' (see inset). Besides sheathed phage tail-like particles (★), ring-like structures (R) can be observed, which probably represent short segments of the tubes seen in axial view. PTA. Magnification of inset: 714.000x (LOTZ and MAYER, unpublished)

Table 4. Particles resembling sheathless phage tails[a]

Host	Outer Diameter	Length	References
Clostridium botulinum	10	100	UEDA and TAKAGI, 1972
Pseudomonas aeruginosa	9	100-120[b]	TAKEYA et al., 1969
Rhizobium spec.	8	200	GISSMANN, 1974
Serratia marcescens	10	150[b]	TRAUB, 1972
Streoptococcus faecalis	7	100	BRADLEY, 1967

[a]Measurements in nanometer units.
[b]Most frequent length.

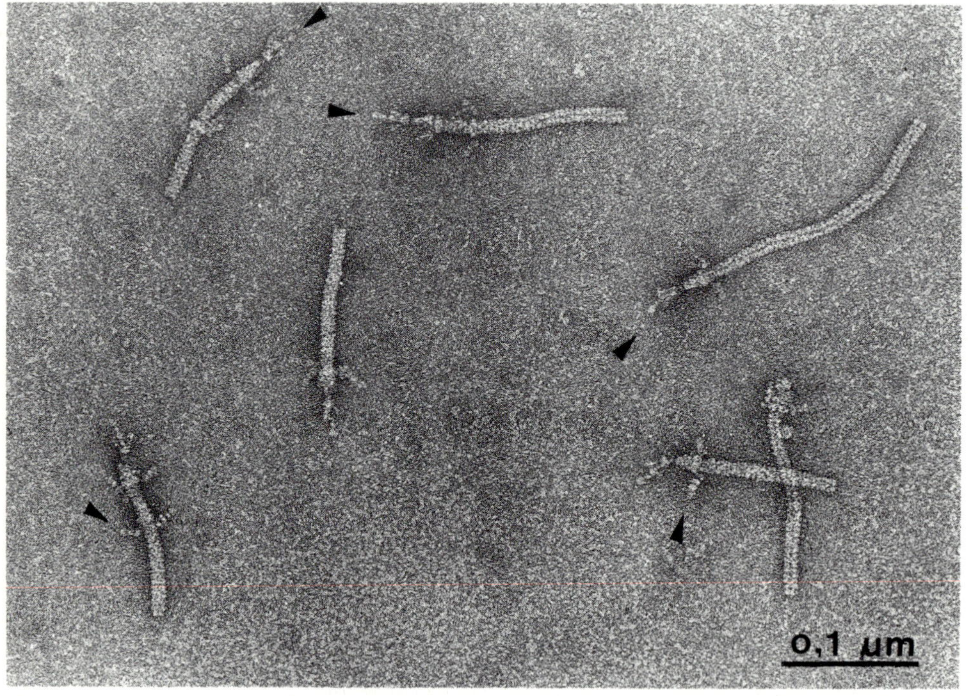

Fig. 10. Electron micrograph of sheathless particles produced by *Serratia marcescens*. Arrows point to fiber-like appendages. PTA (TRAUB and ACKER, unpublished)

("pyocin 28"; TAKEYA et al., 1969), and *S. marcescens* (TRAUB and ACKER, unpublished) were found to vary greatly in length (between 20 and 40 nm in the case of pyocin 28).

Many rods possess one "square" and one pointed end. The particles from *P. aeruginosa* and *Clostridium botulinum* (UEDA and TAKAGI, 1972) carry a fine fiber at their pointed end, which has a length of 20 nm in the case of pyocin 28. These rods are morphologically similar to the long, flexible tail of *E. coli* bacteriophage Lambda, which also carries a fine fiber at its free end (KELLENBERGER and EDGAR, 1971). Fiber-like appendages were also found whith the particles from *Streptococcus faecalis* (BRADLEY, 1967), *S. marcescens* (Fig. 10; TRAUB and ACKER, unpublished), and *Rhizobium* strain 16-2 ("bacteriocin 16-2"; GISSMANN, 1974).

B. Chemical Analysis of Particles

1. Search for a Nucleic Acid Component

BRADLEY and DEWAR (1966) discussed the possibility that the cores of particles with extended sheaths may be filled by some substance such as a short piece of nucleic acid. Alternatively, it has been assumed that the cores of extended particles are empty and are only open to entering staining solution at one end (the opposite end to the base plate; LOTZ and MAYER, 1972). In this case the stain would not penetrate very far into the particles. Contraction of the sheath may result in a removal of the "plug" from the core-end in contact with the base plate (SIMON and ANDERSON, 1967b). The staining solution could now enter the core from both ends and would penetrate it completeley.

Chemical analysis of the sheathed particles produced by *P. aeruginosa, Clostridium botulinum, Proteus vulgaris,* and by *S. grandis* indicates that these structures may consist almost entirely of protein and do not contain nucleic acid.

Preparation of purified pyocin R particles from *P. aeruginosa* showed positive reactions for protein, while the sugar content was negligible (below 0,5 %). The nitrogen content was determined to be 15 ± 0,5 %, which is consistent with a structure entirely made of protein (KAGEYAMA and EGAMI, 1962; KAGEYAMA, 1964). Similarily, the sheathed tail-like particles ("boticin P") produced by *C. botulinum* were found to be composed of 98,8 % protein and 0,4 % carbohydrate. Nucleic acid and lipids were not present in the purified preparation (LAU et al., 1974).

COETZEE and co-workers (1968) determined the composition of particles ("bacteriocin 45") produced by *P. vulgaris*, which were partially purified by differential centrifugation. The composition reported was as follows: 640 μg/ml protein, 0,0 μg/ml DNA-phosphorous, 0,48 μg/ml RNA-phosphorous, and 412 μg/ml for total sugars. The finding of sugars and RNA was ascribed to contaminations by cell-wall material and bacterial ribosomes, respectively.

Earlier reports on the chemical composition of rhapidosomes from *S. grandis* had indicated that these consist of about 70 % protein and 30 % RNA with 40 % - 90 % of the ribose residues methylated at the 2'-hydroxyl (CORRELL and LEWIN, 1964). However, the results of DELK and DEKKER (1969 and 1972) and those of PRICE and ROTTMANN (1970) indicate that the purified particles do not contain any nucleic acid component. Based on phosphate analyses,

Table 5. Amino acid composition of the sheaths from rhapidosomes (DELK and DEKKER, 1972), pyocin R (YUI-FURIHATA, 1972), and INCO particles (PFISTER, 1974)

Amino acid[a]	rhapidosome sheath	pyocin R sheath	INCO sheath
Alanine	11.2 ± 0.3	15.73	16.40
Arginine	3.3 ± 0.1	5.51	5.87
Aspartic acid	12.8 ± 0.6	12.09	11.19
Half cystine	0.9 ± 0.3		0.44
Glutamic acid	10.3 ± 0.1	7.51	10.01
Glycine	9.2	8.28	8.29
Histidine	0.6 ± 0.1	0.59	1.06
Isoleucine	3.5 ± 0.1	4.75	5.83
Leucine	7.2 ± 0.1	7.49	7.26
Lysine	4.6 ± 0.2	3.31	3.83
Methionine	3.1 ± 0.3	0.96	0.09
Phenylalanine	4.2 ± 0.1	3.81	2.60
Proline	4.6 ± 0.3	3.76	3.19
Serine	6.3 ± 0.2	6.89	6.83
Threonine	6.4 ± 0.3	6.16	5.79
Tryptophan	1.0	1.4	
Tryosine	3.5 ± 0.2	1.38	1.33
Valine	7.4 ± 0.3	8.73	9.92

[a]Values in mole %.

preparations of whole rhapidosome particles contained at most 1 - 6 % nucleic acid relative to protein by weight. Most of this could be accounted for by contaminating cellular nucleic acid.

2. *Chemical Analysis of Sheaths*

YUI (1971) succeeded in degrading the pyocin R particles into contracted sheaths and other smaller subunits by treatment with 0,1 N NaOH. The sheaths were purified by sucrose density gradient centrifugation and then analyzed for their amino acid composition (YUI, 1971; YUI-FURIHATA, 1972).

The results of these experiments, together with the amino acid composition of rhapidosome sheaths (DELK and DEKKER, 1972) and of INCO sheaths (PFISTER, 1974), are summarized in Table 5.

The Table 5 shows a close resemblance in amino acid composition of the three structures. The sheaths contain little (if any) cystein, and only a few residues of tryptophan, methionine, and histidine. In contrast, relatively high contents of aspartic acid and glutamic acid (or their amides), as well as valine, alanine, and glycine were found.

Dissoziation of purified pyocin R sheaths in a phosphate buffer (pH 8,4) containing hydrochloride, followed by SDS polyacrylamide gel electrophoresis, revealed a major sheath-protein of molecular weight 35,000 and three minor components of 32,000, 26,000, and 14,500 respectively (YUI-FURIHATA, 1972).

A densitometric scanning of the SDS gels showed that the major component amounts to 73 %, and the sum of the three minor components together occupies 27 % of the sheath protein. This determination together with the molecular weigth of the whole sheath (7.2 x 10^6 daltons; YUI, 1971) yields an estimate of 150 major subunits and 65 minor subunits per sheath which corresponds to a total of 215 sheath components.

A number of 150 major sheath subunits was also determined by SHINOMIYA (1972b), who dissociated the pyocin R sheath in a Tris-HCl-buffer (pH 6,8) containing 2 % SDS, 10 % glycerol, and 5 % β-mercaptoethanol. By applying the method of SDS-disc electrophoresis of LAEMMLI (1970), the author found at least 6 minor components in addition to the major sheath protein.

In contrast to the sheath of pyocin R, that of INCO particles was found to contain only one type of structural protein which had a molecular weight of 56,000 (PFISTER, 1974). The author dissociated the INCO sheaths with 66 % acetic acid and then dialyzed the solution against a phosphate buffer (pH 7,0) containing 8 M urea, 1 % SDS, and 0,5 % β-mercaptoethanol.

III. Production

A. Genetic Determinants

Little is known about the genetic determinants ("defective prophages") coding for the phage tail-like particles. By assuming an extrachromosomal determinant, a number of authors have tried to eliminate these factors with intercalating agents such as acridine orange (AO), acriflavine (AF), or with UV-irradiation, all without success.

The following bacterial strains have been tested with these curing agents: *C. botulinum* (AO; LAU et al., 1974), *P. aeruginosa* (AO, AF, UV; KAGEYAMA et al., 1964), *Rhizobium spec.* (AO; LOTZ, unpublished), *Serratia marcescens* (AO; TRAUB, 1972), and *Spirillum itersonii* (AO; CLARK-WALKER, 1969).

In the case of *P. aeruginosa*, the resistance to curing agents can be readily explained, since the chromosomal location of the pyocinogenic factors (R, R2, and R3) has been shown (KAGEYAMA,

1970a,b; KAGEYAMA et al., 1973). The factors could be transferred into a fertile strain of *P. aeruginosa* (strain PAO of HOLLOWAY). Genetic analysis by conjugation and transduction revealed that they are all incorporated into the bacterial chromosome at a position of 35 min from the origin to transfer.

B. Induction of Defective Prophages

1. Artificial Induction

The production of phage tail-like particles by many bacterial strains can be induced by treatment of growing bacteria with mitomycin C (1 - 3 μg/ml) or with UV-irradiation (at 254 nm). In most cases the bacteria begin to lyse 60 to 90 min after the inducing treatment, releasing the defective phage particles into the medium.

The following strains were shown to be inducible spontaneously (see next Section), as well as artificially (by the agents mentioned above): *Chromobacterium violaceum* (RUCINSKY and COTA-ROBLES, 1973), *Proteus vulgaris* (ITERSON et al., 1967; TAUBENECK, 1967), *P. aeruginosa* (KAGEYAMA and EGAMI, 1962; KAGEYAMA, 1964; TAKEYA et al., 1967), *Rhizobium spec.* (GISSMANN, 1974; LOTZ and MAYER, 1972), *Serratia marcescens* (TRAUB, 1972), and *Spirillum itersonii* (CLARK-WALKER, 1969).

Besides the mentioned agents, phleomycin was shown to induce the pyocin R-production of *P. aeruginosa*, althouth less effectively than mitomycin C (IKEDA and EGAMI, 1966); and nalidixic acid was shown to induce the production of the phage tail-like particles of *S. marcescens* (TRAUB et al., 1973). It was further reported that acriflavine induced the production of the tail-like particles of *C. violaceum* (RUCINSKY and COTA-ROBLES, 1973).

LIU et al. (1969) isolated a temperature-sensitive mutant (RT-1) of *P. aeruginosa* strain R, which grows normally at 28°C, but lyses after about 4 h of growth at 38°C. The cell lysate contained pyocin R activity (= bactericidal pyocin R particles) approximately 100 times above that of the low-temperature cultures; this corresponded to a yield close to that of mitomycin C-induced wild-type cultures. It is possible that *P. aeruginosa* RT-1 harbors a heat-inducible defective prophage.

The heat-induction became irreversible after about 2 h of incubation at 38°C, and a later shift back to 28°C did not prevent subsequent lysis of the cells. The number of colony-formers in the heat-induced culture also started to decrease after 2 h, indicating that the induction of the defective prophage was lethal for the bacteria. This has been similarly shown for the wild-type *P. aeruginosa* strain, which lost its viability shortly after prophage-induction with mitomycin C (IKEDA et al., 1964).

LIU et al. (1969) measured the incorporation of ^{32}P-inorganic phosphate into RNA and DNA of *P. aeruginosa* R (= wild-type) and the mutant RT-1 shifted to 38°C. At this temperature only the

DNA synthesis of the heat-inducible mutant was markedly inhibited. In a similar experiment, using wild-type strain R, IKEDA et al. (1964) had shown that induction of the defective prophage after UV-irradiation was accompanied by an almost complete inhibition of DNA-synthesis, while RNA-synthesis was only slightly impaired.

2. *Spontaneous Induction*

The spontaneous induction of the defective-phage-synthesis has been described for various bacterial strains, some of which have been mentioned in the previous section. Due to the inability of the synthesized particles to reproduce, it is difficult to determine the frequency of spontaneous induction. If the particles have bactericidal activity, lacunae-counting is possible. A lacuna is a tiny clear spot brought about by a single bacteriocin-synthesizing bacterium and is formed on an agar plate seeded with a lawn of sensitive bacteria (OZEKI, 1968).

IKEDA et al. (1964) found that the frequency of induction, as measured by lacunae formation, varied between 20 and 60 % for mitomycin C-induced, and between 10^{-1} and 10^{-2} % for spontaneously induced cells of *P. aeruginosa* R. The authors pointed out that, due to the difficulty of lacunae-counting (caused by the low diffusibility of the pyocin R particles), the values obtained may be lowest estimates.

The fraction of bacteria spontaneously producing phage tail-like particles is higher in the case of *Rhizobium* strain 16-3 (PFISTER and LOTZ, 1974). About 35 % of the cells form large lacunae when plated on nutrient agar plates with sensitive indicator bacteria, while the other 65 % form colonies (Fig. 11a, see arrows). The lacunae do not contain plaque-formers. The colonies are surrounded by relatively broad, ring-like inhibition zones, which are produced by the bactericidal INCO particles diffusing from a growing colony into the surrounding agar.

Due to the high frequency of spontaneous induction, the particles must be produced in relatively large quantities within the growing colonies. The lacunae were not observed, when mutants of the producer strain were plated, which do not synthesize bactericidal INCO particles; in this case the colonies were not surrounded by inhibition zones (Fig. 11b).

Spontaneous prophage induction of some defectively lysogenic strains may depend on the phase of growth and can occur at an increased rate, when bacterical cultures reach the early stationary phase. It has been reported that the rhapidosomes of *S. grandis* were produced in large quantities (about 10^{10}/ml) at the end of exponential growth (LEWIN, 1963). Cells of such cultures lost their viability and lysed.

Similar observations have been reported for cultures of *C. violaceum* approaching and continuing into the stationary phase (RUCINSKY et al., 1972; RUCINSKY and COTA-ROBLES, 1973). The frequency of sphaeroplasts, cell ghosts, and intact cells packed with rosette-

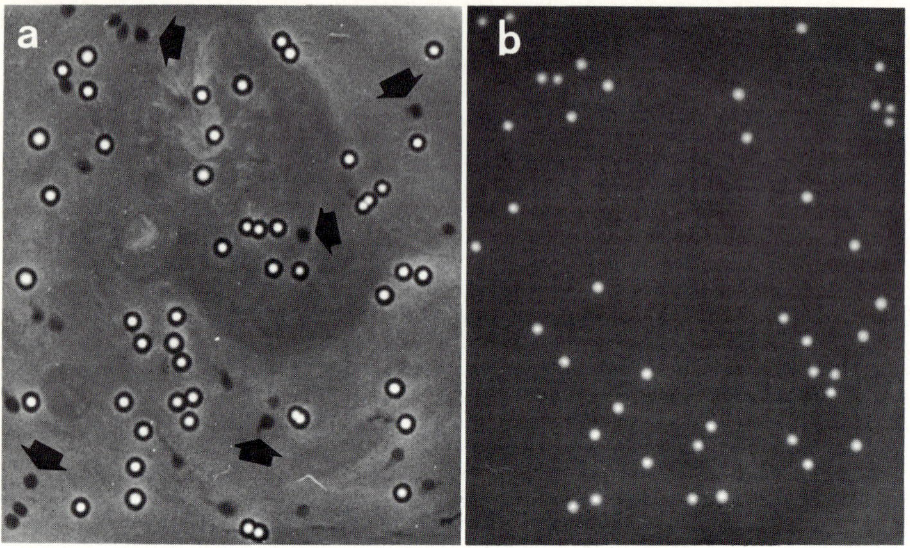

Fig. 11. Colonies of wild-type *Rhizobium* strain 16-3 (a) and of a mutant of this strain (b). Only the wild-type produces the bactericidal INCO particles. After inactivation of the bacteria with chloroform vapor, the colonies were overlayered with soft agar containing cells of a bacteriocin-sensitve *Rhizobium* strain. Following an incubation of the indicator bacteria, inhibition zones could be observed around the colonies of the wild-type strain (a), but not around those of its mutant (b). Moreover, only the agar plate inocculated with the wild-type bacteria showed large lacunae (see arrows) (PFISTER and LOTZ, 1974)

like aggregates of phage tail-like particles increased markedly (see following Section). Finally, PATE and co-workers (1967) observed long, polysheath-like tubes only in thin section of *C. columnaris* cells, which had reached the early stationary phase, when autolysis of the bacteria was in progress.

C. Biosynthesis of Structural Proteins and Maturation of Particles

After induction of *P. aeruginosa* wild-type by UV (KAGEYAMA and EGAMI, 1962) or mitomycin C (IKEDA et al., 1964), pyocin R particles are synthesized *de novo*, and there is probably no pyocin R-precursor in the cells before this treatment.

A detailed study of pyocin R protein synthesis after mitomycin C-induction has been reported by SHINOMIYA (1972a). In one experiment (Fig. 12) the biosynthesis of pyocin R-specific proteins was followed by measuring the incorporation of radioactively labeled ^{14}C-phenylalanine into proteins precipitated by pyocin-specific antiserum. At the same time, the radioactivity incorporated into the acid-insoluble fraction (for determination of total protein) and the intracellular pyocin R bactericidal activity were measured. Some of the results of this experiment were described as follows:

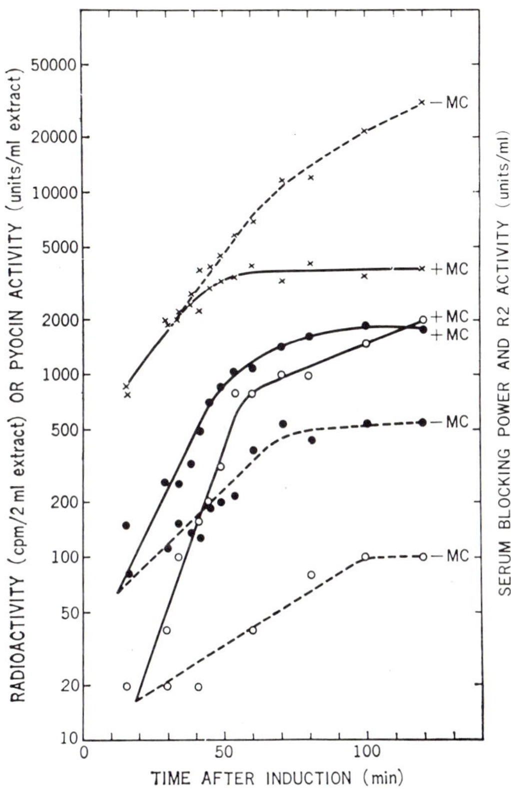

Fig. 12. Synthesis of pyocin R proteins after mitomycin C-induction of *Pseudomonas aeruginosa* P 15. A P15 culture was divided into two parts. Mitomycin C (MC) was added only to one tube. Simultaneously, ^{14}C-1-phenylalanine was added to each tube. Extracts were prepared at intervals with lysozyme-EDTA method and then centrifuged at low speed; supernatants were saved. Radioactivity in the acid-insoluble fraction and in the precipitate formed with pyocin R-specific antiserum were measured. Symbols: ● = radioactivity in pyocin R protein; x = radioactivity in total protein; o = pyocin R activity (SHINOMIYA, 1972a)

1) The synthesis of pyocin proteins started 10 to 15 min after mitomycin C addition; the pyocin activity appeared after 20 to 30 min.

2) The mitomycin C-induced synthesis of pyocin initially proceeded at an exponential rate (until 45 - 50 min for total pyocin proteins and until 55 - 60 min for active pyocins).

3) The proportion of pyocin proteins synthesized compared to total proteins was 2 to 5 % in non-induced cultures and 30 - 40 % after induction with mitomycin C.

The delayed appearance of killing activity after the emergence of the pyocin antigen indicates that the structural proteins by themselves have no bactericidal activity, but have first to undergo a transformation to the active form. It was shown by SHINOMIYA (1972a) that the latter is identical with the *mature* pyocin R particles. The sucrose gradient centrifugation of cell extracts from bacterial cultures pulse-labeled with ^{14}C-phenylalanine, showed a flow of the radioactivity associated with pyocin protein into mature particles.

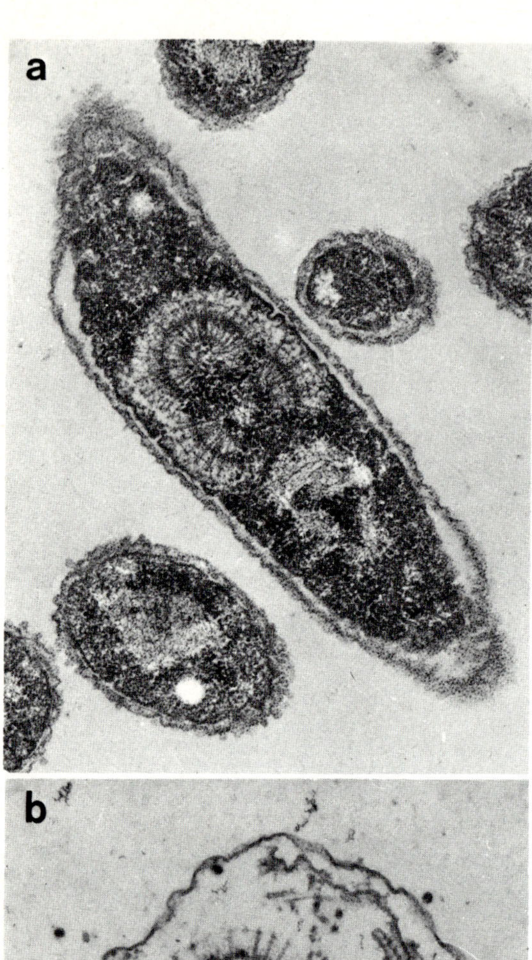

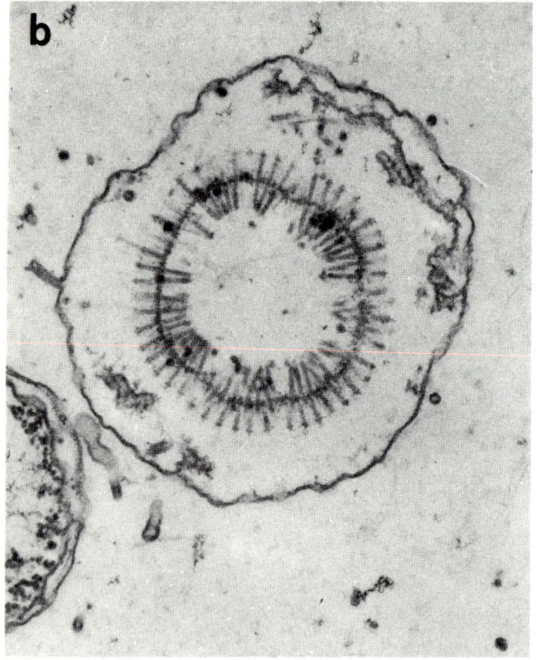

Fig. 13. Electron micrographs of rosette-like aggregates of tail-like particles in cells of *Chromobacterium violaceum*. (a) Longitudinal section through a cell showing the central location of two rosettes; 47.500x. (b) Thin section of ghosted cell with well-defined particle-rosette; 46.700x (RUCINSKY et al., 1972)

Due to their inability to reproduce, it is difficult to estimate the number of phage tail-like particles per induced cell. KAGEYAMA et al. (1964) estimated a burst size of pyocin R particles of about 200. In contrast, the burst size of INCO particles of *Rhizobium* strain 16-3 after spontaneous induction is higher, because a relatively large lacuna was produced by a single induced bacterium on a lawn of sensitive bacteria (Fig. 11a). It has been estimated that the size of such a lacuna corresponds to about 10^4 sensitive bacteria killed (PFISTER and LOTZ, 1974).

The plate shown in Fig. 11a was inocculated with small, stationary phase cells of *Rhizobium* strain 16-3, which can form filaments during growth. It is assumed that the large burst size of INCO particles leading to lacuna-formation is due to the lysis of a filament originating from a 16-3 cell, which experienced spontaneous prophage induction. In agreement with this idea, the bactericidal activity leading to lacuna-formation can be demonstrated on the nutrient agar plate only after a time-lapse corresponding to about 7 cell generations, during which the filaments were probably formed.

RUCINSKY et al. (1972, 1973) studied the organization of the maturing tail-like particles produced by *C. violaceum*. In thin sections of spontaneously induced cells, rosette-like aggregates of such particles (obviously in the extended state) were observed (Fig. 13). The rosettes always appeared to discplace considerable cytoplasm and to be in close contact to the cell membrane. It is interesting that the tails of intracellular T2 and T4 bacteriophages of *E. coli* were also found associated with the plasma membrane (SIMON, 1969). This led SIMON to suggest that the membrane plays a vital role in the *in vivo* assembly of the T-even base plates.

The intracellular organization of phage tail-like particles was also examined after mitomycin C-induction of *P. vulgaris* (de KLERK et al., 1974). In this case, no clear association of the particles with the cell membrane could be observed. Instead of the rosettes, the cells contained striated bands composed of the tail-like particles aggregated in a lateral fashion.

The maturation of the base plate is known to be a prerequisite for further assembly of T4 phage tails (see EISERLING and DICKSON, 1972, for a review). This raises the question as to the mode of maturation of the (sheathed) rhapidosomes of *S. grandis* and of the rhapidosome-like structures produced by *Spirulina*, because the particles do not possess a base plate (Section II.A.1.). This lack may explain why the sheath subunits of the rhapidosomes and rhapidosome-like particles apparently assemble only into contracted sheaths.

IV. Bactericidal Activity

A. General Properties of Bactericidal Particles

A number of phage tail-like particles has been shown to harbor bactericidal activity. Such particles with a contractile sheath are produced by the following bacterial strains: *C. botulinum* ("boticin"; LAU et al., 1974), *Proteus mirabilis* (TAUBENECK, 1963), *Proteus vulgaris* ("bacteriocin 45"; COETZEE et al., 1968), *P. aeruginosa* ("pyocin R"; KAGEYAMA and EGAMI, 1962), *Rhizobium spec.* ("INCO", LOTZ and MAYER, 1972), and *S. marcescens* (TRAUB, 1972).

Some of the strains were shown to produce sheathless particles with bactericidal activity: *P. aeruginosa* ("pyocin 28"; TAKEYA et al., 1967), *Rhizobium spec.* ("bacteriocin 16-2"; GISSMANN, 1974), and *Serratia marcescens* (TRAUB, 1972).

The question has often arisen, as to whether or not the phage-like particles with bactericidal activity are "true bacteriocins". As pointed out by BRADLEY (1967), the answer depends on the definition of a bacteriocin. NOMURA (1967) included the pyocin R-type particles in his review on bacteriocins and defined the latter as "bactericidal substances, apparently protein in nature, which are synthesized by certain strains of bacteria and are active against some other strains of the same or closely related species". In contrast, REEVES (1972) exluded complex antibacterial agents which resemble bacteriophages or their tails from the group of bacteriocins.

BRADLEY (1967) proposed the division of bacteriocins into two basic groups, namely the "low" and the "high" molecular weight forms. For example, the tail-like pyocin R particles have a molecular weight of about 1×10^7 (KAGEYAMA, 1964), whereas the molecular weights of S-type pyocins are much lower (e.g. about 1×10^5 in the case of pyocin S produced by *P. aeruginosa* P28; ITO et al., 1970). In contrast to the low molecular weight bacteriocins, the phage-like bacteriocins are sedimentable in the ultracentrifuge (e.g. at $10^5 \times g$) and can be resolved with the electron microscope.

The known phage tail-like bacteriocins are moreover resistant to DNase, RNase, and ultraviolet light (at 254 nm), but sensitive to heat (e.g. 10 min at 60° - 70°C). In contrast to the "low" molecular weigth forms (NOMURA, 1967; REEVES, 1972), most of the phage tail-like bacteriocins are resistant to trypsin (for exceptions see Section IV.E.).

B. Receptors

It is generally accepted that defective phages with bactericidal activity, like intact bacteriophages, exert their action after adsorbing to specific receptors distributed over the cell surface of sensitive bacteria. The receptor substances have been isolated and studied in the case of *P. vulgaris* (SMIT et al., 1969) and

P. aeruginosa (IKEDA and EGAMI, 1969 and 1973; IKEDA and NISHI, 1973).

SMIT et al. (1969) isolated the lipopolysaccharide (LPS) component from a *P. vulgaris* strain, which was susceptible to the sheathed "bacteriocin 45" particles produced by a second *P. vulgaris* strain. It has been shown that the particles adsorbed to the LPS-fraction. This resulted in the contraction of their sheaths and in the elimination of their bactericidal activity. In contrast, the particles did not adsorb to the LPS-fraction of two bacteriocin 45-resistant mutants of the original strain. Their sheaths remained relaxed and their bactericidal activity was retained. A comparison of the LPS sugars of sensitive and resistant strains showed that the resistant mutants, in contrast to the sensitive indicator strain, lacked glucuronic acid. This indicates that this component is involved in determining the receptor specifity of the LPS-fraction. The following sugar components were found in the LPS-fraction of sensitive as well as resistant strains: glucose, glucosamine, heptose, 2-keto-3-deoxyoctonate.

IKEDA and EGAMI (1969, 1973) analyzed the pyocin R receptor substance of *P. aeruginosa*. As demonstrated for *P. vulgaris*, they could show that the LPS of the pyocin R-sensitive, but not of the resistant strain, contained receptor activity. The LPS-fraction from *P. aeruginosa* P.14 has been dissociated (by heat-treatment in the presence of sodium deoxycholate) into an amino sugar-rich fraction and the LPS subunits without receptor activity. Both components were separated from each other by Sephadex G-100 gel filtration. The carbohydrate moiety had a molecular weight of 1,300 - 1,500 (after acid treatment for the elimination of lipid) and contained glucose, rhamnose, heptose, galactosamine, 2-keto-3-deoxy sugar acid, and phosphate.

The LPS subunits had a molecular weight of 12,000 - 16,000 and contained glucose, rhamnose, heptose, galactosamine, glucosamine, fucosamine, quinovosamine, and also 2-keto-3-deoxy sugar acid. In the absence of sodium deoxycholate the LPS subunits reassociated and recovered their receptor activity, indicating that they are the chemical entity of the receptor. Since the receptor activity of reassociated subunits decreased rapidly, it was concluded that they are normally stabilized by the amino sugar-rich fraction.

Most of the receptor activity of pyocin R-sensitve *P. aeruginosa* cells could be separated from the sphaeroplast membrane fraction by treating the bacteria with lysozyme in the presence of EDTA in hypertonic medium (IKEDA and NISHI, 1973). The results suggest that the pyocin R receptor is located in the bacterial cell wall and that the cytoplasmic membrane is free of receptor activity.

In most cases the producers of phage tail-like bacteriocins proved insensitive to the action of the homologous particles and were shown to lack the receptor activity for their adsorption (for an exception, see Section IV.E.). For example, the particles produced by *P. aeruginosa* (HIGERD et al., 1969; ITO et al., 1970; KAGEYAMA et al., 1964), *P. mirabilis* (TAUBENECK, 1967) and *Rhizobium spec.* (GISSMANN, 1974; LOTZ and MAYER, 1972) do not adsorb to the cells of their respective producer strains. From the results of TAKEYA and co-workers (1967 and 1969) it is apparent that *P. aeruginosa* 28, producing the sheathless pyocin 28 rods, is also resistant to the homologous particles. Recently, resistance to the homologous bacteriocin was also reported for *C. botulinum* producing the boticin P particles (LAU et al., 1974).

As pointed out by GARRO and MARMUR (1970), the insensitivity of a producer strain to the homologous particles may be due to their having been selected for the absence of specific receptors. Alternatively, the authors suggested that the defective prophage may code for a cell-wall component that would prevent attachment of the particles.

C. Adsorption

Adsorption of pyocin R to isolated receptor material led to the inactivation of the bactericidal activity (IKEDA and NISHI, 1973). The reaction was dependent on salt concentration: 0.1 to 0.2 M NaCl was required for optimal inactivation of the pyocin R activity. The reaction proved also dependent on temperature and occurred only slowly at 0°C in comparison to 37°C. KAGEYAMA et al. (1964) had suggested earlier that a reversible step of pyocin R adsorption existed, and that at 0°C the reaction does not pass beyond this step.

It is likely that the sheathed tail-like particles initially have to be in the extended state for the adsorption to and killing of sensitive cells. HIGERD and co-workers (1969) have observed that the freezing of a suspension of pyocin R particles led to sheath contraction and rendered the particles incapable of adsorption. In another experiment the authors demonstrated a direct correlation between the decrease of bactericidal activity of a pyocin R suspension and the conversion of extended particles into their contracted form (by addition of 0.5 M magnesium chloride).

These findings may explain the resistance to freezing of sheathless particles from *S. marcescens* (TRAUB, 1972). Consequently, it should be expected that the sheathless pyocin 28 rods are also resistant to freezing. The contracted state of the sheath of rhapidosomes from *S. grandis*, in addition to the absence of a base plate, may be pertinent to the apparent lack of bactericidal activity demonstrated for these particles (DELK and DEKKER, 1972).

The sheathed particles derived from different bacterial strains adsorb to sensitive bacteria with their base plate oriented toward the cell surface (GOVAN, 1974; HIGERD et al., 1969; LOTZ and MAYER, 1972; SMIT et al., 1969; TAUBENECK, 1969). This adsorption is usually followed by the contraction of the sheath, resulting in an exposure of the inner core (Fig. 14, see arrow).

Very probably, the adsorbing particles attach to bacteria with their short fibers. Accordingly, the distance between cell surface

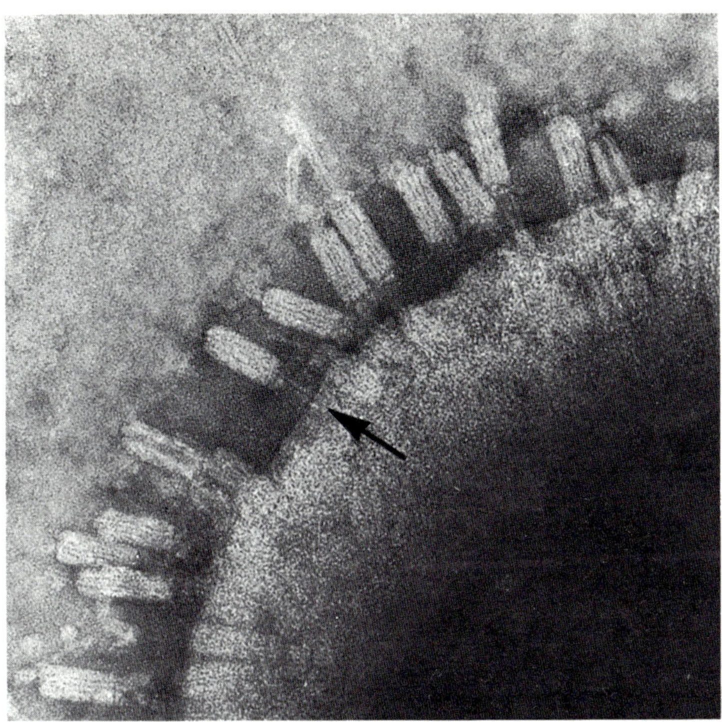

Fig. 14. Electron micrograph of INCO particles of *Rhizobium* strain 16-3 adsorbed to the surface of a sensitive indicator cell. The sheaths of the particles are contracted. URAC; 185.000x. Arrow points to particle with well-defined core (LOTZ and MAYER, 1972)

and base plate of the adsorbed INCO particles shown in Fig. 14 should correspond to the fiber-length (32 nm), if the fibers were orientated vertically towards the cell surface. Measurements taken from this micrograph indicated an average distance of about 29 nm (LOTZ and MAYER, 1972). Artifacts in negative staining, as described by SIMON and ANDERSON (1967a), or a fiber-position not fully vertical, might have resulted in measuring a distance of slightly less than the expected value.

The long fibers of T-even bacteriophages show a characteristic bend near their center, which enables the phage particles to move (passively) towards the cell surface, once the long fibers have attached to the bacterial receptors (SIMON and ANDERSON, 1967a). However, such bends cannot be discerned on fibers of, for example, pyocin R or of INCO particles. In contrast, YUI-FURIHATA (1972) has presented a model of pyocin R which shows fibers with sharp central bends. In addition, the fibers were unusually long in the model, measuring about 70 nm.

The sheathless rods, such as pyocin 28 of *P. aeruginosa* , very probably also attach to the cell surface with their terminal fiber (or fiber-like appendages). In support of this assumption

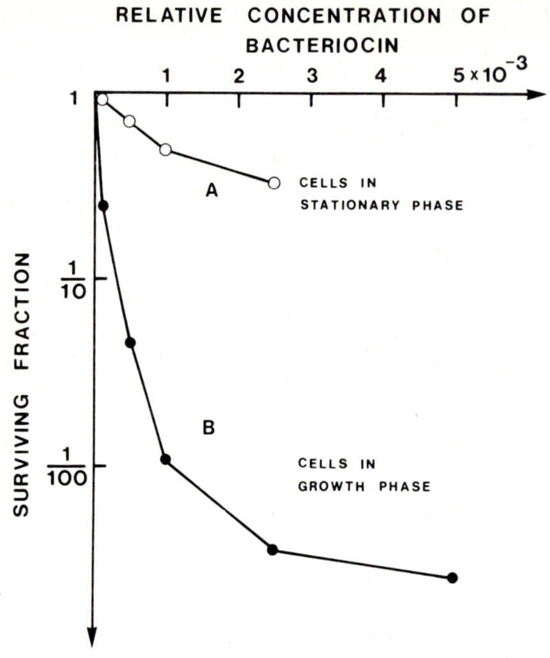

Fig. 15. Survival of INCO-sensitive cells in stationary (A) and early growth phase (B) after treatment with bactericidal INCO particles. Cultures A and B (at a cell titer of $2x10^8$/ml) were treated for 15 min respectively with increasing concentrations of the bacteriocin and then diluted at least 10^{-4} in buffer before they were plated for colony formers (LOTZ and MAYER, 1972)

it has been observed that the free end of the adsorbed rods is always square and does not carry the fiber(s) (GISSMANN, 1974; TAKEYA et al., 1969).

D. Mode of Action

1. Effects on Sensitive Cells

In the case of pyocin R and of INCO particles, the relation between the logarithm of bacterial survival and the concentration of added particles was linear for about the first two $\log_{10}$ of survival (KAGEYAMA et al., 1964; LOTZ and MAYER, 1972; see Fig. 15). The inactivation of sensitive cells by pyocin R is probably a single-hit process, where a single killing unit corresponds to only a few particles (KAGEYAMA et al., 1964; KAZIRO and TANAKA, 1965a).

At higher pyocin R or INCO concentrations the slopes of the survival curves decreased. In the case of pyocin R, it was mentioned that this decrease may partially be due to the presence of pyocin R-resistant mutants. On the other hand, it has been shown that the cells of *Rhizobium* strain 16-2, which survived higher concentrations of INCO particles, were nevertheless of wild-type sensitivity (LOTZ and MAYER, 1972). This indicates that the relative insensitivity of cells at higher bacteriocin concentrations is not due to genetic, but rather physiological differences within the cell population. Similar problems have been discussed in the case of some low molecular weight bacteriocins (see review by

REEVES, 1972), and it has been proposed that the decrease in slope at higher bacteriocin concentrations may be due to a heterogeneity in the number of active receptor sites per cell.

A strong influence of the physiological state on the pyocin R- or INCO-sensitivity is indicated by the finding that growing cells of *P. aeruginosa* or *Rhizobium* are much more sensitive to the mentioned agents than bacteria in the stationary phase (Fig. 15).

It has been shown by KAZIRO and TANAKA (1965a) that addition of pyocin R to growing cells of *P. aeruginosa* results in an immediate cessation of the bacterial DNA-, RNA-, and protein synthesis. Furthermore, there is probably no turnover of either nucleic acids or protein in the pyocin R-treated cells, since an incorporation of ^{14}C-adenine or ^{14}C-phenylalanine into cold or hot TCA-insoluble fractions, respectively, has not been observed. Inhibition of DNA synthesis was not accompanied by a degradation of DNA into small, TCA-soluble fragments, and the sedimentation coefficients of DNA from untreated and DNA from pyocin R-treated cells were similar (19.0 S and 18.0 S, respectively; KAZIRO and TANAKA, 1965b).

After pyocin R treatment, the rate of ^{32}P-incorporation into TCA-soluble nucleosides decreased to about half that of the untreated control (KAZIRO and TANAKA, 1965a). Since the respiration of treated cells proceeded at an almost constant rate, for a considerable period of time after pyocin R addition, the authors concluded that the primary effect of pyocin R action was unlikely to involve an inhibition of the bacterial energy metabolism.

Another investigation of the pyocin R-mediated inhibition of protein synthesis indicated an effect on the bacterial ribosomes (KAZIRO et al., 1964; KAZIRO and TANAKA, 1965b). Extracts of pyocin R-treated cells were unable to support the poly-U-dependent incorporation of ^{14}C-phenylalanine into the acid-insoluble fraction. Moreover, the ribosomes of these cells showed a marked alteration in their sedimentation pattern, indicative of their degradation.

The action of the sheathless pyocin 28 rods on sensitive cells has been examined by electron microscopy (OHNISHI et al., 1971). While untreated cells showed the fibrous nucleoplasm with low electron density surrounded by a clear margin of cytoplasm, the nucleoplasm of pyocin 28-treated bacteria was dispersed throughout the cells with no clear boundary between it and the cytoplasm. The mechanism leading to this morphological change is unknown.

The authors discussed the possibility that the pyocin 28 adsorption may affect bacterial DNA bound to the cytoplasmic membrane, resulting in the dispersion of the DNA.

2. Mechanism Leading to Inactivation

It is conceivable that tail-like particles either kill sensitive cells by their specific contact with the bacterial surface or, alternatively, by injecting a bactericidal substance (KAZIRO and TANAKA, 1965b).

Killing by injection: The hypothetical component to be injected probably does not consist of nucleic acid, since pyocin R particles most likely do not contain DNA or RNA (see Section II.B.1.). Moreover, the bactericidal activity of particles from different sources proved UV-resistant (Section IV.A.), a property not to be expected, if nucleic acid was the bactericidal principle. Although pyocin R can inhibit the protein synthesis of sensitive cells (previous Section), it was not possible to demonstrate the presence or the accumulation of an inhibitor of *in vitro* protein synthesis, in extracts of pyocin R-treated cells (KAZIRO and TANAKA, 1965b). The authors, therefore, concluded that pyocin R particles remain at the surface of sensitive cells and do not inject such a substance.

Killing by contact: If sensitive cells are killed by a specific contact with a phage tail-like bacteriocin, the question arises as to which component of the latter may be responsible for the effect.

With the sheathed particles the experimental approach to this problem could prove difficult, because their structural integrity appears to be a prerequisite for their bactericidal activity. SHINOMIYA (1972a) has shown that only the fully assembled form of pyocin R, but not its intracellular precursors, are active (Section III.C.). Similar problems were also referred to by DUCKWORTH (1970) when discussing the killing action of nucleic acid-free "ghosts" of T-even *E. coli* bacteriophages. The author reported that several attempts to identify the bactericidal components by fractionation of the ghosts were unsuccessful.

In contrast, the following two observations indicate that the sheathless particles, such as pyocin 28 of *P. aeruginosa* or bacteriocin 16-2 of *Rhizobium*, need not be structurally intact in order to act bactericidally: (1) TAKEYA et al. (1969) showed that the pyocin 28 activity was scarcely affected by ultrasonic treatment (10 kcyc./sec, 200 W) for 10 min, and only slightly reduced after 30 min. (2) The original bacteriocin 16-2 activity was not diminished significantly by CsCl density gradient centrifugation, although this procedure led to a dissociation of the rods into ring-like particles (GISSMANN, 1974).

This resistance of the pyocin 28 and bacteriocin 16-2 activity could be explained, if only the pointed end of the rods harbored a bactericidal component, and if the latter was able to inactive sensitive bacteria by specifically adsorbing to a cellular receptor. The question therefore arises, whether such a component can be isolated from the particles and shown to exert the bactericidal activity common to the structurally intact rods.

The mode of action of the phage tail-like particles will be discussed further in Section V.A.3., where it will be compared with the bactericidal action of the more extensively studied nucleic acid-free "ghosts" of T-even *E. coli* bacteriophages.

E. Vibriocin: An Exception to the Rule?

In 1968, JAYAWARDENE and FARKAS-HIMSLEY presented evidence for the particulate and phage tail-like nature of vibriocin, a bacteriocin produced by *Vibrio comma* (strain ATCC 9168). Since this bacteriocin differs markedly from the known phage tail-like bacteriocins described above, it will be discussed separately.

1. General Properties

Vibriocin differs from most phage tail-like bacteriocins in its sensitivity to trypsin, its heat-resistance, and its stability at extreme pH-values.

Within 7 to 10 min after its adsorption to sensitive cells, the bactericidal activity of vibriocin can be inhibited by incubating the bacteria with trypsin (FARKAS-HIMSLEY and SEYFRIED, 1963; JAYAWARDENE and FARKAS-HIMSLEY, 1970). The vibriocin activity proved resistant to heat-treatment at 56°C for at least 3 h or to boiling for 1 min, but was inactivated completely by boiling for 10 min or more. The vibriocin activity was stable within a pH-range of 1,3 to 12,5 (30 min of treatment at 30°C; FARKAS-HIMSLEY and SEYFRIED, 1963).

In contrast, the bactericidal activity of pyocin R was stable only between pH 7 and 9 (KAGEYAMA and EGAMI, 1962), and it was completely destroyed by a heat-treatment of 70°C for 10 min (HIGERD et al., 1967).

Vibriocin furthermore differs from other tail-like bacteriocins (Section IV.B.), because it adsorbs to the (insensitive) producer strain (JAYAWARDENE and FARKAS-HIMSLEY, 1969).

2. Production and Purification

Vibriocin is not produced spontaneously by *V. comma*, but its synthesis can be induced with mitomycin C. The mitomycin C-treated bacteria lysed after 90 to 120 min, releasing the vibriocin (JAYAWARDENE and FARKAS-HIMSLEY, 1968 and 1969). This crude lysate, after a 10-fold concentration by lyophilization, was purified on an Sephadex G-200 column. The filtration resulted in two peaks of which the first one overlapped with the peak for vibriocin activity. The active fractions were pooled and centrifuged at 150,000x g for 6 h. The sediment contained about 90 % of the total vibriocin activity. The resuspended pellet contained many particles resembling sheathed phage tails. Such particles were neither observed in the supernatant of the high-speed centrifugation, nor in the second (inactive) peak of the gel-filtration. It was therefore concluded that these particles are the bactericidal agent. They have a lenght of about 110 nm. Their sheaths were mostly in the contracted state, having an outer diameter of approximately 24 nm (JAYAWARDENE and FARKAS-HIMSLEY, 1968 and 1969).

3. *Mode of Action*

It is likely that the vibriocin action affects membrane functions of sensitive cells. JAYAWARDENE and FARKAS-HIMSLEY (1970) have shown that there is extensive leakage of previously accumulated $^{42}K^{+}$ and of UV-absorbing material from the sensitive bacteria after adsorption of the vibriocin. In addition, vibriocin almost immediately inhibited the transport of metabolic precursors for DNA- and RNA-synthesis (KROL and FARKAS-HIMSLEY, 1972).

Vibriocin also affects the macromolecular syntheses of sensitive cells. Within 5 min after its adsorption, the bacterial DNA synthesis stopped, RNA synthesis was strongly inhibited, and protein synthesis continued at about one half the normal rate (JAYAWARDENE and FARKAS-HIMSLEY, 1970). In contrast to the action of pyocin R (Section IV.D.1.) vibriocin-treatment led to a degradation of the bacterial DNA, as shown by changes in the acid-precipitable radioactivity from ^{3}H-thymidine-prelabeled DNA.

In conclusion, the mentioned properties of vibriocin are reminiscent of some low molecular weigth bacteriocins (for reviews of the latter: see NOMURA, 1967 and REEVES, 1972). A further characterization of the active principle of vibriocin should prove of interest and may explain why it differs so markedly from the described phage tail-like bacteriocins.

Of the other tail-like bacteriocins, only boticin P of *C. botulinum* (LAU et al., 1974) has been described as trypsin-sensitive. The boticin activity was completely destroyed by treating the agent with trypsin for 60 min in 0,1 M phosphate buffer at pH 8,0. The samples were adjusted to pH 7,2 before testing for bacteriocin activity. The data presented do not completely exclude an inactivation of boticin P, not caused by the trypsin, but by the 1 h of incubation at pH 8 during the enzyme treatment: the authors have shown that this bacteriocin is sensitive to pH; it is stable within a pH-range of 6,5 to 7,5.

V. Evolution

A. Relationship to Intact Bacteriophages

1. *Structural and Serological Relationships*

Morphological similarities between the described particles and the tails of intact bacteriophages have been discussed by numerous authors. An interesting similarity exists between the contractile tail of *P. aeruginosa* phage PB-1 and sheathed phage tail-like particles such as pyocin R (BRADLEY, 1967; BRADLEY and ROBERTSON, 1968). Like pyocin R, phage PB-1 carries short tail fibers attached to its base plate. The PB-1 fibers are folded up against the extended sheath, while particles with contracted sheaths have their fibers splayed out from the base plate. A similar orientation of fibers has been postulated for the extended, respectively contracted INCO particles of *Rhizobium spec.* (LOTZ and MAYER, 1972).

In the case of bacteriophages T4 of *E. coli* (MOODY, 1967a and b) and G of *Bacillus megaterium* (DONELLI et al., 1972) sheath contraction has been discussed in terms of a rearrangement of the structural sheath subunits. This would lead to a partial intercalation of the subunits from neighboring sheath annuli. Preliminary optical diffraction studies indicate that sheath contraction of the mentioned INCO particles may occur in a similar fashion (PFISTER, 1974). More extensive studies of this type are needed in order to evaluate the structural relationships between the defective particles and known intact bacteriophages.

Similarities also exist in the amino acid composition of the sheath subunits of several defective phages (e.g. pyocin R, INCO particles, and rhapidosomes from *S. grandis*) and of intact *E. coli* bacteriophage T2 (SARKAR et al., 1964; see also Section II.B.2.).

It has been shown by HOMMA and co-workers (1967 and 1968) that serological relations exist between R-type pyocins ($A1_{mc}$, R_{mc}, and R_{sp}) and intact temperate bacteriophages of *P. aeruginosa*. Out of 10 temperate phages investigated, three (Aø2(2), Aø11(11), and Aø14(14)) showed a cross-reaction with antisera produced against any of the three pyocins. Conversely, antisera against these phage strains showed cross-reactions with the three pyocins. Temperate bacteriophage Aø11(11) was found to possess a contractile sheath similar to that of the pyocin R-type particles.

ITO and KAGEYAMA (1970) also described an intact temperate phage (PS3) of *P. aeruginosa* which was immunologically and morphologically related to R-type pyocins (R, R2, R3, and R4). Like bacteriophage Aø11(11), phage PS3 also has a contractile sheath. From the fact that antiserum produced against the pyocin R-type particles was able to neutralize phage PS3, it has been concluded that both types of particles have certain components in common. Such a component probably resides in structures other than the sheath (e.g. the base plate or the fibers), because the antiserum against isolated pyocin R-type sheaths, while inactivating pyocin-action, did not markedly inactivate the phage.

2. *Common Receptor Sites*

ITO and KAGEYAMA (1970) have analyzed the receptor sites for phage PS3 and for different R-type pyocins of *P. aeruginosa*. Mutants of *P. aeruginosa* resistant to either one of these agents were selected and tested for cross-resistance. In this way, a model concerning the arrangement of the receptor sites, along a common backbone of the cell-wall was established (Fig. 16). The results indicate that the receptor site for phage PS3 may be identical with that for pyocin R4 and is closely related to that for pyocin R2. Most bacterial mutants resistant to pyocin R showed resistance to the other agents, while only a few remained sensitive. This result could be explained by assuming a branching of the pyocin R-receptor, off the common receptor backbone, as shown in Fig. 16.

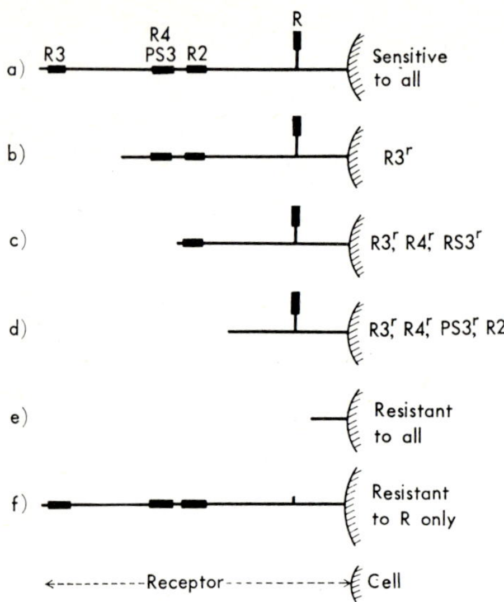

Fig. 16. Schematic representation of the receptors for different R-type pyocins and for phage PS3 of *Pseudomonas aeruginosa* strain P15-16 (ITO and KAGEYAMA, 1970)

3. Common Mode of Action

The effects of pyocin R on susceptible cells (Section IV.D.1.) resembles that of the nucleic acid-free "ghosts" of the virulent bacteriophage T4 of *E. coli* (for a review on the action of T-even ghosts, see: DUCKWORTH, 1970). In both cases a specific attachment of the particles to the bacteria leads to an inhibition of the host macromolecular syntheses, but leaves the energy-generating metabolism (partially) intact. No immediate degradation of the host chromosome has been observed in either case.

The inhibition of protein synthesis in T2 ghost-treated *E. coli* cells is due to a defect in the bacterial ribosomes (FABRICANT and KENNELL, 1970; NUGENT and KENNELL, 1972). SIMON and KENNELL (1974) have shown recently that the 16S rRNA of the 30S ribosomal subunit is being degraded due to the ghost-treatment. It should be interesting to find out whether the ribosomes of pyocin R-treated *P. aeruginosa* cells, which are also inactivated, show similar defects.

Pyocin R particles, as well as T-even ghosts, were shown to induce leakage of cellular material into the medium, indicating changes in the membrane properties of the bacteria. In the case of pyocin R, KAGEYAMA et al. (1964) have shown that UV-absorbing (at 260 nm) material is released from *P. aeruginosa* cells in amounts nearly proportional to the concentration of the pyocin added.

As pointed out by NOMURA (1967), the significance of the leakage induced by pyocin R is not clear, because some lytic enzyme is synthesized simultaneously with the pyocin R particles (KAGEYAMA et al., 1964). The observed leakage could therefore result from the action of such an enzyme, adsorbed to the particles, rather than from the pyocin itself. On the other hand, DUCKWORTH (1969) has demonstrated that the product of the T4 lysozyme gene, plays no role in the biological activity of phage ghosts.

A leakage of cellular contents has also been demonstrated after infection of bacteria with intact bacteriophages (see review by DUCKWORTH, 1970). SHAPIRA et al. (1974) have shown that the adsorption to *E. coli* cells of either intact phage T4 or T4-ghosts caused a leakage of intracellular potassium with identical efflux-rates during the first minute. If protein synthesis was allowed during the following minutes, the leak was "sealed" and the intracellular concentration of potassium restored in phage-, but not in ghost-treated cells.

From these observations, the authors concluded that bacteriophage T4, as well as its ghosts, may inflict the same type of initial damage to the bacterial permeability barrier. This damage would normally be repaired by a phage-specific process, which occurred after phage infection. As pointed out by SHAPIRA and co-workers, the question arises whether this repair process is directly related to the genetic exclusion of superinfecting T4 phage genomes and to the "immunity" to T4 ghosts of T4-infected bacteria.

Superinfection exclusion (CHILDS, 1973; MUFTI, 1972) and immunity to ghosts (DUCKWORTH, 1971; VALLÉE et al., 1972) require the expression of the *imm* gene of bacteriophage T4 (CHILDS, 1973; CORNETT and VALLÉE, 1973; VALLÉE and CORNETT, 1972). The *imm* gene product is thought to act in a stoichiometric rather than a catalytic fashion (OKAMOTO, 1973; VALLÉE and CORNETT, 1973). Recently, OKAMOTO and YUTSUDO (1974) have shown that, in addition to the *imm* gene product, the *s* gene product of phage T4 is also required for establishment and maintenance of the immunity of T4-infected cells to T4 ghosts.

VALLÉE and CORNETT (1972 and 1973) have proposed a hypothesis for the mechanism underlying immunity to T4 ghosts, which is based on the electron microscopic characterizations of the *E. coli* cell envelope performed by BAYER and STARKEY. BAYER (1968a) has demonstrated that adhesions exist between the cytoplasmic membrane and the cell-wall of *E. coli* (200 to 400 in the case of *E. coli* B). Adsorbed T1 to T7 phages (BAYER, 1968b) and phage ØX174 (BAYER and STARKEY, 1972) were shown to be preferentially located over such areas. It was concluded that phage adsorption occurs almost exclusively to these areas of the cell-wall, where infection of the cell by the viral nucleic acid probably occurs.

VALLÉE and CORNETT (1972 and 1973) proposed that a cell wall-membrane junction might contain <u>several specific</u> transmission "channels" for the different substances or structures acting on the membrane from the outside of the cell (for example: colicins, different phages, etc.). T4-ghosts, adsorbed to the receptors of the host cell-wall, would transmit their disruptive effect to the cytoplasmic membrane through specific sites within the cell wall-membrane adhesions. It has been suggested by DUCKWORTH (1970) that the inhibitory effects of ghost-adsorption may be due to allosteric changes within the cytoplasmic membrane, which would lead to changes in some membrane functions.

SABET and SCHNAITMAN (1973) have estimated that there are about 220 molecules of the colicin E3 receptor protein per cell, which is close to the number of cell wall-membrane adhesions reported by BAYER (1968a). The authors therefore speculated that the colicin E3 receptor may also be located at the adhesion areas.

VALLÉE and CORNETT (1972 and 1973) further suggested that the T4-immunity reaction hinders the mentioned transmission and injection processes due to a stoichiometric association of the *imm* protein with a component of the adhesion sites. Correspondingly, the *s* gene product, which is thought to act "cooperatively" with the *imm* protein (OKAMOTO and YUTSUDO, 1974) may also interact with the adhesion sites.

It is possible that the adsorption of certain temperate bacteriophages, like that of T4-ghosts, may inflict damage to the bacterial cell which would be repaired by a phage-specific process, once the viral genome had entered its host. If the phage tail-like bacteriocins originated from such phages, their bactericidal activity could be explained: the defective particles would still be able to inflict the "damage-by-adsorption" to their host, but – due to the absence of a functional genome – would be unable to reverse this potentially lethal effect.

To test this hypothesis, it would be interesting to know whether the ghosts of temperate bacteriophages such as phage PS3 or AØ11(11), which are possibly related to the pyocin R-type bacteriocins of *P. aeruginosa* (see Section V.A.1.), harbored bactericidal activity.

If such temperate phages existed, their prophages should be expected to code for two different types of immunity-substances: (a) for a repressor which protects the lysogenic cell by suppressing the "lytic" functions of the prophage and of superinfecting, homologous phage chromosomes (PTASHNE, 1971), and (b) for a substance functionally similar to the *imm*- and *s*-gene products of phage T4, which would protect the lysogenic cell against the damage inflicted by the adsorption of homologous phage particles. The *imm*- or *s*-type substance would not be necessary, where the prophage coded for a lysogenic conversion of the cell wall, leading to a loss of the phage receptor activity of the cell surface.

B. Relation to Low Molecular Weigth Bacteriocins

The similarities between "low" molecular weight bacteriocins and intact bacteriophages have been discussed by a number of authors (FREDERICQ, 1963; IVÁNOVICS, 1962; LURIA, 1964; NOMURA, 1963; TAUBENECK, 1967; for a recent review, see REEVES, 1972).

It has often been mentioned that some low molecular weight bacteriocins may have evolved from bacteriophages (or from phage tail-like bacteriocins) and may represent extreme cases of defective lysogeny. The bacteriocins may consist of base plate components; their adsorption to specific phage receptors would lead to an inactivation of sensitive bacteria. For example, it has been shown recently that the low molecular weight colicin M and bacteriophage T5 of *E. coli* bind to the same receptor protein, which is located in the outer membrane of the bacteria. The protein consists of a single polypeptide chain of molecular weight 85.000 (BRAUN et al., 1973; BRAUN and WOLFF, 1973).

On the other hand, in the case of pyocin R, the mentioned hypothesis is not supported by the available data. No serological relationship exists between two low molecular weight bacteriocins (A2 and A3) and three R-type pyocins ($A1_{mc}$, R_{mc}, R_{sp}) as well as 10 kinds of temperate phages, all of *P. aeruginosa* (HOMMA and SHIONOYA, 1967). It has also been shown that four R-type pyocins (R, R_2, R_3, R_4) differed serologically and in their receptor specifity from the low molecular weight pyocin S (ITO et al., 1970).

Because only little is known about the mechanism of action of phage tail-like bacteriocins, a comparison with the action of their low molecular weight couter-parts is scarcely possible. Interesting parallels appear to exist between the mode of action of T2-ghosts (previous Section) and that of colicin E3 (BOWMAN et al., 1973). Both agents inactivate the 30S ribosomal subunit and induce alterations in its 16S rRNA.

C. Origin

Defective prophages may have arisen in nature by progressive addition or reduction of phage-specific genetic information (BRADLEY, 1967; GARRO and MARMUR, 1970). In general, only the loss of genetic information has been discussed in the literature. Such a loss could result from spontaneously occurring point- or deletion-mutations of prophage genes essential for phage viability (e.g. for plaque-forming ability). The lethal genetic defects would remain "silent" as long as gene functions involved in lytic phage growth remained repressed. In this case, the prophage DNA would be passively replicated as part of the host chromosome.

After induction of lysogenic cells, phage genomes with deletions may be generated by rarely occurring "illegitimate cross-overs" between the prophage and a segment of the bacterial chromosome adjacent to the prophage locus; (this mechanism has been described for the formation of specialized transducing phages, such as phage Lambda of *E. coli*; CAMPBELL, 1968 and 1971; FRANKLIN, 1971; OZEKI and IKEDA, 1968).

The prophage abnormally excised in this fashion from the bacterial chromosome would be covalently linked at one end to a short segment of bacterial DNA and would have lost a corresponding fragment of its own DNA from the other end; the latter would have been "left behind" in the bacterial chromosome. The resulting defective virus progeny may still possess all functions necessary for the establishment of lysogeny or, alternatively, may be dependent on corresponding functions of an intact homologous "helper phage" , co-infecting the host simultaneously with the defective phage.

The origin of the tail-like particles from intact bacteriophages is indicated where phage head-like structures are synthesized in addition to the tails. This is the case with *Listeria monocytogenes* strain 1896x (BRADLEY and DEWAR, 1966). The authors reported that numerous empty tail-less heads were produced by *L. monocytogenes* in addition to the sheathed tail-like particles, and that in a few cases the heads were found attached to tails. Furthermore, RUCINSKY et al. (1972) occasionally observed empty heads attached to the sheathed particles produced by *Chromobacterium violaceum*.

The genetic information of a defective prophage may be reduced further by mutation. For example, a spontaneous mutant of defectively lysogenic strain D-52 of *P. mirabilis* was shown to produce only polysheath and polycore-like tubes, whereas the wild-type also synthesized the sheathed tail-like particles (TAUBENECK, 1967 and 1969). In addition, nitrosoguanidine-induced mutants of wild-type strain D-52 were isolated, which produced only polycore-like tubes.

An interesting example for the loss of genetic information of a defective prophage was recently reported for defective phage PBSX of *Bacillus subtilis*. PBSX particles can act bactericidally and consist of a sheathed tail and a head which contains only the DNA of their host (see review by GARRO and MARMUR, 1970). THURM and GARRO (1974) isolated and mapped a number of PBSX prophage mutants. The PBSX-specific mutations were shown to be clustered between *arg* C and *met* C on the host chromosome. One of the mutants proved to be defective for phage head-formation and, following induction, produced only tail structures. The tails still exhibited the killing activity of the whole phage.

It is conceivable that with increasing defectivity of the prophages it may become increasingly difficult to identify the produced particles as "phage-like". For example, newly isolated strain 16-20 of *Rhizobium* was found to produce only non-bactericidal particles which resemble "flat discs" (Fig. 17). The particles can be seen primarily in axial view and have a central hole. Their outer diameter (20 nm) is similar to that of the contracted sheath of INCO particles (24 nm) produced by *Rhizobium* strain 16-3. When positioned vertically on the supporting carbon film, the discs should be visible as "white lines" in the film of negative stain. Such lines were observed on the micrograph shown in Fig. 17 and set in brackets. The lines have a length of about 23 nm and a width of 7 nm. Curved projections can be seen around the periphery of some of the particles. The average angle between two adjacent projections was 30°, indicating a 12-fold rotation symmetry of the particles. It is possible that the discs consist of sheath subunits, which may be unable to polymerize into longer sheath- or polysheath-like tubes.

D. Function

The widespread occurrence of defective prophages in different bacterial species has been taken as an indication for some essential role in normal cellular function (GARRO and MARMUR, 1970). It is interesting in this context that a wide distribution of closely related R-type pyocins was found among the *Pseudomonads*:

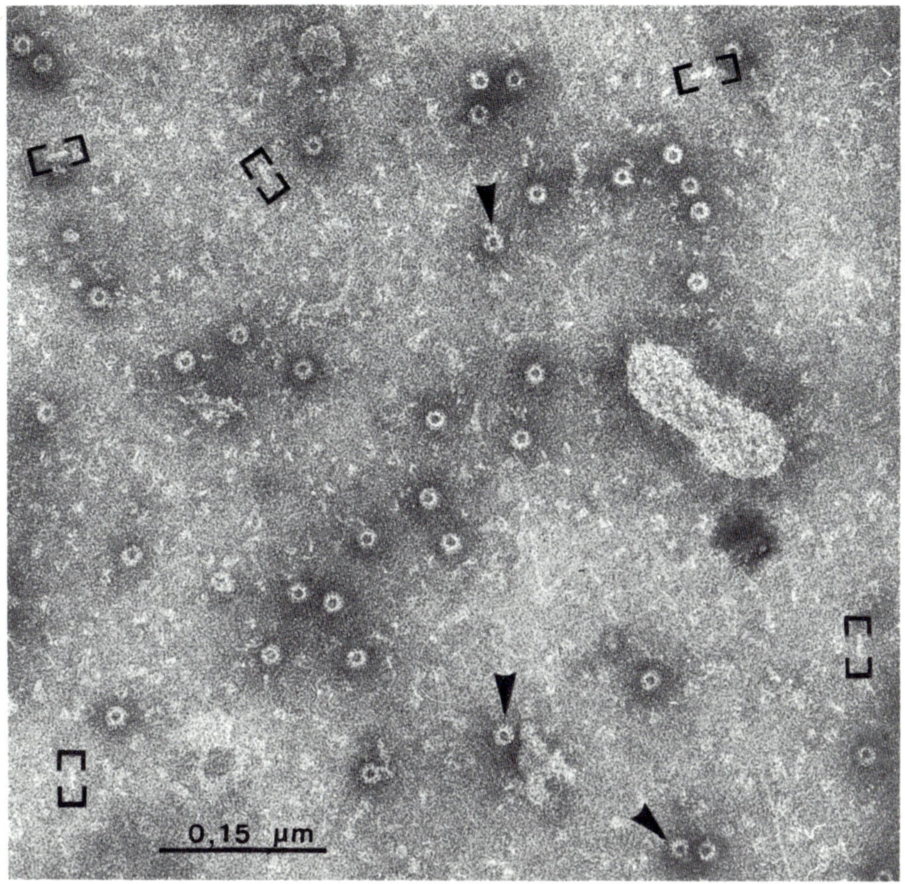

Fig. 17. Electron micrograph of 'flat disks' produced by strain 16-20 of *Rhizobium spec.* Curved projections can be discerned around the periphery of some of the disks (see arrows). The "white lines" in brackets probably represent disks positioned vertically on the supporting carbon film. PTA

pyocin R and R4 are produced by strains originally isolated by J.A. HOMMA in Tokyo, and the R2- and R3-producing strains were isolated in Australia (ITO et al., 1970).

As discussed for the producers of low molecular weight bacteriocins (REEVES, 1972), it may be argued that the production of the bactericidal particles confers a selective advantage to the (bacteriocin resistant!) defective lysogens. In a natural habitat the particles would kill sensitive bacteria of closely related strains, which would otherwise compete with the bacteriocinogenic strain for the "natural resources".

Defective prophages, as their intact counter-parts, may help to protect the bacterial cell against virus infection. (1) It is possible that some of the defective prophages code for a conversion of the bacterial cell envelope (Section IV.B.). Due to conversion the respective strain would be resistant to a number of potentially lethal bacteriophages or bacteriocins (for a recent review on lysogenic conversion, see: BARKSDALE and ARDEN, 1974). (2) Where phage adsorption is not prevented, the invading foreign viral nucleic acid may be broken down by prophage-coded restriction enzymes (see reviews by ARBER, 1971, and ARBER and LINN, 1969). (3) If the infecting DNA was not restricted and homologous to the prophage DNA (a possibility not unlikely for lysogens in a natural habitat), its transcription into mRNA would be repressed by prophage-coded repressor proteins (PTASHNE, 1971).

Finally, as mentioned in the previous Section, in some of the naturally occurring defective prophages a portion of the original phage genome may have been replaced by a chromosomal fragment from the previous bacterial host. It is possible that the genes located on the transduced chromosomal fragment confer a selective advantage to the defectively lysogenic strain.

Probably not all of the defective prophages are of selective advantage; some of them even appear to have a harmful effect on their host. For example, a correlation may exist between the autolysis of a culture of *S. grandis* bacteria reaching the stationary phase of growth (LEWIN, 1963) and the spontaneous induction of the rhapidosome prophage: large numbers of (non-bactericidal) rhapidosome particles are liberated by the lysing cells.

VI. Conclusions

With respect to structural complexity, chemical composition, receptor specifity, and - possibly - mode of action, the phage tail-like particles are similar to the tails of intact bacteriophages.

Many of the defective lysogens have been shown to be inducible and to be inactivated upon induction. In addition, it has been shown that the cells of *P. aeruginosa* are free of pyocin R-specific precursors before induction. As with intact lysogens (for recent reviews, see: CALENDAR, 1970; DOVE, 1971; ECHOLS, 1971; HERSKOWITZ, 1973) the induction probably involves a turn-on of "lytic" functions of the defective prophage, which are otherwise repressed.

Little is as yet known about the genetic determinants coding for the phage tail-like particles; in the case of the pyocin R-prophage their chromosomal location has been demonstrated. As discussed by GARRO and MARMUR (1970), naturally occurring defective prophages, like intact ones, may also exist as extrachromosomal plasmids. For example, the genome of temperate *E. coli* phage P1 (IKEDA and TOMIZAWA, 1968), and of two mutants of phage Lambda (λdv, MATSUBARA and KAISER, 1968; λN^-, SIGNER, 1969, and LIEB, 1970) are carried as plasmids.

As pointed out by HAYES (1968), lysogeny is now known to be so prevalent and widespread among the strains of different bacterial species that it must be regarded as the normal, rather than the exceptional, state. If, as it appears, the naturally occurring defective phages originated (and originate) from intact prophages, defective lysogeny is probably as widespread in nature as its intact counter-part.

The study of naturally occurring defective lysogeny would be facilitated, if the intact "ancestors" of some defective prophages could be isolated. As yet, revertants to wild-type genotype have not been described in the literature. This may indicate the presence of deletions in the defective genomes or, alternatively, the reversion frequencies are small enough to have remained undetected.

Reversion frequencies for defective mutants of phage Lambda are typically in the order of 10^{-6} per generation (CAMPBELL, 1968). The number of revertants is therefore vastly less than the amount of phage particles found in a culture of intact lysogens. Revertants to wild-type would be detected by their ability to form plaques on an indicator. Since defective particles with bactericidal activity can inhibit phage growth (LOTZ and MAYER, 1972), an attempt to separate possible revertants from the mass of mutant particles may prove essential before testing for plaque-formers.

Although a number of phage tail-like particles from different bacterial species has been shown to act bactericidally, little is known about their mode of action. The inactivation of bacteria may be due to a "damage-by-adsorption", rather than to the injection of a bactericidal substance.

Since the primary target of the bactericidal action may be located on the inner membrane, it would be interesting to find out whether (1) the particles, like a number of *E. coli* phages, adsorb to areas of adhesion between the cell-wall and the cytoplasmic membrane and (2), whether mutants can be isolated from sensitive strains, which still have the receptors for particle-adsorption, but which are tolerant to the bactericidal action. In this case changes in some membrane properties should be expected.

Because of their relative structural simplicity, the (non-contractile) pyocin 28-type particles appear well-suited for an attempt to isolate from them the (hypothetical) bactericidal component. If the terminal fiber of pyocin 28 constitutes the bactericidal principle, specific adsorption of isolated fibers to sensitive bacteria should lead to an inactivation of the cells.

Acknowledgments

I wish to thank the authors who kindly provided electron micrographs and diagrams for the review. I would also like to thank the following for helpful comments: Ann S. DELK, A.J. GARRO, M. KAGEYAMA, H. PFISTER, and U. TAUBENECK. I am indepted to R. SCHMITT and F. MAYER for a critical reading of the manuscript, to F.C. CANNON and J.E. BERINGER for correcting my English, and to Dörthe SKRAWEK and Beatrix GÖRG for excellent technical assistance. The cited papers of the author have been supported by the Deutsche Forschungsgemeinschaft. Literature survey has been completed with few exceptions to December 1974.

References

ACKERMANN, H.W., BROCHU, G.: Particulate bacteriocins, p. 629. Handbook of Microbiology (Ed. A.I. LASKIN and H.A. LECHEVALIER), vol. I. Cleveland: CRC Press 1973.

ACKERMANN, H.W., GAUVREAU, L.: Phages défectifs chez *Chromobacterium*. Zbl. Bakt. Hyg., I. Abt. Orig. A, 221, 196 (1972).

ADHIKARI, P.C., CHATTERJEE, S.N.: Rhapidosomes in *Vibrio* species. Can. J. Microbiol. 18, 541 (1972).

AMAKO, K., YASUNAKA, K., TAKEYA, K.: Relationship between rhapidosome and pyocin in *Pseudomonas fluorescens*. J. Gen. Microbiol. 62, 107 (1970).

ARBER, W.: Host-controlled variation. In: The bacteriophage Lambda (Ed. A.D. HERSHEY), p. 83. Cold Spring Harbor Lab. 1971.

ARBER, W., LINN, S.: DNA modification and restriction. Ann. Rev. Biochem. 38, 467 (1969).

BARKSDALE, L., ARDEN, S.B.: Persisting bacteriophage infections, lysogeny, and phage conversions. Ann. Rev. Microbiol. 28, 265 (1974).

BAYER, M.E.: Areas of adhesion between wall and membrane of *Escherichia coli*. J. Gen. Microbiol. 53, 395 (1968a).

BAYER, M.E.: Adsorption of bacteriophages to adhesions between wall and membrane of *Escherichia coli*. J. Virol. 2, 346 (1968b).

BAYER, M.E., STARKEY, T.W.: The adsorption of bacteriophage øX174 and its interaction with *Escherichia coli*; a kinetic and morphological study. Virology 49, 236 (1972).

BOWMAN, C.M., SIDIKARO, J., NOMURA, M.: Mode of action of colicin E3. In: Chemistry and functions of colicins (Ed. L.P. HAGER), p. 87. New York and London: Academic Press, Inc. 1973.

BRADLEY, D.E.: The structure and infective process of a *Pseudomonas aeruginosa* bacteriophage containing ribonucleic acid. J. Gen. Microbiol. 45, 83 (1966).

BRADLEY, D.E.: Ultrastructure of bacteriophages and bacteriocins. Bacteriol. Rev. 31, 230 (1967).

BRADLEY, D.E., DEWAR, C.A.: The structure of phage-like objects associated with non-induced bacteriocinogenic bacteria. J. Gen. Microbiol. 45, 399 (1966).

BRADLEY, D.E., ROBERTSON, D.: The structure and infective process of a contractile *Pseudomonas aeruginosa* bacteriophage. J. Gen. Virol. 3, 247 (1968).

BRANDIS, H., ŠMARDA, J.: Bacteriocine und bacteriocinähnliche Substanzen. Jena: VEB Gustav Fischer Verlag 1971.

BRAUN, V., SCHALLER, K., WOLFF, H.: A common receptor protein for phage T5 and Colicin M in the outer membrane of *Escherichia coli* B. Biochim. Biophys. Acta 323, 87 (1973).

BRAUN, V., WOLFF, H.: Characterization of the receptor protein for phage T5 and Colicin M in the outer membrane of *E. coli* B. FEBS Letters 34, 77 (1973).

CALENDAR, R.: The regulation of phage development. Ann. Rev. Microbiol. 24, 241 (1970).

CAMPBELL, A.M.: Techniques for studying defective bacteriophages, p. 279. In: Methods in Virology (Eds. K. MARAMOROSCH and H. KOPROWSKI), vol IV. New York: Academic Press Inc. 1968.

CAMPBELL, A.: Genetic structure. In: The bacteriophage Lambda (Ed. A.D. HERSHEY), p. 13. Cold Spring Harbor Lab. 1971.

CHANG, H.Y.Y., ALLEN, M.M.: The isolation of Rhapidosomes from the blue-green alga, *Spirulina*. J. Gen. Microbiol. 18, 121 (1974).

CHAPMAN, G., HILLIER, J., JOHNSON, F.H.: Observations on the bacteriophagy of *Erwinia carotovora*. J. Bacteriol. 61, 261 (1951).

CHILDS, J.D.: Superinfection exclusion by incomplete genomes of bacteriophage T4. J. Virol. 11, 1 (1973).

CLARK-WALKER, G.D.: Association of microcyst formation in *Spirillum itersonii* with the spontaneous induction of a defective bacteriophage. J. Bacteriol. 97, 885 (1969).

COETZEE, H.L., de KLERK, H.C., COETZEE, J.N.: Bacteriophage-tail-like particles associated with intraspecies killing of *Proteus vulgaris*. J. Gen. Virol. 2, 29 (1968).

CORNETT, J.B., VALLÉE, M.: The map position of the immunity (imm) gene of bacteriophage T4. Virology 51, 506 (1973).

CORRELL, D.L., LEWIN, R.A.: Rod-shaped ribonucleoprotein particles from *Saprospira*. Can. J. Microbiol. 10, 63 (1964).

DE KLERK, H.C., COETZEE, J.N., LECATSAS, G.: Intracellular organization of bacteriocin particles in *Proteus vulgaris*. Arch. Microbiol. 98, 271 (1974).

DELK, A.S., DEKKER, C.A.: Rhapidosomes: Absence of a highly 2'-O-methylated RNA component. Science 166, 1646 (1969).

DELK, A.S., DEKKER, C.A.: Characterization of Rhapidosomes of *Saprospira grandis*. J. Mol. Biol. 64, 287 (1972).

DONELLI, G., GUGLIELMI, F., PAOLETTI, L.: Structure and physico-chemical properties of bacteriophage G. I. Arrangement of protein subunits and contraction process of tail sheath. J. Mol. Biol. 71, 113 (1972).

DOVE, W.F.: Biological inferences. In: The bacteriophage Lambda (Ed. A.D. HERSHEY), p. 297. Cold Spring Harbor Lab. 1971.

DUCKWORTH, D.H.: Role of lysozyme in the biological activity of bacteriophage ghosts. J. Virol. 3, 92 (1969).

DUCKWORTH, D.H.: Biological activity of bacteriophage ghosts and "take-over" of host functions by bacteriophage. Bacteriol. Rev. 34, 344 (1970).

DUCKWORTH, D.H.: Inhibition of T4 bacteriophage multiplication by superinfecting ghosts and development of tolerance after bacteriophage infection. J. Virol. 7, 8 (1971).

ECHOLS, H.: Lysogeny: viral repression and site-specific recombination. Ann. Rev. Biochem. 40, 827 (1971).

EISERLING, F.A., DICKSON, R.C.: Assembly of viruses. Ann. Rev. Biochem. 41, 467 (1972).

FABRICANT, R., KENNELL, D.: Inhibition of host protein synthesis during infection of *Escherichia coli* by bacteriophage T4. III. Inhibition by ghosts. J. Virol. 6, 772 (1970).

FARKAS-HIMSLEY, H., KORMENDY, A., JAYAWARDENE, A.: Electron microscopy of *Vibrio comma* during vibriocin production. Cytobios. 3, 97 (1971).

FARKAS-HIMSLEY, H., SEYFRIED, P.L.: Lethal biosynthesis of a bacteriocin, vibriocin, by *V. comma*. I. Conditions affecting its production and detection. Can. J. Microbiol. 9, 329 (1963).

FRANKLIN, N.C.: Illegitimate Recombination. In: The bacteriophage Lambda (Ed. A.D. HERSHEY), p. 175. Cold Spring Harbor Lab. 1971.

FREDERICQ, P.: On the nature of colicinogenic factors: a review. J. Theoret. Biol. 4, 159 (1963).

GARRO, A.J., MARMUR, J.: Defective bacteriophages. J. Cell Physiol. 76, 253 (1970).

GISSMANN, L.: A new bacteriocin from a strain of *Rhizobium* resembling a flexible bacteriophage tail. Thesis for diploma, University of Erlangen, Germany (1974); (in German).

GOVAN, J.R.W.: Studies on the pyocins of *Pseudomonas aeruginosa*: morphology and mode of action of contractile pyocins. J. Gen. Microbiol. 80, 1 (1974).

GRÄF, W.: Bewegungsorganellen bei Myxobakterien. Arch. Hyg. 149, 518 (1965).

GUMPERT, J., TAUBENECK, U.: "Mikrotubuli" bei *Proteus mirabilis* als Produkte defekter Lysogenie. Z. Allg. Mikrobiol. 8, 101 (1968).

HAYES, W.: The genetics of bacteria and their viruses. Oxford and Edinburgh: Blackwell 1968.

HERSKOWITZ, I.: Control of gene expression in bacteriophage Lambda. Ann. Rev. Genetics 7, 289 (1973).

HIGERD, T.B., BAECHLER, C.A., BERK, R.S.: *In vitro* and *in vivo* characterization of pyocin. J. Bacteriol. 93, 1976 (1967).

HIGERD, T.B., BAECHLER, C.A., BERK, R.S.: Morphological studies on relaxed and contracted forms of purified pyocin particles. J. Bacteriol. 98, 1378 (1969).

HOMMA, J.Y., GOTO, S., SHIONOYA, H.: Relationship between pyocine and temperate phage of *Pseudomonas aeruginosa*. II. Isolation of pyocines from strain P 1-III and their characteristics. Japan. J. Exp. Med. 37, 373 (1967).

HOMMA, J.Y., SHIONOYA, H.: Relationship between pyocine and temperate phage of *Pseudomonas aeruginosa*. III. Serological relationship between pyocines and temperate phages. Japan. J. Exp. Med. 37, 395 (1967).

HOMMA, J.Y., WATABE, H., TANABE, Y.: The temperate phages having serological relationship with the "phage bound pyocine". Japan. J. Exp. Med. 38, 213 (1968).

HUMMELER, K., ANDERSON, T.F., BROWN, R.A.: Identification of poliovirus particles of different antigenicity by specific agglutination as seen in the electron microscope. Virology 16, 84 (1962).

IIDA, H., INOUE, K.: Rhapidosomes of *Clostridium botulinum* type E. Japan. J. Microbiol. 12, 353 (1968).

IKEDA, H., TOMIZAWA, J.: Prophage P1, an extrachromosomal replication unit. Cold Spring Harbor Symp. Quant. Biol. 33, 791 (1968).

IKEDA, K., EGAMI, F.: Effects of antibiotics and antimetabolites on the induced formation of pyocin R. Z. Allg. Mikrobiol. 6, 219 (1966).

IKEDA, K., EGAMI, F.: Receptor substance for pyocin R. I. Partial purification and chemical properties. J. Biochem. (Tokyo) 65, 603 (1969).

IKEDA, K., EGAMI, F.: Lipopolysaccharide of *Pseudomonas aeruginosa* with special reference to pyocin R receptor activity. J. Gen. Appl. Microbiol. 19, 115 (1973).

IKEDA, K., KAGEYAMA, M., EGAMI, F.: Studies of a pyocin. II. Mode of production of the pyocin. J. Biochem. (Tokyo) 55, 54 (1964).

IKEDA, K., NISHI, Y.: Interaction between pyocin R and pyocin R receptor. J. Gen. Appl. Microbiol. 19, 209 (1973).

INOUE, K., IIDA, H.: Bacteriophages of *Clostridium botulinum*. J. Virol. 2, 537 (1968).

ISHII, S.I., NISHI, Y., EGAMI, F.: The fine structure of a pyocin. J. Mol. Biol. 13, 428 (1965).

ITERSON, W. VAN, HOENIGER, J.F.M., NIJMAN VAN ZANTEN, E.: A "microtubule" in a bacterium. J. Cell. Biol. 32, 1 (1967).

ITO, S., KAGEYAMA, M.: Relationship between pyocins and a bacteriophage in *Pseudomonas aeruginosa*. J. Gen. Appl. Microbiol. 16, 231 (1970).

ITO, S., KAGEYAMA, M., EGAMI, F.: Isolation and characterization of pyocins from several strains of *Pseudomonas aeruginosa*. J. Gen. Appl. Microbiol. 16, 205 (1970).

IVÂNOVICS, G.: Bacteriocins and bacteriocin-like substances. Bacteriol. Rev. 26, 108 (1962).

JAYAWARDENE, A., FARKAS-HIMSLEY, H.: Particulate nature of vibriocin: a bacteriocin from *Vibrio comma*. Nature 219, 79 (1968).

JAYAWARDENE, A., FARKAS-HIMSLEY, H.: Vibriocin: a bacteriocin from *Vibrio comma*. I. Production, purification, morphology, and immunological studies. Microbios 1B, 87 (1969).

JAYAWARDENE, A., FARKAS-HIMSLEY, H.: Mode of action of vibriocin. J. Bacteriol. 102, 382 (1970).

KAGEYAMA, M.: Studies of a pyocin. I. Physical and chemical properties. J. Biochem. (Tokyo) 55, 49 (1964).

KAGEYAMA, M.: Genetic mapping of a bacteriocinogenic factor in *Pseudomonas aeruginosa*. I. Mapping of pyocin R2 factor by conjugation. J. Gen. Appl. Microbiol. 16, 523 (1970a).

KAGEYAMA, M.: Genetic mapping of a bacteriocinogenic factor in *Pseudomonas aeruginosa*. II. Mapping of pyocin R2 factor by transduction with phage F116. J. Gen. Appl. Microbiol. 16, 531 (1970b).

KAGEYAMA, M., EGAMI, F.: On the purification and some properties of a pyocin, a bacteriocin produced by *Pseudomonas aeruginosa*. Life Sciences 9, 471 (1962).

KAGEYAMA, M., IKEDA, K., EGAMI, F.: Studies of a pyocin. III. Biological properties of the pyocin. J. Biochem. (Tokyo) 55, 59 (1964).

KAGEYAMA, M., SHINOMIYA, T., OHSUMI, M.: Pyocins or defective phages in *Pseudomonas aeruginosa*. Fed. Proc. 32, 491 (1973).

KAZIRO, Y., TANAKA, M., SHIMAZONO, N.: Mode of action of pyocin: Inactivation of ribosomes in supporting poly U directed incorporation of phenylalanine. Biochem. Biophys. Res. Commun. 17, 624 (1964).

KAZIRO, Y., TANAKA, M.: Studies on the mode of action of pyocin. I. Inhibition of macromolecular synthesis in sensitive cells. J. Biochem. (Tokyo) 57, 689 (1965a).

KAZIRO, Y., TANAKA, M.: Studies on the mode of action of pyocin. II. Inactivation of ribosomes. J. Biochem. (Tokyo) 58, 357 (1965b).

KELLENBERGER, E., BOY DE LA TOUR, E.: On the fine structure of normal and "polymerized" tail sheath of phage T4. J. Ultrastruct. Res. 11, 545 (1964).

KELLENBERGER, E., EDGAR, R.S.: Structure and assembly of phage particles. In: The bacteriophage Lambda (Ed. A.D. HERSHEY), p. 271. Cold Spring Harbor Lab. 1971.

KLUG, A., BERGER, J.E.: An optical method for the analysis of periodicties in electron micrographs, and some observations on the mechanism of negative staining. J. Mol. Biol. 10, 565 (1964).

KLUG, A., DE ROSIER, D.J.: Optical filtering of electron micrographs: reconstruction of one-sided images. Nature 212, 29 (1966).

KROL, P.M., FARKAS-HIMSLEY, H.: Mode of action of vibriocin: effects on membrane permeability and transport. Microbios 6, 199 (1972).

LAEMMLI, U.K.: Cleavage of structural proteins during the assembly of the head of bacteriophage T4. Nature 227, 680 (1970).

LANG, D., McDONALD, T.O., GARDNER, E.W.: Electron microscopy of particles associated with a bacteriocinogenic *Vibrio cholerae* strain. J. Bacteriol. 95, 708 (1968).

LAU, A.H.S., HAWIRKO, R.Z., CHOW, C.T.: Purification and properties of boticin P produced by *Clostridium botulinum*. Can. J. Microbiol. 20, 385 (1974).

LEWIN, R.A.: Rod-sheaped particles in *Saprospira*. Nature 198, 103 (1963).

LEWIN, R.A., KIETHE, J.: Formation of rhapidosomes in *Saprospira*. Can. J. Microbiol. 11, 935 (1965).

LIEB, M.: λ mutants which persist as plasmids. J. Virol. 6, 218 (1970).

LIU, H.J., KAZIRO, Y., HORIUCHI, T.: A temperature-sensitive induction of pyocin R synthesis in a mutant strain of *Pseudomonas aeruginosa* R. Japan. J. Exp. Med. 39, 371 (1969).

LOTZ, W., MAYER, F.: Isolation and characterization of a bacteriophage tail-like bacteriocin from a strain of *Rhizobium*. J. Virol. 9, 160 (1972).

LURIA, S.E.: On the mechanism of action of colicines. Ann. Inst. Pasteur 107 (Suppl. 5), 67 (1964).

MATSUBARA, K., KAISER, A.D.: λdv: an autonomously replicating DNA fragment. Cold Spring Harbor Symp. Quant. Biol. 33, 769 (1968).

MOODY, M.F.: Structure of the sheath of bacteriophage T4. I. Structure of the contracted sheath and polysheath. J. Mol. Biol. 25, 167 (1967a).

MOODY, M.F.: Structure of the sheath of bacteriophage T4. II. Rearrangement of the sheath subunits during contraction. J. Mol. Biol. 25, 201 (1967b).

MUFTI, S.: A bacteriophage T4 mutant defective in protection against superinfecting phage. J. Gen. Virol. 17, 119 (1972).

NOMURA, M.: Mode of action of colicines. Cold Spring Harbor Symp. Quant. Biol. 28, 315 (1963).

NOMURA, M.: Colicins and related bacteriocins. Ann. Rev. Microbiol. 21, 257 (1967).

NUGENT, K., KENNELL, D.: Polypeptide synthesis by extracts from *Escherichia coli* treated with T2 ghosts. J. Virol. 10, 1199 (1972).

OHNISHI, Y., TAKADE, A., TAKEYA, K.: Morphological changes in *Pseudomonas aeruginosa* treated with rod-shaped pyocin 28. Japan. J. Microbiol. 15, 201 (1971).

OKAMOTO, K.: Role of T4 phage-directed protein in the establishment of resistance to T4 ghosts. Virology 56, 595 (1973).

OKAMOTO, K., YUTSUDO, M.: Participation of the *s* gene product of phage T4 in the establishment of resistance to T4 ghosts. Virology 58, 369 (1974).
OZEKI, H.: Methods for the study of colicine and colicinogeny, p. 565. In: Methods in virology (Eds. K. MARAMOROSCH and H. KOPROWSKI), vol. IV. New York: Academic Press Inc. 1968.
OZEKI, H., IKEDA, H.: Transduction mechanisms. Ann. Rev. Genetics 2, 245 (1968).
PATE, J.L., JOHNSON, J.L., ORDAL, E.J.: The fine structure of *Chondrococcus columnaris*. II. Structure and formation of rhapidosomes. J. Cell Biol. 35, 15 (1967).
PFISTER, H.: Characterization of the substructures of a bacteriophage tail-like bacteriocin from a strain of *Rhizobium*. Doctoral dissertation, University of Erlangen, Germany (1974); (in German).
PFISTER, H., LOTZ, W.: Spontaninduktion eines defekten Prophagen von *Rhizobium lupini*. Zbl. Bakt. Hyg., I. Abt. Orig. A 228, 179 (1974).
PRICE, A.R., ROTTMAN, F.: Nucleic acids from *S. grandis*: the absence of 2'-o-methyl RNA. Biochim. Biophys. Acta 199, 288 (1970).
PTASHNE, M.: Repressor and its action. In: The bacteriophage Lambda (Ed. A.D. HERSHEY), p. 221. Cold Spring Harbor Lab. 1971.
RAUTENSTEIN, Y.I., MOSKALENKO, L.N.: Defective lysogeny among several cultures of *Rhizobium meliloti* nodule bacteria. Mikrobiologija 39, 507 (1970). (in Russian)
REEVES, P.: The bacteriocins. (Molecular biology, biochemistry, and biophysics, vol. 11). Berlin-Heidelberg-New York: Springer 1972.
REICHENBACH, H.: Die wahre Natur der *Myxobakterien*-"Rhapidosomen". Arch. Microbiol. 56, 371 (1967).
REICHLE, R.E., LEWIN, R.A.: Purification and structure of rhapidosomes. Can. J. Microbiol. 14, 211 (1968).
RUCINSKY, T.E., COTA-ROBLES, E.H.: The intracellular organization of bacteriophage tail-like particles in cells of *Chromobacterium violaceum* following mitomycin C treatment. J. Ultrastruct. Res. 43, 260 (1973).
RUCINSKY, T.E., GREGORY, J.P., COTA-ROBLES, E.H.: Organization of bacteriophage tail-like particles in cells of *Chromobacterium violaceum*. J. Bacteriol. 110, 754 (1972).
SABET, S.F., SCHNAITMAN, C.A.: Chemistry of the colicin E receptor. In: Chemistry and functions of colicins (Ed. L.P. HAGER), p. 59. New York and London: Academic Press 1973.
SARKAR, N., SARKAR, S., KOZLOFF, L.M.: Tail components of T2 bacteriophage. I. Properties of the isolated contractile tail sheath. Biochemistry 3, 511 (1964).
SHAPIRA, A., GIBERMAN, E., KOHN, A.: Recoverable potassium fluxes variations following adsorption of T4 phage and their ghosts on *Escherichia coli* B. J. Gen. Virol. 23, 159 (1974).
SHINOMIYA, T.: Studies on biosynthesis and morphogenesis of R-type pyocins of *Pseudomonas aeruginosa*. II. Biosynthesis of antigenic proteins and their assembly into pyocin particles in mitomycin C-induced cells. J. Biochem. (Tokyo) 72, 39 (1972a).
SHINOMIYA, T.: Studies on biosynthesis and morphogenesis of R-type pyocins of *Pseudomonas aeruginosa*. III. Subunits of pyocin R and their precipitability by anti-pyocin R serum. J. Biochem. (Tokyo) 72, 499 (1972b).

SIGNER, E.R.: Plasmid formation: a new mode of lysogeny by phage λ. Nature 223, 158 (1969).
SIMON, L.D.: The infection of *Escherichia coli* by T2 and T4 bacteriophages as seen in the electron microscope. III. Membrane-associated intracellular bacteriophages. Virology 38, 285 (1969).
SIMON, L.D., ANDERSON, T.F.: The infection of *Escherichia coli* by T2 and T4 bacteriophages as seen in the electron microscope. I. Attachment and penetration. Virology 32, 279 (1967a).
SIMON, L.D., ANDERSON, T.F.: The infection of *Escherichia coli* by T2 and T4 bacteriophages as seen in the electron microscope. II. Structure and function of the baseplate. Virology 32, 298 (1967b).
SIMON, M., KENNELL, D.: Defective 30S ribosomal subunits after infection of *Escherichia coli* by T2 ghosts. J. Virol. 14, 1310 (1974).
SMIT, J.A., HUGO, N., DE KLERK, H.C.: A receptor for a *Proteus vulgaris* bacteriocin. J. Gen. Virol. 5, 33 (1969).
TAKEYA, K., MINAMISHIMA, Y., AMAKO, K., OHNISHI, Y.: A small rod-shaped pyocin. Virology 31, 166 (1967).
TAKEYA, K., MINAMISHIMA, Y., OHNISHI, Y, AMAKO, K.: Rod-shaped pyocin 28. J. Gen. Virol. 4, 145 (1969).
TAUBENECK, U.: Über die Produktion biologisch aktiver Phagenschwänze durch einen defekt lysogenen *Proteus mirabilis*-Stamm. Z. Naturforsch. 18b, 989 (1963).
TAUBENECK, U.: Über inkomplette Bakteriophagen aus defekt lysogenen *Proteus mirabilis*-Stämmen. Biol. Zbl. 86 (Suppl.), 45 (1967).
TAUBENECK, U.: Virusproteine in Bakterienzellen. Z. Allg. Mikrobiol. 9, 315 (1969).
THURM, P., GARRO, A.J.: Structural and genetic analysis of the defective *Bacillus subtilis* phage PBSX. (submitted for publication, 1974).
TIKHONENKO, A.S.: Ultrastructure of bacterial viruses. New York, London: Plenum Press 1970.
TRAUB, W.H.: Studies on group A bacteriocins of *Serratia marcescens*: Preliminary characterization of two subgroups of bacteriocins. Zbl. Bakt. Hyg. I. Abt. Orig. A. 222, 232 (1972).
TRAUB, W.H., KLEBER, I., SCHABER, I.: Induction of group A bacteriocins of *Serratia marcescens* by nalidixic acid. Nature New Biol. 245, 144 (1973).
UEDA, M., TAKAGI, A.: Rhapidosomes in *Chlostridium botulinum*. J. Gen. Appl. Microbiol. 18, 81 (1972).
VALLÉE, M., CORNETT, J.B.: A new gene of bacteriophage T4 determining immunity against superinfecting ghosts and phage in T4-infected *Escherichia coli*. Virology 48, 777 (1972).
VALLÉE, M., CORNETT, J.B.: The immunity reaction of bacteriophage T4: a non-catalytic reaction. Virology 53, 441 (1973).
VALLEE, M., CORNETT, J.B., BERNSTEIN, H.: The action of bacteriophage T4 ghosts on *Escherichia coli* and the immunity to this action developed in cells preinfected with T4. Virology 48, 766 (1972).
YAMAMOTO, T.: Presence of rhapidosomes in various species of bacteria and their morphological characteristics. J. Bacteriol. 94, 1746 (1967).
YUI, C.: Structure of pyocin R. I. Isolation of sheath from pyocin R by alkali treatment and its properties. J. Biochem. (Tokyo) 69, 101 (1971).
YUI-FURIHATA, C.: Structure of pyocin R. II. Subunits of sheath. J. Biochem. (Tokyo) 72, 1 (1972).

The Genesis of Multicellular Organization and the Control of Gene Expression in *Dictyostelium discoideum*

Maurice Sussman

I. Introduction

In recent years, the merger of biochemistry and genetics into a single, unified, experimental approach has provided insights into the ways in which gene expression may be regulated: *via* control of transcription, the processing of the transcripts, and their transport and entrance into polysomal complexes; by control of translation and by processing of the polypeptide products; by regulation of protein conformation and activity through interactions with small effector molecules or direct protein-protein interactions; by compartmentalization of proteins, polysomal complexes, etc. in vesicles and reticulated structures. The point has been reached now where one can account in detail for transient phenotypic modulations in bacteria, such as enzyme induction and repression, and even complex developmental programs in phages and smaller animal viruses. The details of m-RNA synthesis, processing and transport in Eucaryotes are rapidly emerging. Hormone-induced enzyme modulations in mammalian cells, programs of cytodifferentiation in hematopoietic cells, pancreatic acinar and islet cells, myoblasts, chondroblasts, lens tissue, etc. are now being described at the molecular-genetic level.

But one aspect of gene expression, that which is involved in the genesis of multicellular organization during the development of plants, animals and colonial protista such as *Dictyostelium discoideum*, remains almost as great a mystery as when the fundamental problems were posed by the great classical morphogenetists like WILSON, CHILD, DRIESCH, HARRISON et al.

Three interrelated, universal properties of such systems generate the bulk of these problems:

a) What a cell does in a developing multicellular assembly depends on where it is.

   That is, a cell can determine within rather close limits whether it is at the top or bottom, front or back, inside or outside of a multicellular assembly and then elect to follow a compatible program of cytodifferentiation. If its position is changed early enough, it can jettison one program and elect another.

b) What a cell does in a developing multicellular assembly depends on who its neighbors are, how many they are and what they are doing.

   That is, the development of a cell within the assembly is triggered and modulated by the activities of its nearest neighbors, of neighboring cell clusters and the assembly as a whole. For example, a large cluster of neutral fold cells can do what a small cluster cannot (GROBSTEIN and ZWILLING, 1953); a pancreatic salivary, or kidney epithelial rudiment at the right stage in the presence of associated mesenchemal cells at the right stage can do what it cannot do in their absence (GROBSTEIN, 1964). The development of a cell in the precise middle of a coelenterate bud will depend on whether the bud is programmed to develop into the hydroid form or the Medusa and if into the hydroid form, whether a feeding polyp or a defensive polyp or a gonoid.

c) In a developing multicellular assembly, not all cells capable of doing something actually do it.

Thus the vertebrate heart invariably appears at once circumscribed site within a field of cells all of which are potentially able to form a heart. The outlying cells can realize this potential only if the normal heart rudiment is removed (DE HAAN, 1968). Such cell groups termed morphogenetic fields by CHILD (1941) operate in the formation of the eye, limb, gut and many other organs of vertebrates and invertebrates and in colonial protista. CHILD recognized the two generic properties of such fields, polarity which determines the position of the rudiment within the field and dominance which ensures that only one rudiment will appear.

All of these phenomena provoke the same basic questions. What molecules convey to a cell the information[1] regarding its position within the multicellular assembly and within a morphogenetic field? What molecules inform a cell about the activities of its neighbors and of the assembly as a whole? What are the cellular receptors and how is the information transduced into instructions which ultimately trigger and modulate programs of gene expression? Recently serious interest in the theoretical and empyrical aspects of these phenomena has been renewed. We think that *Dictyostelium* is an ideal system for this purpose: the general properties described above are present in paradigmal form stripped of nonessential complications.

The formation of the organized, multicellular fruiting body occurs only after growth and cell division have stopped and can be conducted rapidly and conveniently in large numbers and under defined experimental conditions (SUSSMAN, 1966). The development of methods for the selection of diploid heterozygotes and haploid segregants (KATZ and SUSSMAN, 1972) has already led to the identification of 13 markers and 5 linkage groups (there are seven chromosomes in the haplophase). The remainder of this article will be devoted to a description of three experimental opportunities which I think will be especially valuable in providing answers to the above questions.

---

[1] Already graced by the felicitous phrase "*positional information*", by LEWIS WOLPERT (1971).

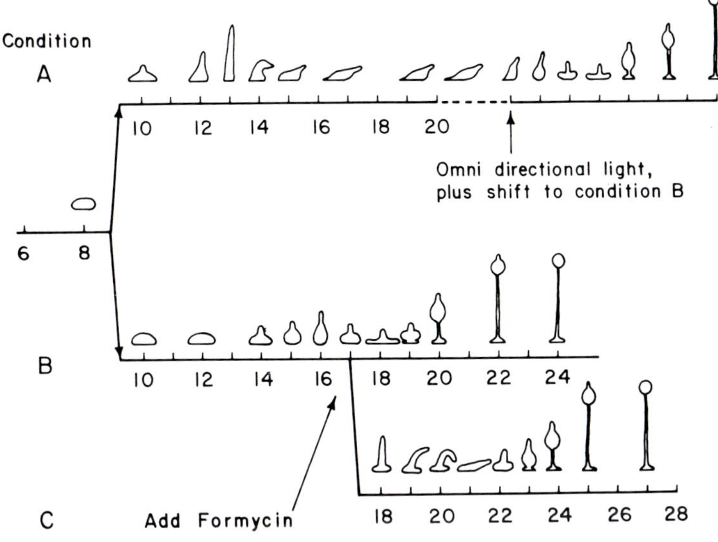

Fig. 1. Alternative morphogenetic pathways in *D. discoideum*. Cells are harvested from growth plates, washed and dispensed in aliquots of 10^8 cells on 2" Millipore or Whatman Filter circles on absorbent support pads inside Petri dishes. After 8 h they form ca 1,000 equal-sized aggregates. To induce the aggregates to develop into migrating slugs, the support pads are saturated with water or a weak (0.01 M) phosphate buffer solution and the plates are incubated in the dark or in a weak horizontal light gradient (condition A). To induce aggregates to develop directly into fruiting bodies, the support pads are saturated with a solution (LPS) containing salts at relatively high concentrations and 0.05 M phosphate buffer pH 6.5. Absorbent pads cemented to the dish covers are saturated with 1 M phosphate pH 6.0 to absorb volatile alkaline metabolites (condition B). Condition C involves addition of 1 mM Formycin B to the support pad fluid at zero time or any time up to 17 h (NEWELL, TELSER, and SUSSMAN, 1969; BRACKENBURY et al., 1974)

II. The Migrating Slug

In *Dictyostelium discoideum*, vegetative cells which have entered the stationary growth phase collect together into organized multicellular aggregates (Fig. 1). Influenced by several environmental parameters (pH, salt concentration, humidity, light and the concentration of a metabolite(s) produced prior to and during aggregation) each aggregate elects either to construct a fruiting body directly at the site of aggregation or to transform into a migrating slug and move away (NEWELL, TELSER, and SUSSMAN, 1969). The slug can migrate over the substratum for hours or days (SLIFKIN and BONNER, 1952) but if exposed to omnidirectional light and/or shifted to conditions (low pH, high ionic strenght, low humidity, etc.) that favor fruiting, it immediately stops migrating and constructs a fruiting body over a 7-h period as shown in Fig. 2.

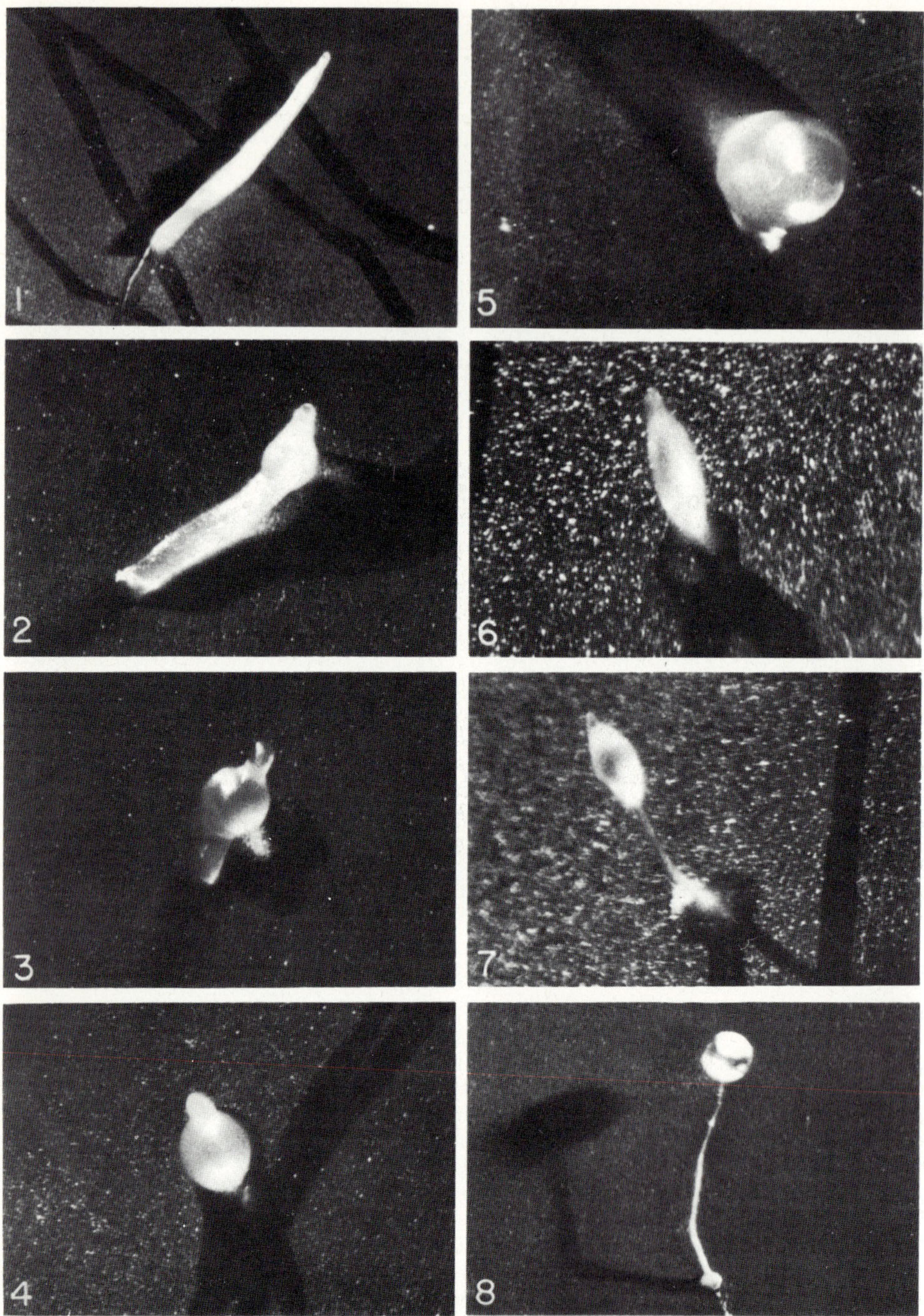

Fig. 2. Photographs of a migrating slug induced to fruit by exposure to omnidirectional light and a shift to condition B (NEWELL, TELSER and SUSSMAN, 1969)

Committment to fruit construction: If at any time prior to the 17-h stage of fruit construction, filters bearing the aggregates are shifted from condition B (favoring fruiting) to condition A (favoring slug migration), the aggregates stop fruit construction and are transformed into migrating slugs. However, between 16 and 17 h of fruit construction, the aggregates become comitted to that pathway and continue to follow it despite the shift to condition A.

The effect of Formycin B: Formycin B is an analog of adenosine and inosine.

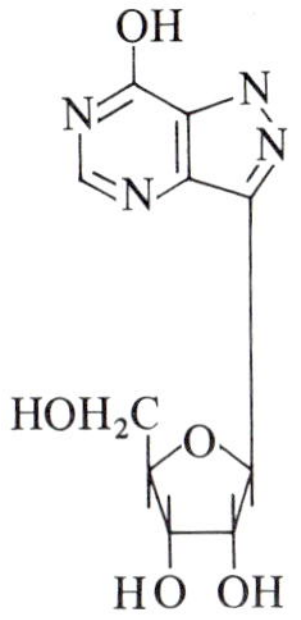

Fig. 1C shows the pathway followed by aggregates which are allowed to develop under conditions which favor fruiting but which are exposed to Formycin B starting from 17 h (BRACKENBURY et al., 1974). Instead of the normal Mexican Hat shape at 18 h the aggregate assumes a deviant form with a greatly expanded apex and reduced base. The apex progressively assumes the shape of a slug while the basal portion continues to regress until the whole aggregate is transformed into a migrating slug. This condition is very transitory. Almost immediately, the slug reverts to a shape resembling an early Mexican Hat (the 17-h stage of fruit construction) and thereafter follows the normal fruiting sequence. Addition of Formycin B prior to 17 h does not alter this response. Addition after 17 h is ineffective. Slugs exposed to Formycin B under conditions favoring slug migration ignore it. But if exposed to omnidirectional light and/or switched to conditions favoring fruiting while in the presence of Formycin B the slugs stop migrating and enter the fruiting mode in normal fashion. When they reach the 17 h (formycin-sensitive) stage they revert to slugs in the characteristic manner shown in Fig. 1C and then fruit. Thus the formycin-induced decision is clearly different from the earlier decision which is induced by high pH, low ionic strength, etc.

In summary, Formycin B induces aggregates to transform into migrating slugs but it does so in an environment that otherwise favors fruiting body construction and acts at a time when the aggregate has already reached an advanced stage of fruit construction. Once formed, the formycin-induced slug responds to external conditions (omnidirectional light, pH, etc.) exactly as does the slug which is induced earlier by high pH, low ionic strength, etc.

Functional and structural organization of the slug: In the dark the slug migrates randomly. In a horizontal light gradient it moves directly toward the source (BONNER et al., 1950), turning if necessary[2].

A recent study (POFF and LOOMIS, 1973) indicated that the anterior 5 - 10 % of the slug controls this phototactic behavior. The convex apical surface acts as a lens. Light appears to cause a higher rate of migration and the performance of the slug in the light gradients can be explained by an assymetrically accelerated migration of cells in the tip induced by differential illumination. The photoreceptor system appears to include a photoreceptor pigment, possibly a flavin, and a photo-responsive pigment which has been purified and characterized as a b-type cytochrome (POFF and BUTLER, 1974). "Blind" mutant strains have been isolated whose slugs are not photoactic.

Classical experiments of RAPER (1941) showed that the slug is a morphogenetic field. Experiments in which tips were removed, replaced, rotated and grafted showed that the tip controls the migratory process and establishes the polarity and domincance relationships within this field. Recent experiments (LOOMIS, 1972) raise the possibility that gradients in the thickness of the slug sheath[3] may account for some (but not all) of these field properties.

The cells at the apical tip also differ morphologically from those in the posterior in several respects. These include: differential staining properties (BONNER, CHIQUOINE, and KOLDERIE, 1955), specific organelles (GREGG and BADMAN, 1970; HOHL and HAMAMOTO, 1969), vesicular and laminar cytoplasmic structures (MILLER, QUANCE, and ASHWORTH, 1969; MAEDA and TAKEUCHI, 1969), differences in density (MILLER, QUANCE, and ASHWORTH, 1969; MAEDA, SUGITA, and TAKEUCHI, 1973).

Positional information in the slug: The relative positions of cells in the slug (and in the developing fruiting body) are determined by the order in which they originally entered the multicellular aggregate (BONNER, 1944) and are maintained during slug migration (RAPER, 1940). Cells which entered first occupy the apical tip of the aggregate and later of the slug, and those

---

[2] The complex responses of *D. discoideum* aggregates to ligth, pH, etc. must certainly have significant selective value. Since the vegetative cells grow by feeding on bacteria in the soil, they would be most abundant in the levels immediately below the surface, where most bacteria are found. Aggregates formed there after exhaustion of the food supply might routinely transform into slugs and move to the surface, either randomly or directed by a light gradient. Once at the surface the slug would be exposed to omnidirectional light (plus low humidity and high salt concentrations) and would therefore stop migrating and start fruiting. These suppositions are testable but have not yet been examined empirically in ecological studies.

[3] The slug synthesizes and moves inside a thin sheath (that looks like a miniature dialysis tube) which is spun out behind.

that entered afterward occupy progressively more posterial positions. The ultimate developmental fates of the cells in the fruiting body are determined by these positions. Thus cells at the apical tip (the anterior 25 - 30 %) of the slug will comprise the stalk of the mature fruiting body. The next 60 % will comprise the spores and the posterior 10 - 15 % will form the basal disc.

Qualitative differences in protein composition between cells in different parts of the slug have been demonstrated. For example, the enzyme UDP galactose: polysaccharide transferase catalyzes the incorporation of galactose into a mucopolysaccharide which is synthesized during fruit construction and is uniquely associated with the spores (WHITE and SUSSMAN, 1963). Migrating slugs were fragmented and assayed for this enzyme at a time when it had already attained its peak level of specific activity. All of the activity was associated with those cells destined to form the spores of the mature fruit. No activity was detected in those cells destined to become the stalk. In contrast, the enzyme UDP glucose pyrophosphorylase, both at its peak level and during its accumulation, was found to be distributed uniformly in all parts of the migrating slug and in all portions taken from developing fruits (NEWELL, ELLINGSON, and SUSSMAN, 1968).

GREGG (1965) and TAKEUCHI (1963) have demonstrated that spore-specific and stalk-specific antigens are already detectible (by fluorescent antibodies) in the migrating slug and are confined to the precise regions that will supply the spores and the stalk cells of the mature fruit.

When slugs are cut into two segments, both of the separated segments ultimately construct complete fruiting bodies (RAPER, 1941). This requires a redetermination of developmental fates. Thus in a segment comprising the posterior half of the original slug, that portion constituting the new apex (which normally would have developed into spores) will now form the stalk. In a segment comprising the front third of the original slug, the majority of cells will now develop into spores although if untouched they would have become stalk cells. Examination of the segments at various stages of development using fluorescent antibodies showed that spore-specific and stalk-specific antigens disappeared and appeared within the segments in patterns consistent with the altered developmental fates of the component cells.

<u>Programs of enzyme accumulation and disappearance in migrating slugs and developing fruits:</u> The cells which make up the mature fruiting body are very different in composition from what they were as vegetative amoebae at the end of exponential growth. Turnover studies indicate that at least 80 % of the RNA and 90 % of the protein present in the cells of the fruiting body are synthesized during the period of fruit construction (FRANKE and SUSSMAN, 1973; COCUCCI and SUSSMAN, 1970; WRIGHT and ANDERSON, 1960).

Data from several laboratories indicate that this wholesale turnover includes a significant differential change in protein composition. Thus a considerable number of enzyme activities have been shown to increase dramatically at specific stages of fruit construction and of these, many then disappear partly or completely (SUSSMAN and SUSSMAN, 1969; LOOMIS, 1969; NEWELL, 1971; SUSSMAN and NEWELL, 1972; FIRTEL and BRACKENBURY, 1972; FIRTEL and BONNER, 1972; GARROD and ASHWORTH, 1973). Even catalytic activities that show little or no changes appear to involve replacements of one isozymic form by another (PONG and LOOMIS, 1973).

These patterns of accumulation and disappearance are altered in morphogenetically deranged mutant strains in a manner which is consistent with the nature of morphogenetic derangement and the stage at which it occurs. The accumulations of activity require concurrent protein synthesis and prior RNA synthesis. For one of these enzymes, there is direct experimental support for the conclusion that the accumulation is the result of a 20-fold increase in the differential rate of synthesis (see below).

The patterns of accumulation and disappearance of at least four of these enzymes have been shown to depend upon which morphogenetic alternative is elected, slug migration or fruit construction. The enzymes are: UDP glucose pyrophosphorylase (ASHWORTH and SUSSMAN, 1967; NEWELL and SUSSMAN, 1969) which catalyzes the synthesis of UDP glucose; UDP galactose-4 epimerase (TELSER and SUSSMAN, 1971) which converts UDP glucose into UDP galactose; UPD galactose: polysaccharide transferase (SUSSMAN and OSBORN, 1964) which transfers galactose from UDP galactose to a mucopolysaccharide that is associated uniquely with the spores of the fruiting body; trehalose-6-P synthetase (ROTH and SUSSMAN, 1968) which catalyzes the synthesis of trehalose-6-P from UDP glucose and glucose-6-P. Fig. 3 summarizes the results schematically. The experimental points have been eliminated from the curves for pictorial clarity but can be seen in the original reports (BRACKENBURY et al., 1974; NEWELL et al., 1972; ELLINGSON et al., 1971). The performance of each enzyme is differently affected as seen in Fig. 3, by the choice of morphogenetic pathways.

It should, however, be noted here that in a series of publications, WRIGHT (1966), WRIGHT et al. (1968, 1972) have raised objections to the conclusion that, during fruit construction, enzyme levels do change significantly and that substantial differential changes in protein composition do occur. Two of the enzymes described above were among those studied. Despite previous statements to the contrary (WRIGHT et al., 1960; 1967; 1968) WRIGHT et al. agree (WRIGHT, 1966) that UDP glucose pyrophosphorylase activity does accumulate significantly during fruit construction and that the enzyme is synthesized *de novo* as originally reported (FRANKE and SUSSMAN, 1971). But they have now concluded that the increase in activity is due to a differential decrease in turnover of the protein rather than an increase in the rate of synthesis. However, the data of FRANKE and SUSSMAN (1973) demonstrate that the proportion of labeled amino acids incorporated into this protein *vs* all others increases about 20-fold throughout the period in which the enzyme actually accumulates. Moreover, the results of both FRANKE and SUSSMAN (1973) and GUSTAFSON and WRIGHT (1973)

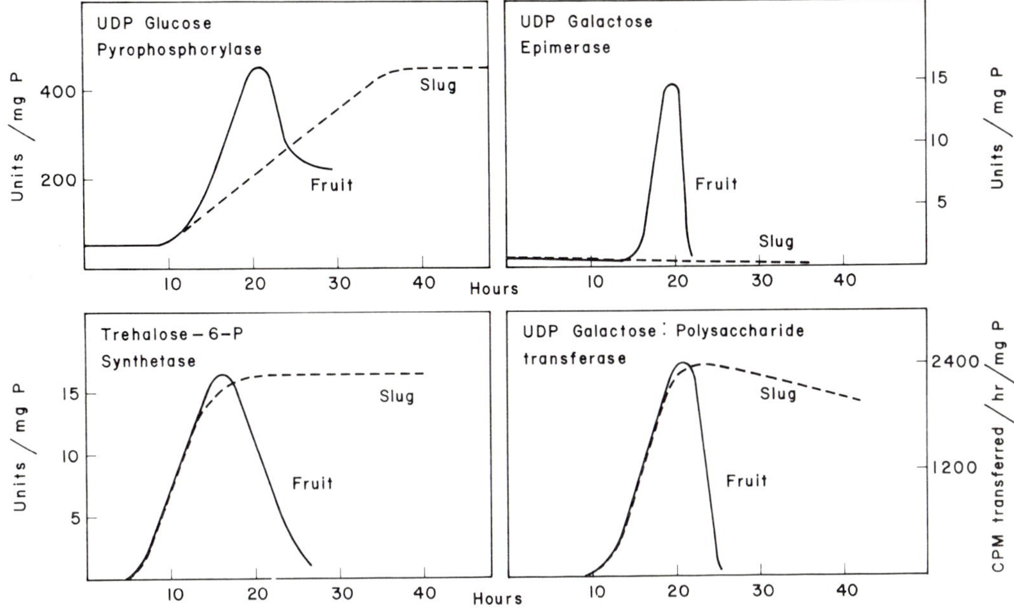

Fig. 3. Patterns of enzyme accumulation and disappearance in cell aggregates which develop into migrating slugs and in those that develop directly into fruiting bodies. The experimental data from which these curves were drawn are found elsewhere (NEWELL and SUSSMAN, 1970; ELLINGSON et al., 1971; BRACKENBURY et al., 1974)

in comparing incorporation rates during short and long pulses fail to yield evidence of high turnover or of changes in turnover rates. Hence, the accumulation of this enzyme appears to be the direct result of a differential increase in the rate of its synthesis. This increase is in turn the result of a prior period of genetic transcription (ROTH, ASHWORTH and SUSSMAN, 1968).

KILLICK and WRIGHT (1972) recently reported that previous measurements of trehalose-6-phosphate synthetase activity (SUSSMAN and NEWELL, 1972; ROTH and SUSSMAN, 1968) were erroneous because the enzyme is extremely cold-sensitive and because the negligible activity of cell extracts from an early stage of fruit construction can be "unmasked" i.e. increased many-fold by ammonium sulfate fractionation. The experiments have been repeated according to the published descriptions by S. ALEXANDER in our laboratory. He finds that the enzyme is not cold-sensitive but like most enzymes is increasingly stable at lower temperatures and significant activity is not lost in extracts which are treated and assayed as previously reported. Despite many attempts he has failed to confirm the unmasking in extracts of cells grown in axenic medium, or in association with *Aerobacter aerogenes* or with *Escherichia coli*. He finds that the pattern of accumulation and disappearance is precisely as reported by ROTH et al. (1968), confirmed by NEWELL et al. (1972) and recently extended to the axenic strain of *D. discoideum* by ASHWORTH et al. (1973). We are unable to account for the discrepancies in the experimental results.

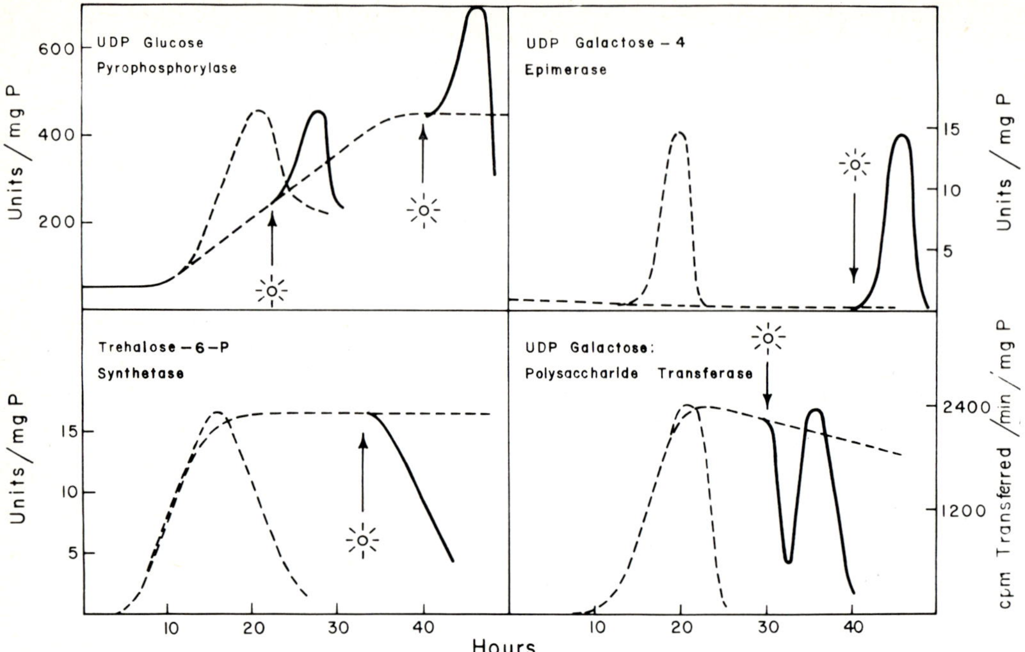

Fig. 4. Patterns of enzyme accumulation and disappearance in migrating slugs which have been induced to stop migrating and construct fruiting bodies (solid lines). The dotted lines included for reference, are taken from Fig. 3. The arrows designate the times at which the slugs were exposed to omnidirectional light and shifted to conditions that favor fruiting. The experimental data from which these curves were drawn are found elsewhere (NEWELL and SUSSMAN, 1970; ELLINGSON et al., 1971; BRACKENBURY et al., 1974)

The patterns also change dramatically in migrating slugs which are induced to stop migrating and to start fruit construction by exposure to omnidirectional light and a shift to conditions of low pH, high salt, etc. Thus, when migrating slugs are induced to fruit before they have finished the normal round of UDPG pyrophosphorylase synthesis they complete this round rapidly and reach the same peak as do aggregates which fruit directly (Fig. 4). However, if they are allowed to migrate long enough to have finished a complete round and are then induced to construct fruiting bodies, they initiate a second complete round of pyrophosphorylase synthesis!

The responses of the other three enzymes do not depend upon the time at which the slug is induced to stop migrating and start fruiting. Thus a normal though delayed round of epimerase accumulation is initiated. No additional T-6-P synthetase activity accumulates and the amount already present is lost at the appropriate stage of fruit construction exactly as in aggregates that develop directly into fruits. During the first 2 h after the induction of fruiting, the previously accumulated UDPG galactose: polysaccharide transferase disappears dramatically and a second complete round of accumulation is initiated. Previous experiments

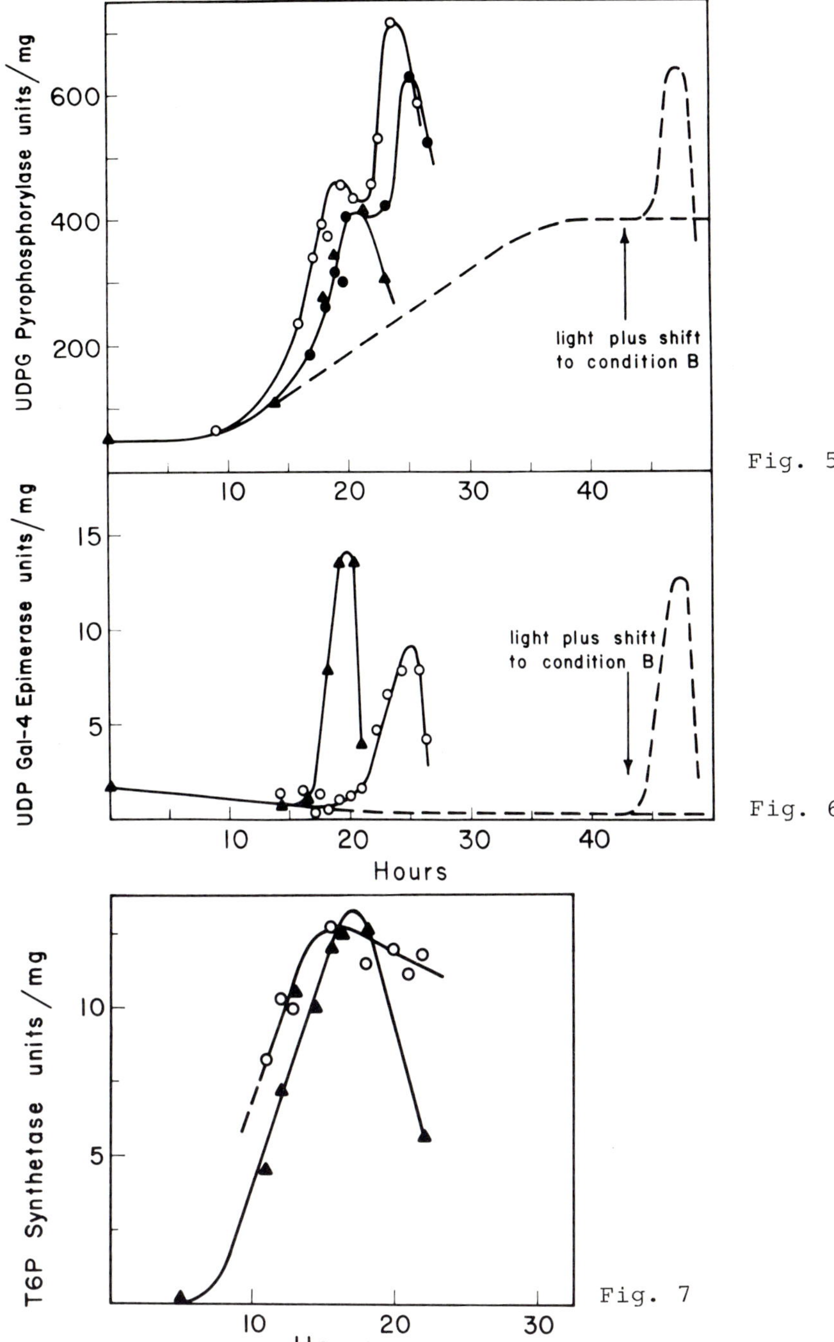

Figs. 5-7. Accumulation and disappearance of UDP glucose pyrophosphorylase, UDP galactose-4 epimerase and trehalose-6-P synthetase in the presence of Formycin B (BRACKENBURY et al., 1974). Cells were dispensed on filter circles and incubated in condition B (shown in Fig. 1): (o) exposed to 1 mM Formycin B starting from zero time; (●) exposed to Formycin B starting at 17 h by shifting the filters to new pads; (▲) untreated control

have shown that the transferase and epimerase are both preferentially released by the cells into the extracellular space at a late stage of fruiting body construction (SUSSMAN and LOVGREN, 1965; TELSER and SUSSMAN, 1971). It is thought that these enzymes are sequestered within vesicles in which the mucopoly saccharide is synthesized and that during spore formation the vesicles fuse with the cell membrane and evert, thereby providing the spore with an outer coat of mucopolysaccharide and releasing the enzymes (TELSER and SUSSMAN, 1971). Recent electron micrographic studies support this view (GREGG and BADMAN, 1970; HOHL and HAMAMOTO, 1969). Perhaps the loss of transferase by slugs which are induced to fruit results from the extrusion by the prespore cells of the vesicles that had previously accumulated.

When aggregates were induced by Formycin B to become slugs at a later stage and then to resume the fruiting program, the patterns of enzyme accumulation and disappearance were consistent with those described above. Thus, cells exposed to Formycin B starting from zero time showed the pattern of UDP glucose pyrophosphorylase accumulation characteristic of the fruiting program (because aggregation occurs 1 - 2 h earlier in the presence of Formycin B, the period of enzyme accumulation was shifted forward accordingly). The subsequent initiation of a second round of pyrophosphorylase accumulation coincided with the reversion of the Formycin-induced slugs to the fruiting program. Aggregates exposed to Formycin B from 17 h responded similarly (Fig. 5). In cells exposed to Formycin B from zero time UDP galactose-4 epimerase did not accumulate at the usual time but only after the slugs had reverted to the fruiting program and reached the customary morphogenetic stage (Fig. 6). Trehalose-6-P synthetase accumulated as usual but its disappearance was delayed by the transformation to migrating slugs (Fig. 7).

Two other enzymes, N-acetyl glucosaminidase and alanine transaminase were examined. Both accumulate very early, even before the first signs of cell aggregation (LOOMIS, 1969; FIRTEL and BRACKENBURY, 1972) and the levels remain constant thereafter. These patterns were unaffected by the presence of Formycin even when present at zero time (BRACKENBURY et al., 1974). Thus they reinforce the conclusion that the patterns of accumulation and disappearance are keyed to specific morphogenetic events.

Summary

1) A variety of proteins accumulate and some disappear during morphogenesis in *D. discoideum*. These events appear to involve a program of gene expression including differential transcription and translation.

2) Within any given cell this program is influenced directly by its relative position within the multicellular assembly.

3) Each pattern of accumulation and disappearance is affected in a specific manner by the choice of morphogenetic pathways which the cell aggregate elects and it keeps perfectly in step with the flow of morphogenetic events along the pathway.

4) Depending on the morphogenetic circumstances, a particular enzyme may accumulate rapidly or slowly, now or later, or not at all. For some enzymes, there can only be one round of accumulation; for others, there can be two successive rounds. Each such round requires a prior period of RNA synthesis, that is blocked by d-actinomycin. In each round, whether fast or slow, immediate or delayed, initial or additional, a quantal amount of activity characteristic of each enzyme accumulates. This pattern of regulation has been termed *Quantal Control* (SUSSMAN and NEWELL, 1972).

III. Cell Association as an Initiator of Gene Expression

Several hours after the cells have entered the stationary growth phase they become sticky. If shaken gently in buffer they form loose amorphous clumps at first and then tight organized aggregates with a thin covering like the cortical layer of a migrating slug or developing fruit. If deposited on solid substratum, such structures immediately transform into slugs (BONNER, 1950; GERISCH, 1961). The cell adhesion is species-specific; there appear to be restricted sites on the cell membrane for such connections (BEUG et al., 1973; SHAFFER, 1963). The acquisition of adhesivity is accompanied by the appearance of new antigenic determinants located at the cell surface (SONNEBORN et al., 1963; GERISCH et al., 1969; BEUG et al., 1970). Exposure of vegetative cells to the purified homologous γ-globulin inhibits aggregation (GERISCH et al., 1969; BEUG et al., 1970) without impairing the chemotactic responsiveness of the cells (SONNEBORN et al., 1963; GERISCH et al., 1969; BEUG et al., 1970).

Cell aggregates can be harvested from filters even after 16 - 18-h development and be reduced to a dispersed cell suspension by vigorous repeated pipettings. If washed and redeposited on fresh filters at the original cell density, the cells reaggregate within minutes, and by 2 - 3 h have recapitulated the morphogenetic sequence that took as much as 16 - 18 h the first time (Fig. 8). Then they continue the program of fruit construction at the normal rate. If made to reaggregate in the presence of cycloheximide or d-actinomycin they simply recapitulate previous morphogenesis but do not develop beyond the stage at which they were dispersed (NEWELL, FRANKE, and SUSSMAN, 1972).

Fig. 9 illustrates the course of enzyme accumulation during disaggregation, reaggregation and subsequent development. Data are given for the pyrophosphorylase and the epimerase (the other two enzymes showed similar patterns). In each case the accumulation of enzyme activity was immediately stopped by disaggregation and was not resumed until after the cells had reaggregated. Then a new, complete round of accumulation was initiated regardless of how much enzyme had previously accumulated during the first round. If the aggregates were dispersed a second time, once again enzyme accumulation stopped until after the cells had reaggregated and then a third complete round of accumulation ensued! In the case of the epimerase, the previously accumulated activity was dra-

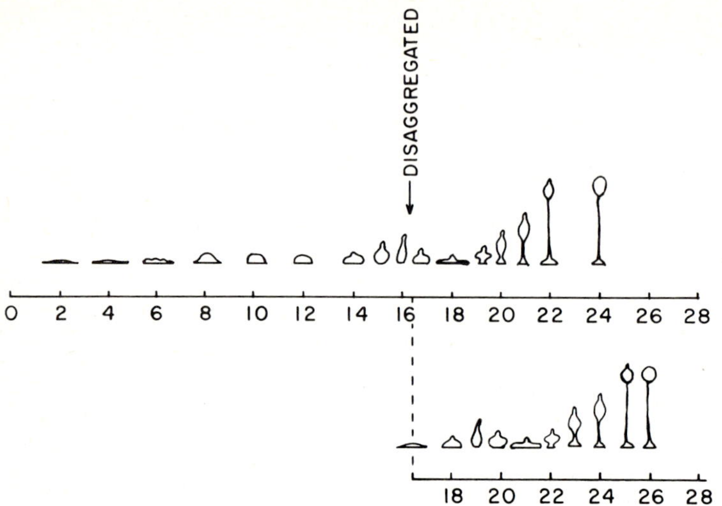

Fig. 8. The reaggregation and subsequent development of disaggregated cells (NEWELL, LONGLANDS, and SUSSMAN, 1971). Aggregates were harvested from filters and dispersed by repeated vigorous pipettings to a suspension of largely single cells with some doublets and triplets but no clumps. After one washing in the centrifuge the cells were resuspended and aliquots of 10^8 cells were deposited on fresh filters

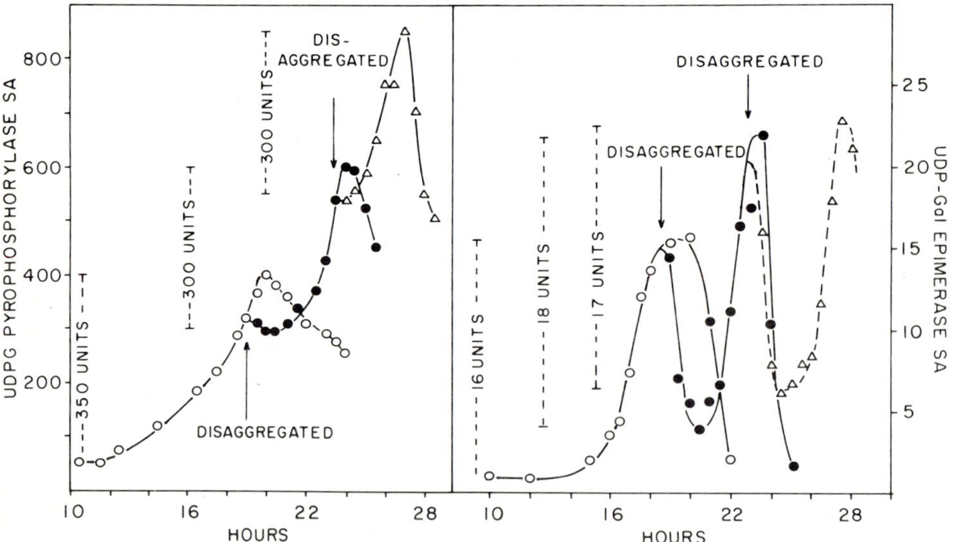

Fig. 9. Enzyme accumulation during two successive disaggregations and reaggregations (NEWELL, FRANKE, and SUSSMAN, 1972). At each arrow, developing aggregates were dispersed and the cells redeposited over fresh filters at the original cell density

matically lost immediately after each disaggregation and during the subsequent reaggregation. This may be related to the fact that this enzyme is normally preferentially released by the cells at the last stages of fruit construction (TELSER and SUSSMAN, 1971).

In the presence of d-actinomycin the cells reaggregated and recapitulated previous development but did not develop further, nor did they initiate additional rounds of enzyme accumulation. They merely retained the levels previously reached. If the disaggregated cells were redeposited at a very low density so that they could not reaggregate, they also failed to initiate additional rounds of enzyme accumulation.

As mentioned previously, N-acetyl glucosaminidase accumulates very early in the developmental sequence (LOOMIS, 1969). Its appearance is triggered long before the onset of aggregation and accumulates normally in most aggregateless mutants. Hence, one would not expect the pattern of accumulation of this enzyme to be altered by disaggregation and subsequent reaggregation. In fact it is not.

From these results it is possible to conclude that the formation of the multicellular aggregate triggers a differential transcription of the genome which results in the synthesis and accumulation of a set of enzymes[4]. Disaggregation is a signal to abandon the program of synthesis, an eminently sensible decision on the part of the disaggregated cell since it cannot be sure if in the next moment it will meet other disaggregated cells and resume morphogenesis or meet bacteria, begin feeding and reenter the cell growth and division cycle. When such cells do reaggregate, the same program of gene expression is retriggered. The accumulation of quantal levels of the enzymes follows automatically regardless of the amounts previously accumulated and keeps perfectly in step with subsequent morphogenesis just as it did the first time.

IV. c-AMP, c-AMPsites, and the Morphogenetic Fields of *D. discoideum*

Aggregation and chemotaxis: Some hours after stationary phase cells are dispensed on a solid substratum they begin to assume elongated shapes and orient themselves radially (Fig. 10). This wave of orientation spreads outward. Meanwhile the excited cells move in toward the center, forming ramified streams as they do so. When they reach the center they enter the growing, slightly conical multicellular aggregate. The number of aggregative centers and the number of cells that enter them are complex functions of several parameters, to wit:

---

[4] FIRTEL and BONNER (1972) have shown that a fifth enzyme, glycogen phosphorylase displays the same quantal pattern during reaggregation and subsequent development.

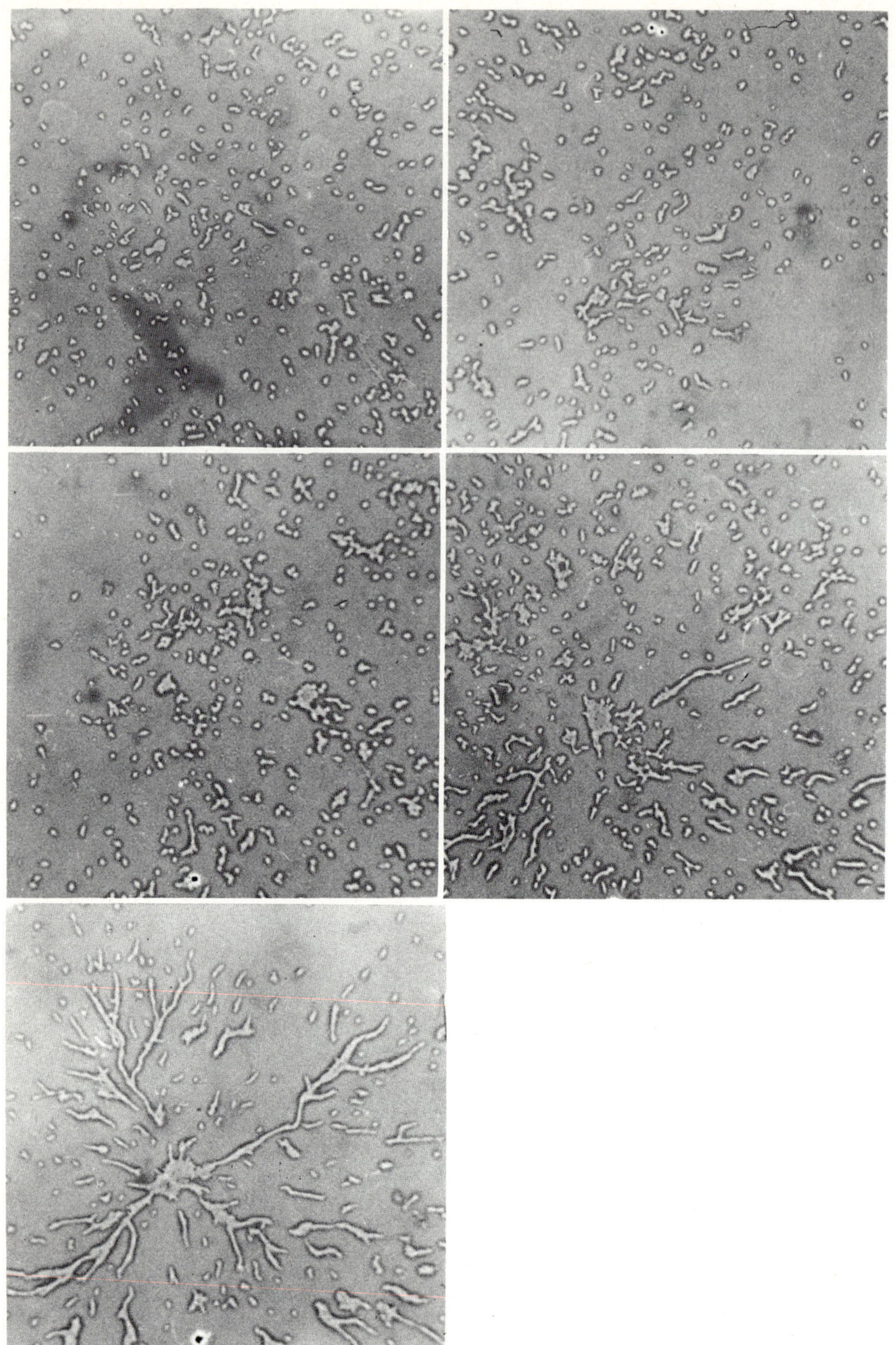

Fig. 10. Early stages of aggregation in *D. discoideum* (SUSSMAN and SUSSMAN,1961)

1) Population density. Population samples containing the same number of cells are dispersed on a solid substratum at different population densities (SUSSMAN and NOEL, 1952; SUSSMAN and SUSSMAN, 1961). There is found to be a threshold density below which no aggregative centers form and an optimal density at which a maximum number of aggregative centers appear, leaving practically no cells unaggregated. At still higher densities the centers are fewer but bigger[5] (since the potential center forming agencies are close enough together to compete with each other).

2) Cell number. Population samples containing different numbers of cells are dispensed on a solid substratum at the optimal density. It is found that the more cells there are the greater the number of aggregative centers in direct proportion (SUSSMAN and NOEL, 1952). By simultaneously increasing both cell number and cell density (above the optimal), one observes a constant "aggregation territory", i.e. a single aggregate forms within a constant area no matter how many more cells are present and how high the density is (BONNER and DODD, 1962).

3) Genetic constitution. Differences in aggregative performance as described above are observable between mutants and wild type, haploid and diploid populations, and mixtures thereof (SUSSMAN and SUSSMAN, 1961).

4) Nature of the substratum. pH, divalent metals, specific compounds like histidine and adenine influence the sensitivity of the responding cells to the chemotactic signal (SUSSMAN and SUSSMAN, 1961).

Mixtures of small numbers of wild-type and large numbers of aggregateless mutant cells show that single wild-type cells can initiate the formation of an aggregative center (SUSSMAN and SUSSMAN, 1961). Where the response is made difficult (by elimination of salts) the initiators can be shown to consist of a class of single especially large and active cells (SUSSMAN and SUSSMAN, 1961). When the response is made easier (by including salts) other cells can serve (KONIJN and RAPER, 1961). When single cells are separated with a micromanipulator from an aggregate and are added to a population of cells which have just stopped growing, any cell of the former can initiate a center among the latter (SUSSMAN and SUSSMAN, 1961). Cell clumps can also be initiators of aggregation (and perhaps generally are).

Periodicity in aggregating populations: Cells that aggregate at fairly high population densities move in toward the center in concerted, periodic fashion (GERISCH et al., 1966). As shown by time-lapse cinematography, excited cells suddenly coalesce into rings and these appear and disappear with a regular periodicity of about 5 min. A mutant strain with greatly reduced c-AMP phosphodiesterase activity produces aggregation patterns much

---

[5]Depending on the initial population density the wild-type aggregates can contain as few as 50 - 100 cells or as many as 10^5 cells.

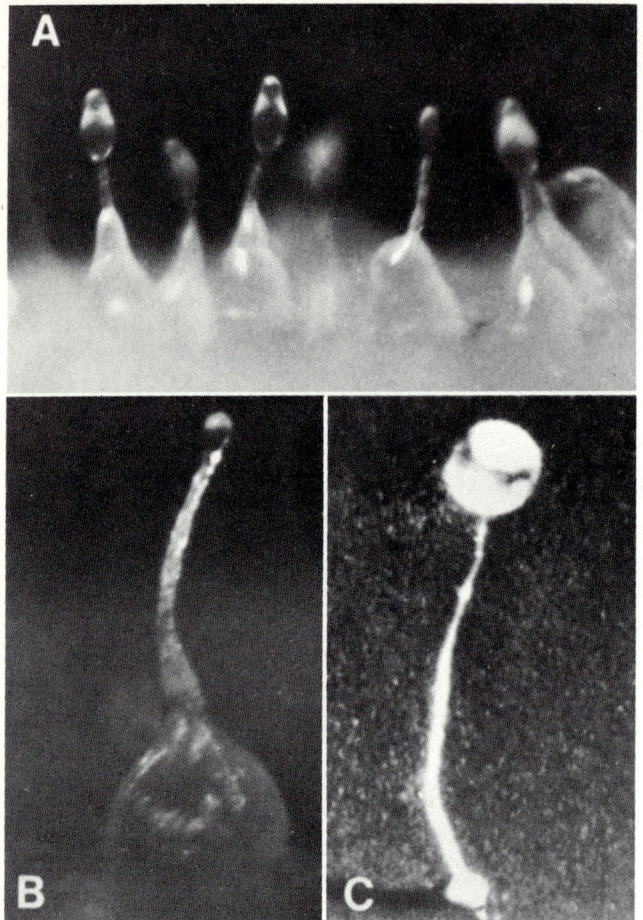

Fig. 11A-C. The effect of c-AMP on fruit construction. Filters bearing cell aggregates after 16-h incubation were shifted to fresh pads containing 3 mM c-AMP in LPS (see Legend to Fig. 1) and incubated until no further morphogenetic changes occurred. (A) Filter shifted after 16-h development. (B) Filter shifted between 16-18 h. (C) Control

larger in area than the wild type. The mutant cells show concerted movements but the periods are substantially longer than 5 min (RIEDEL and GERISCH, 1971).

Cells taken from preaggregative stages have been suspended in buffer and shaken gently at densities around 10^7 cells/ml. In this condition they show concerted, periodic alterations in volume and shape (detected by light scattering) and changes in redox potential (GERISCH, 1973). On solid substratum, cells at this stage can be induced to form regular radial aggregates around an artificial source of c-AMP if it is delivered with a constant periodicity (ROBERTSON et al., 1972).

Chemotactic activity of c-AMP: Many years ago the oriented movements of aggregating cells of *D. discoideum* were shown to stem from a chemotactic attraction exerted by the more central cells upon the more peripheral ones and it was found that when a cell is excited by the chemotactic agent, it produces (BONNER, 1947). In recent years 3', 5'-cyclic AMP (c-AMP) has been shown to possess chemotactic activity (KONIJN et al., 1967). Further more the chemotactic activities of extracellular fluids obtained from *D. discoideum* aggregates could be accounted for by the concentration of c-AMP.

Involvement of c-AMP in other morphogenetic fields: If filters bearing aggregates at the 16-h stage of fruit construction are shifted to fresh pads saturated with LPS containing 1 - 6 mM c-AMP, normal development is observed up to the Mexican Hat stage. Then the cells at the apex carry out their normal activity, i.e. they construct a cellulose-ensheathed lower stalk by the usual concerted upward movements while the rest of the cells remain at the base (some of them transform into normal spores). Hence, the presence of c-AMP in the substratum disturbs the normal polarity relationships (NESTLE and SUSSMAN, 1972). Fig. 11 shows photographs of these upside-down fruiting bodies.

If filters bearing migrating slugs are shifted to absorbent pads containing c-AMP and exposed to omnidirectional light, the slugs immediately stop migrating and they break up into a number of mini slugs (at some concentrations of c-AMP the cells at the apex of the slug disperse and fan out over the surfact in a delta-like pattern). From each mini slug one or more upside-down fruiting bodies develop (NESTLE and SUSSMAN, 1972). Thus, exposure to exogenous c-AMP disrupts polarity and dominance relationships in the slugs. Fig. 12 shows photographs of these events.

Enzymes involved in c-AMP metabolism: Crude extracts of *D. discoideum* contain at least four enzyme activities related to c-AMP metabolism. These reactions can be combined in the metabolic net work shown in Fig. 13 which may conceivably operate *in vivo*. The remarkable regulatory properties of two of the enzymes can in principle explain the pulsed production of and the periodic response to c-AMP during aggregation.

The extracts contain an adenylate cyclase which may be membrane-bound but is easily solubilized. It has a molecular weight of 100,000 Daltons and can be purified ca 100-fold from crude extract by collecting the 30,000 X G supernatant and separating it on an agarose column (ROSSOMANDO and SUSSMAN, 1973). A second enzyme converts ATP directly to 5'-AMP by elimination of pyrophosphate (ROSSOMANDO and SUSSMAN, 1973). It appears in the same fraction as the cyclase after 100-fold purification but can be separated from it. This reaction has previously been reported in bull semen (HEPPEL and HILMOE, 1953) and snake venom (JOHNSON et al., 1953) and was called "α, β ATPase" and "PP-forming ATPase". We have given it a more conventional name, ATP pyrophosphate hydrolase. Two forms of a third enzyme, c-AMP phosphodiesterase,

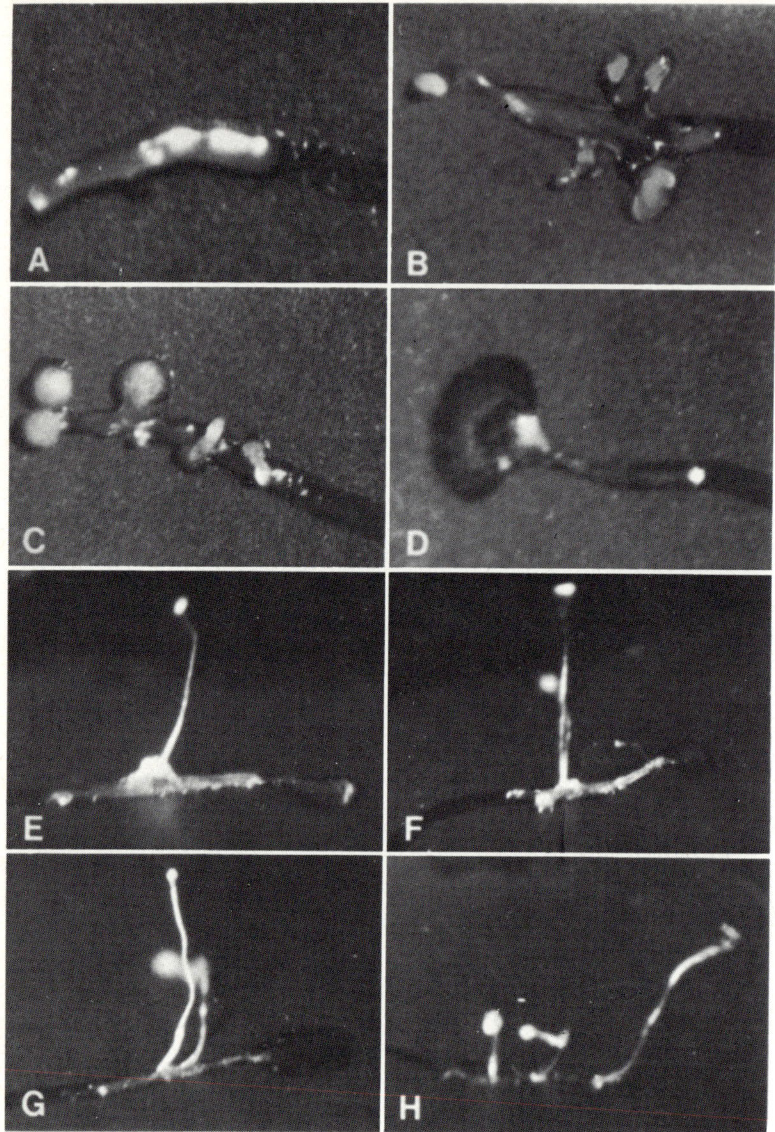

Fig. 12A-H. Effect of c-AMP on migrating slugs (NESTLE and SUSSMAN, 1972). Washed cells were dispensed in a thin line at one end of a 2 % agar-water substratum in a rectangular plastic container. Black Millipore Filters were placed on the agar in the middle of the box. The box was covered with aluminium foil and a small hole was cut opposite the line of cells. The cells aggregated and formed slugs after 16 h and the slugs migrated directly toward the light source at a mean rate of 2 mm/h. After a period calculated to allow the slugs to migrate on to the filters the boxes were opened and the filters were shifted to pads saturated with LPS containing 6 mM c-AMP (B, C), 0.6 mM (D), 3 mM (E, F), 1.5 mM (G), and 0.15 mM (H). (A) shows a slug prior to treatment

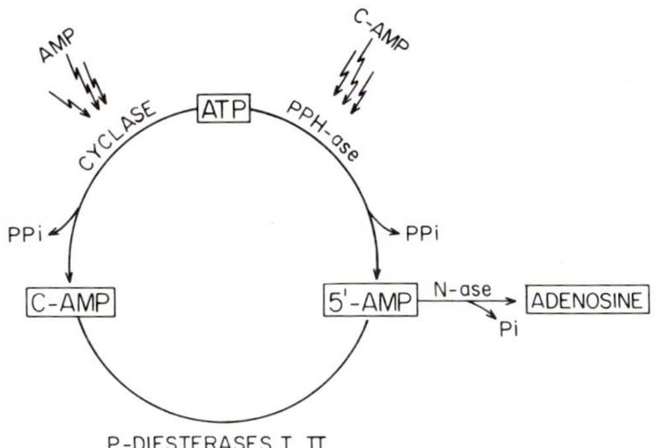

Fig. 13. A possible network of enzymes in c-AMP metabolism in *D. discoideum*. See text for details

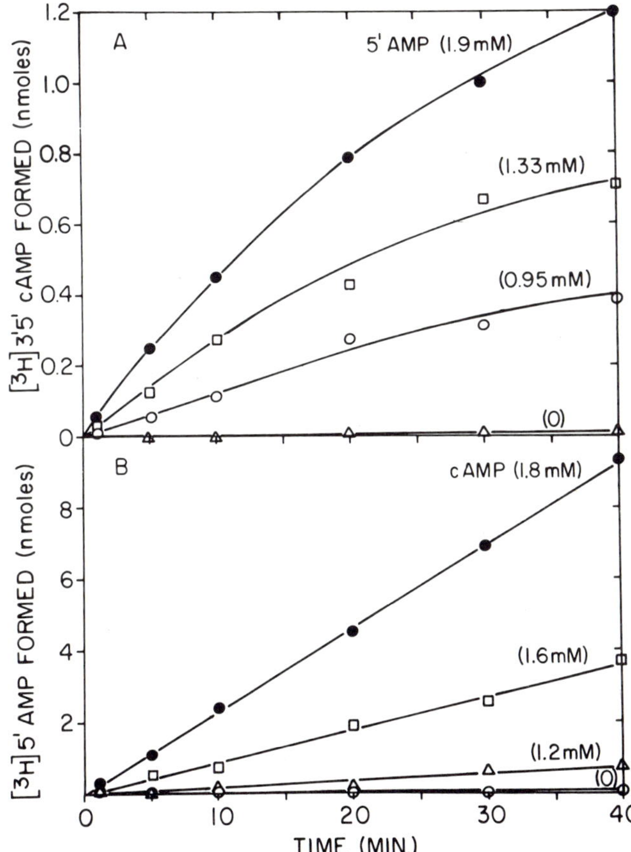

Fig. 14A and B. Time course of the reaction catalyzed by (A) adenylate cyclase at different concentrations of 5'-AMP and (B) ATP pyrophosphate hydrolase at different concentrations of c-AMP (ROSSOMANDO and SUSSMAN, 1973)

exist. One form is cell-bound and the other is largely extracellular, with molecular weights of 130,000 and 65,000 Daltons and with different Km values (CHASSY, 1972; GERISCH et al.,

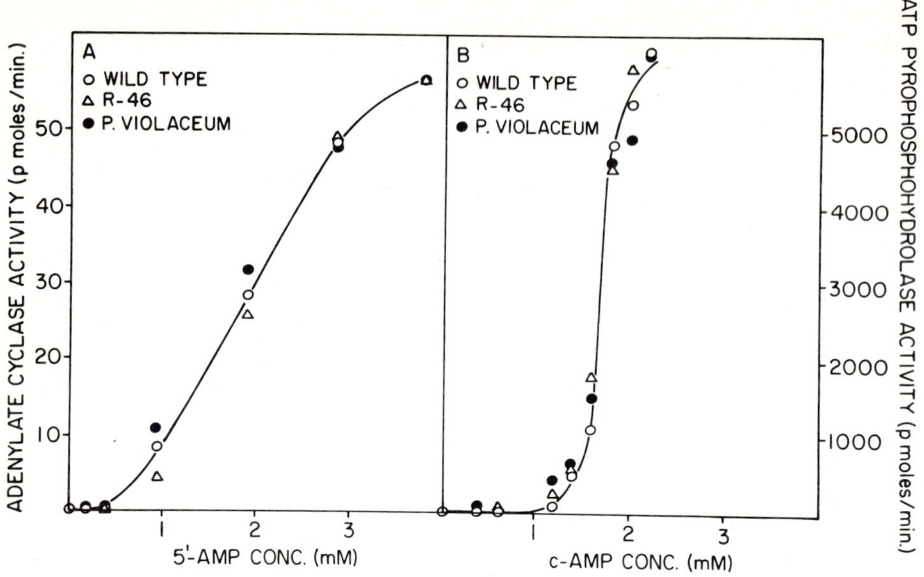

Fig. 15 A and B. Activities as functions of activator concentrations (ROSSOMANDO and SUSSMAN, 1973). Purified preparations were employed from *D. discoideum* wild type (o), a mutant R-46 with low phosphodiesterase activity (Δ), and another species polysphondylium violaceum (●). Data points of the latter two were normalized to the first

1972; PANNBACKER and BRAVARD, 1972). Finally a 5'-Nucleotidase has been demonstrated which can convert 5'-AMP to adenosine (ROSSOMANDO and SUSSMAN, 1973). Though all 4 enzymes appear in the 30,000 X g supernatant → they are also associated with a purified membrane vesicle preparation and can be released by treatment with triton X-100 (ROSSOMANDO, personal communication).

The remarkable regulatory properties of the adenylate cyclase and the ATP pyrophosphate hydrolase are shown in Fig. 14. The cyclase is activated by 5'-AMP, the product of the hydrolase, the hydrolase is activated by c-AMP the product of the cyclase. Thus, the two enzymes provide a positive feedback loop. As seen in Fig. 15 the activations show cooperativity, that of the pyrophosphate hydrolase being so extreme as to make its activation almost a step function.

The activations are specific. Thus, only 3'5' c-AMP activates the pyrophosphate hydrolase. 2',3' - cAMP and 3',5' c-GMP are completely inactive. The adenylatecyclase is not activated by 2',3'-cAMP. 3'5'-cIMP is a competitive inhibitor (see Fig. 17). 3'-5' cGMP is about one third as active as 3',5'-cAMP and 3, 5' deoxy c-AMP is fully as active. The latter finding has interesting implications. Recent results indicate that c-AMP may be involved in the control of Eucaryotic cell division (HADDEN et al., 1972).

The ability of a deoxynucleotide to activate dictyostelium adenylate cyclase raises the question of its extension to other cyclases and the possibility that it may act as a feedback control in cell division.

Theories of chemotaxis and aggregation: The network shown in Fig. 13 could conceivably account for two puzzling features of the chemotactic system in *D. discoideum*. They are:

1) A relay operates in which cells excited by the chemotactic agent then produce it and thereby attract other cells. The production appears to be pulsed.

2) Concentric rings of cells responding cells move periodically toward the aggregative center in a concerted pattern with a characteristic frequency and pathlength, both alterable by mutation.

If the regulatory loop does operated *in vivo*, the first puzzling feature can be satisfactorly explained. Thus, if a cell were exposed to a pulse of exogenous c-AMP, its pyrophosphate hydrolase, situated at the cell surface, would be activated and would produce 5'-AMP. This would in turn activate the cyclase to produce c-AMP. The intracellular concentration of c-AMP would quickly rise and then it might be released by a time or concentration dependent mechanism into the extracellular space. Any c-AMP remaining in the cell would by hydrolyzed by the cell-bound phosphodiesterase to 5'-AMP and the latter would be converted to adenosine by the 5'-nucleotidase, thereby bringing the system back to an inactive state. Thus a cell exposed to c-AMP would in turn synthesize and release c-AMP. If the exposure were in the form of a pulse periodically delivered, the synthesis and release would also be periodic. The presence of extracellular phosphodiesterase would help to preserve the pulsed character of the excreted c-AMP.

The regulatory properties of the pyrophosphate hydrolase can also account in principle for the second puzzling feature. As seen in Fig. 15, the activation is extraordinarily cooperative so that the enzyme would function as an on-off switch *in vivo*. The threshold concentration of c-AMP needed for activation is extremely high. Such a threshold could be exceeded only by a group of cells releasing c-AMP in concert. The pyrophosphate hydrolase of their nearest neighbors would be activated and they in turn would accumulate and secrete a pulse of c-AMP. Meanwhile they would respond chemotactically by bunching up. The synchronous production and secretion of c-AMP and the chemotactically induced bunching-up would ensure that the pulse will be concentrated enough to activate the next set of nearest neighbors.

Albert GOLDBETER has recently incorporated these enzymic elements and their regulatory properties into a formal model of c-AMP production (GOLDBETER, personal comm.). A computer program based on this model has generated a cyclical pattern of c-AMP synthesis with a frequency of ca 5 min. The assumption that an "on-off switch" operates i.e. that the activation of the pyrophosphate hydrolase is highly cooperative, turns out to be a crucial one in this program.

In principle, exposure of the cells to exogenous 5'-AMP could also induce c-AMP synthesis. However, in crude extracts, the rate of 5' AMP production *via* the pyrophosphate hydrolase is about 100-fold greater then the rate of c-AMP production by the cyclase, hence, *in vivo*, one would expect exogenous c-AMP rather than 5'-AMP to be the rate-limiting agent. Nevertheless it should be noted that in early experiments (SUSSMAN, LEE, and KERR, 1956) the chemotactic activity of extracellular fluid from nascent aggregates could be approximated by mixing in characteristic proportions two components of that fluid that had been separated by paper chromatography. The R_f values of the components indicates that they must have been c-AMP and 5'-AMP respectively, a conclusion which is supported by the finding that the former was converted to the latter by an enzyme present in the fluid (SUSSMAN, LEE, and KERR, 1956) presumably the phosphodiesterase.

Two groups have proposed generalized theories of chemotactic behavior, one based upon randomly produced discontinuities in the concentration of the chemotactic agent (KELLER and SEGEL, 1970) and another based upon assumed periods of sensitivity and recalcitrance to the agent (COHEN and ROBERTSON, 1971). While both theories contain extremely interesting implications, they have not yet proved to have significant predictive values, though the fault may lie with the experimenters rather than the model builders.

<u>c-AMP, migrating slugs, and still more theories:</u> As noted previously, the migrating slug is a morphogenetic field and the ultimate developmental fates of the component cells depend on their positions within it. These fates can be altered if the slug is cut into segments. Hence, each cell must know not only its distance from the front but also from the back of the slug. These polarity and dominance relationships can be disturbed by exposing the slug to exogenous c-AMP (NESTLE and SUSSMAN, 1972). Moreover, cells in different parts of the slug show dramatic differences in their capacity to synthesize and/or release c-AMP (BONNER, 1949).

The general problem of determining relative position along a longitudinal axis by monitoring gradients of diffusible substances has recently been treated in elegant fashion by WOLPERT (1969) and by CRICK (1970). Both single and double (opposing) gradients have been invoked. An interesting modification is offered by GOODWIN and COHEN (1969). Two periodic waves with different velocities are imagined to be propagated by a boundary pacemaker cell or cell group. Because of the difference in velocities, a phase shift would occur from point to point and this is considered to supply the positional information. However, unless one imagines echoes bouncing back from the rear it is difficult to see how a cell could monitor its position with respect to both the front and the back of the slug.

McMAHON (1973) has recently offered a particularly ingenious and elegant model specifically aimed at *D. discoideum* slugs but generally applicable to other forms. He imagines that the intracellular concentration of a morphogen rather than an extracellular gradient determines developmental fate. This concentration

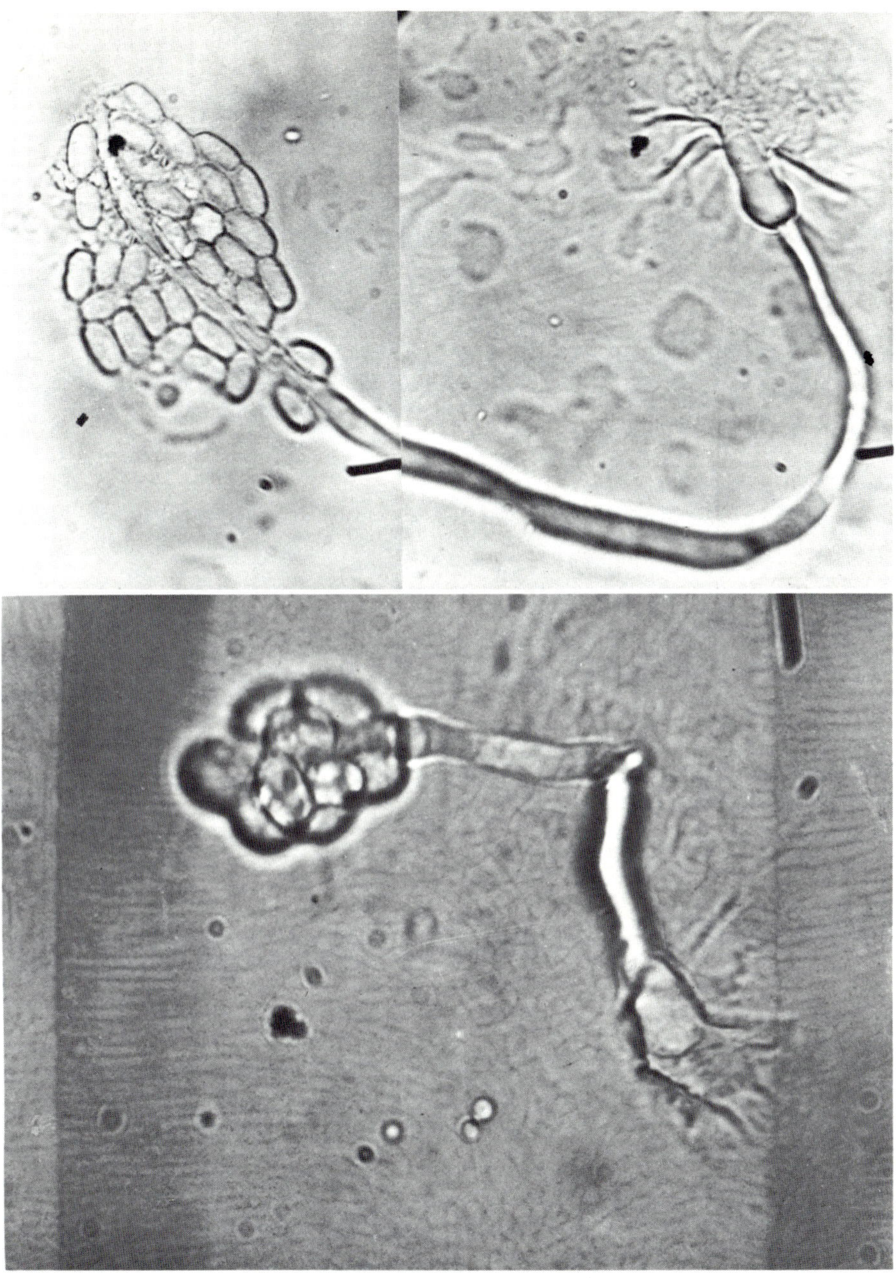

Figs. 16 and 17. Photographs taken *in situ* under immersion oil of fruiting bodies of the mutant Fty-1 (SUSSMAN, 1955)

is imagined to be controlled by macromolecules at the cell surface that can monitor cell contacts in a directional fashion. The control is by negative feedback so that ultimately only two steady-state concentrations of the morphogen are permissible. Thus discrete developmental fates can be distributed precisely along the axis of the slug.

Epilog: Remember the Fruity Mutant!

In considering the kinds of cell interactions that can regulate morphogenesis and cytodifferentiation within the migrating slugs and nascent fruits of *D. discoideum*, there is an unconscious tendency to think in terms of interactions occurring among tens and hundreds of thousands of cells. These are Gaussian Numbers and they lead to Gaussian Thoughts.

It is necessary to emphasize that Poissonian Thoughts may be much more suited to the elucidation of these interactions since aggregates of as little as 10 - 20 cells can produce perfectly proportioned fruiting bodies with terminally differentiated component cells.

As mentioned previously, the sizes of fruiting bodies (which are determined by the number of cells that enter a single aggregate) depend upon the population density of the cells before aggregation. In the wild type, the observed sizes of these aggregates have ranged between 10^5 cells and 50 - 100 cells. The fruity mutant of *D. discoideum* FTy-1 is recognized by the fact that its colonies contain a huge number of very tiny fruits (SUSSMAN, 1955). It turns out that in this strain, the relation between population density and the number and size of the cell aggregates has been changed dramatically. At very high population densities the fruits are much smaller than wild-type fruits would be, but still contain tens of thousands of cells. At very low densities aggregates of a few as 10 - 20 cells can form and give rise to fruiting bodies with normal spores, stalk and basal disc cells. Figs. 16 and 17 show two such fruits containing respectively ca 50 and precisely 12 cells!

By criteria such as proportion of cell types, ratio of spore mass to stalk height, etc. the dimensions of the fruits from the largest to the smallest are constant. The point of this epilog is to emphasize that any theory of morphogenetic regulation in *D. discoideum* must be capable of explaining how the developmental fates of 12 cells can be allocated with the same precision and in the same pattern as the fates of 100,000 cells.

References

ASHWORTH, J.M., SUSSMAN, M.: J. Biol. Chem. 242, 1696 (1967).

BEUG, H., GERISCH, G., KEMPFF, S., RIEDEL, V., CREMER, G.: Exp. Cell. Res. 63, 147 (1970).

BEUG, H., KATZ, F.E., STEIN, A., GERISCH, G.: Proc. Nat. Acad. Sci. (Wash.) 70, 3150 (1973).

BONNER, J.T.: Amer. J. Bot. 31, 175 (1944).

BONNER, J.T.: J. Exp. Zool. 106, 1 (1947).

BONNER, J.T.: J. Exp. Zool. 110, 259 (1949).

BONNER, J.T.: Biol. Bull. 99, 143 (1950).

BONNER, J.T., CHIQUOINE, A.D., KOLDERIE, M.: J. Exp. Zool. 130, 133 (1955).

BONNER, J.T., CLARKE, W.W., NEELY, C.L., SLIFKIN, M.: J. Cell. Comp. Physiol. 36, 149 (1950).

BONNER, J.T., DODD, M.: Biol. Bull. 122, 13 (1962).

BRACKENBURY, R.W., SCHINDLER, J., ALEXANDER, S., SUSSMAN, M.: J. Mol. Biol. 90, 529 (1974).

CHASSY, B.M.: Science 175, 1016 (1972).

CHILD, M.: Problems and patterns of development. Chicago: Univ. of Chicago Press 1941.

COCUCCI, S., SUSSMAN, M.: J. Cell Biol. 545, 399 (1970).

COHEN, M.H., ROBERTSON, A.: J. Theor. Biol. 31, 101 (1971).

CRICK, F.: Nature 225, 420 (1970).

DE HAAN, R.L.: In: 27th Symp. Soc. Dev. Biol. (M. LOCKE, Ed.), p. 208. New York: Academic Press 1968.

ELLINGSON, J.S., TELSER, A., SUSSMAN, M.: Biochim. Biophys. Acta 244, 388 (1971).

FIRTEL, R., BONNER, J.: Dev. Biol. 29, 85 (1972).

FIRTEL, R., BRACKENBURY, R.W.: Dev. Biol. 27, 307 (1972).

FRANKE, J., SUSSMAN, M.: J. Biol. Chem. 246, 6381 (1971).

FRANKE, J., SUSSMAN, M: J. Mol. Biol. 81, 173 (1973).

GARROD, D., ASHWORTH, J.M.: Symp. Soc. for Gen. Microbiol. XXIII, 407 (1973).

GERISCH, G.: Dev. Biol. 3, 685 (1961).

GERISCH, G.: Personal communication (1973).

GERISCH, G., MALCHOW, D., RIEDEL, V., MULLER, E., EVERY, M.: Nature 235, 90 (1972).

GERISCH, G., MALCHOW, D., WILHELMS, H., LÜDERETZ, O.: Europ. J. Bioch. 9, 229 (1969).

GERISCH, G., NORMANN, I., BEUG, H.: Naturwissenschaften 53, 618 (1966).

GOLDBETER, A.: Personal communication.

GOODWIN, B., COHEN, M.H.: J. Theor. Biol. 25, 49 (1969).

GREGG, J.H.: Dev. Biol. 12, 377 (1965).

GREGG, J.H., BADMAN, W.S.: Dev. Biol. 22, 96 (1970).

GROBSTEIN, C.: Science 143, 643 (1964).

GROBSTEIN, C., ZWILLING, E.: J. Exp. Zool. 122, 259 (1953).

GUSTAFSON, G.L., WRIGHT, B.E.: Critical reviews in Microbiology, Vol. 1, p. 453 (1973).

GUSTAFSON, G.L., WRIGHT, B.W.: Biochem. Biophys. Res. Comm. 50, 438 (1973).

HADDEN, J.W., HADDEN, E.M., HADDOX, M.K., GOLDBERG, N.D.: Proc. Nat. Acad. Sci. (Wash.) 69, 3024 (1972).

HEPPEL, L.A., HILMOE, R.J.: J. Biol. Chem. 202, 217 (1953).

HOHL, H.R., HAMAMOTO, S.T.: J. Ultrastructure Res. 26, 442 (1969).
JOHNSON, M., KAYE, M.A.G., HEMS, R., KREBS, H.A.: Biochem. J. 54, 625 (1953).
KATZ, E.R., SUSSMAN, M.: Proc. Nat. Acad. Sci. (Wash.) 69, 495 (1972).
KELLER, E.F., SEGEL, L.A.: J. Theor. Biol. 26, 399 (1970).
KILLICK, K.A., WRIGHT, B.E.: J. Biol. Chem. 247, 2967 (1972).
KONIJN, T., RAPER, K.B.: Dev. Biol. 3, 725 (1961).
KONIJN, T.M., VAN DE MEENE, J.G., BONNER, J.T., BARKLEY, D.S.: Proc. Nat. Acad. Sci. (Wash.) 58, 1152 (1967).
LOOMIS, W.F. Jr.: J. Bacteriol. 97, 1149 (1969).
LOOMIS, W.F. Jr.: Nature 240, 6 (1972).
MAEDA, Y., SUGITA, K., TAKEUCHI, I.: Bot. Mag. 86, 5 (1973).
MAEDA, Y., TAKEUCHI, I.: Development, Growth and Differentiation 11, 232 (1969).
McMAHON, D.: Proc. Nat. Acad. Sci. (Wash.) 70, 3296 (1973).
MILLER, Z.I., QUANCE, J., ASHWORTH, J.M.: Biochem. J. 114, 815 (1969).
NESTLE, M., SUSSMAN, M.: Dev. Biol. 28, 545 (1972).
NEWELL, P.C.: In: Essays in Biochemistry (Eds. P.N. CAMPBELL, F. DICKENS), vol. 7, p. 87. New York: Academic Press 1971.
NEWELL, P.C., ELLINGSON, J.S., SUSSMAN, M.: Biochim. Biophys. Acta 177, 610 (1968).
NEWELL, P.C., FRANKE, J., SUSSMAN, M.: J. Mol. Biol. 63, 373 (1972).
NEWELL, P.C., LONGLANDS, M., SUSSMAN, M.: J. Mol. Biol. 58, 541 (1971).
NEWELL, P.C., SUSSMAN, M.: J. Biol. Chem. 244, 2990 (1969).
NEWELL, P.C., SUSSMAN, M.: J. Mol. Biol. 49, 627 (1970).
NEWELL, P.C., TELSER, A., SUSSMAN, M.: J. Bacteriol. 100, 763 (1969).
PANNBACKER, R., BRAVARD, L.: Science 175, 1014 (1972).
POFF, K.L., BUTLER, W.L.: Nature, in press (1974).
POFF, K.L., LOOMIS, W.F.: Exp. Cell Res. 82, 236 (1973).
PONG, S., LOOMIS, W.F. Jr.: J. Biol. Chem. 248, 4867 (1973).
RAPER, K.B.: J.E. Mitchell Sci. Soc. 56, 241 (1940).
RAPER, K.B.: Growth Symposium III, 41 (1941).
RIEDEL, V., GERISCH, G.: Biochem. Biophys. Res. Comm. 42, 119 (1971).
ROBERTSON, A., DRAGE, D., COHEN, M.H.: Science 175, 333 (1972).
ROSSOMANDO, E.F.: Personal communication.
ROSSOMANDO, E.F., SUSSMAN, M.: Proc. Nat. Acad. Sci. (Wash.) 70, 1254 (1973).
ROTH, R.M., ASHWORTH, J.M., SUSSMAN, M.: Proc. Nat. Acad. Sci. (Wash.) 59, 1235 (1968).
ROTH, R.M., SUSSMAN, M.: J. Biol. Chem. 243, 5081 (1968).
SHAFFER, B.M.: In: Primitive Motile Systems in Cell Biology (Eds. R.D. ALLEN, N. DAMIYA), p. 387. New York: Academic Press 1963.
SLIFKIN, M.K., BONNER, J.T.: Biol. Bull. 102, 273 (1952).
SONNEBORN, D.R., SUSSMAN, M., LEVINE, L.: J. Bacteriol. 87, 1321 (1963).
SUSSMAN, M.: J. Gen. Microbiol. 13, 295 (1955).
SUSSMAN, M.: Methods in Cell Physiology (Ed. D. PRESCOTT), vol. II, p. 397. New York: Academic Press 1966.
SUSSMAN, M., LEE, F., KERR, N.S.: Science 123, 1171 (1956).

SUSSMAN, M., LOVGREN, N.: Exp. Cell Res. 38, 97 (1965).
SUSSMAN, M., NEWELL, P.C.: In: Molecular Genetics and Developmental Biology (Ed. M. SUSSMAN), p. 275. New Jersey: Prentic Hall 1972.
SUSSMAN, M, NOEL, E.: Biol. Bull. 103, 259 (1952).
SUSSMAN, M., OSBORN, M.J.: Proc. Nat. Acad. Sci. (Wash.) 52, 81 (1964).
SUSSMAN, M., SUSSMAN, R.R.: Exp. Cell Res. Suppl. 8, 91 (1961).
SUSSMAN, M., SUSSMAN, R.R.: Symp. Soc. for Gen. Microbiol. XIX, 403 (1969).
SUSSMAN, R.R.: Biochim. Biophys. Acta 149, 407 (1967).
TAKEUCHI, I.: Dev. Biol. 8, 1 (1963).
TELSER, A., SUSSMAN, M.: J. Biol. Chem. 246, 2252 (1971).
WHITE, G.J., SUSSMAN, M.: Biochim. Biophys. Acta 74, 179 (1963).
WOLPERT, L.: J. Theor. Biol. 25, 1 (1969).
WOLPERT, L.: Current Topics in Developmental Biology 6, 183 (1971).
WRIGHT, B.E.: Proc. Nat. Acad. Sci. (Wash.) 46, 798 (1960).
WRIGHT, B.E.: J. Cell Physiol. 72, Suppl. 145 (1968).
WRIGHT, B.E.: Science 153, 830 (1966).
WRIGHT, B.E., ANDERSON, M.: Biochim. Biophys. Acta 43, 62 (1960).
WRIGHT, B.E., DAHLBERG, D.: J. Bacteriol. 95, 983 (1968).
WRIGHT, B.E., PANNBACKER, R.: J. Bacteriol. 93, 1762 (1967).

Isoprenoids and Bacteriorhodopsin in Halobacteria

Dieter Oesterhelt

I. Introduction

Halophilic (salt-loving), thermophilic, acidophilic or psychrophilic organisms are all examples of biological adaption to extreme environmental conditions. Halophilic organisms are distinguished from marine organisms by the requirement of more than 3 % NaCl in their growth medium, extremely halophilic bacteria require more than 15 % NaCl for optimal growth. The genus Halobacteria, object of this article, belongs to the extreme halophiles. A third group comprises the so-called halotolerant organisms which are able to grow in the absence of salt[1] as well as in the presence of high salt concentrations (GIBBONS, 1969). Halophiles occur in nature wherever their special growth conditions are met and may then reach very high cell densities, e. g. in open ponds where salt is produced by evaporation of sea water. The bright red color of such brines is partially due to carotenoids of extreme halophiles, which grow out their biological concurrence (for illustration see OESTERHELT, 1974a).

The interest of microbiologists for halophiles started when reddening of heavily salted products such as fish or hides was found to be caused by these microorganisms. The interest has decreased, however, with the decline of salt as a preservative (GIBBONS, 1969). More recently extreme halophiles, especially Halobacteria, were considered to be useful biological objects by electronmicroscopists and biochemists for several reasons.

Halobacteria show unusual properties. Their ribosomes retain their structural integrity and most of their enzymes will be catalytically active only in high salt. The cells will lyse when they are exposed to pure water. Several reviews on halophilism appeared and cover the literature until 1973 (LARSEN, 1962; LARSEN, 1967; KUSHNER, 1968; GIBBONS, 1969), LARSEN, 1973). Recently retinal was found in *Halobacterium halobium* and related strains. It is bound to a rhodopsin-like protein called bacteriorhodopsin. This is found in a differentiated area of the cell membrane called the purple membrane and mediates light-energy conversion.

It is the aim of this article to give an introduction into this new photoreceptor system. First a brief summarization of some special features of Halobacteria will be presented, followed by

---

[1]salt is used synonymously to NaCl throughout this article

the description of the isoprenoid compounds, expecially carotenoids of these organisms. Then various aspects of bacteriorhodopsin in the purple membrane and its significance for the bacterial cell will be discussed.

II. General Features of Halobacteria

A. Growth Requirements

Halobacteria belong to the order pseudomonales and are described as gram negative rod-shaped bacteria which are highly pleomorphic (BERGEY, 1957, 1974). Some species are motile by means of polar flagellation. All species require more than 15 % salt for optimal growth. Besides Na^+ and Cl^-, which are not replaceable by other ions, smaller amounts of potassium and magnesium are necessary to sustain growth. Potassium is accumulated inside the cells more than a hundredfold (CHRISTIAN and WALTHO, 1962; GINZBURG et al., 1971) and replaces internally Na^+ to a large extent. Magnesium seems to be necessary for the maintenance of the shape of the cells, trace amounts of iron and manganese influence growth and pigmentation respectively (GIBBONS, 1969).

In its natural environment, Halobacteria live on concentrated biological detriments and this may explain why freshly isolated strains show gelatinolytic activity which may be lost during cultivation (GIBBONS, 1957). Usually carbohydrates are not attacked by the cells whereas amino acids are metabolized by all strains (but see TOMLINSON et al., 1974). In general, no significant differences in the metabolic apparatus compared to nonhalophilic aerobes have been found so far. It should be pointed out, however, that systematic studies on the metabolism of Halobacteria have not been reported. In the laboratory the cells are usually grown on protein digests in liquid culture, using sparged air for aeration although synthetic media have been developed (GIBBONS, 1969). Most Halobacteria strains are obligate aerobes, but strictly anaerobe and facultative aerobe strains have been described in the literature (see GIBBONS, 1969). The growth rate will depend on medium, temperature and rate of aeration, the generation time ranging from 4 to 20 h. The temperature optima are around 40° although cells will grow eventually up to 55°. This tendency towards thermophilism may be due to the elevated temperatures in the shallow brines under tropical sunshine, one of the natural habitat of Halobacteria (GIBBONS, 1969; WULFF and KERADJOPOULUS, 1973; OESTERHELT and STOECKENIUS, 1974).

B. Morphology

Examined with the phase-contrast microscope *H. halobium* cells appear as rods of a variable length (3 - 10 μ) and have an approximate diameter of 0,5 μ. Without added magnesium the cells grow in irregular forms. Halobacteria lack the typical components of bacterial cell walls, for instance muramic or diaminopimelic acid (KUSHNER et al., 1964). A wall-like structure on the surface of *H. halobium* cells has been demonstrated by electron microscopy (STOECKENIUS and ROWEN, 1967). It consists of mainly protein in regular array as it can be seen in electron micrographs of freeze-dried cell envelope preparations. The envelope of thin-sectioned cells clearly shows the wall component and discernible from it, the cell membrane. Intracytoplasmic membranes are not found in Halobacteria except in *H. halobium* where membrane-bounded gas vacuoles exist (STOECKENIUS and KUNAU, 1968).

Halobacteria offer two attractive properties to membrane biologists. As expected from the chemical composition of the cell envelope, the rigidity of the cells is very low and weak mechanical forces will break the cells and allow the cytoplasma to be diluted into the surrounding medium. The broken cells reform closed vesicles which can be washed free of cytoplasmic components (OESTERHELT and STOECKENIUS, 1974). They can be used to study transport processes in a cell free system such as proton translocation (see p. 153), ion and amino acid transport similar to the system described first by KABACK in *E. coli* (KABACK, 1971; LOMBARDI, REEVES, and KABACK, 1973). Exposure of whole cells or cell envelopes to pure water results in lysis, i.e. disintegration of cell wall components and the cell membrane itself (see LARSEN, 1967; STOECKENIUS and ROWEN, 1967; STOECKENIUS and KUNAU, 1968). A method for fractionation of the various envelope components has been developed (OESTERHELT and STOECKENIUS, 1974) and the protein and carbohydrate composition of the cell envelope of *H. salinarium* has been reported (MESCHER et al., 1974).

C. Cell Constituents

Halobacteria do not synthesize fatty acids under their growth conditions. The only, invariable side chain of their lipids is dihydrophytol, the C_{20}-saturated isoprenoid alcohol (Fig. 1). It is bound to the α- and β-hydroxyl group of the glycerol moiety of the cellular lipids *via* an ether linkage. KATES and his

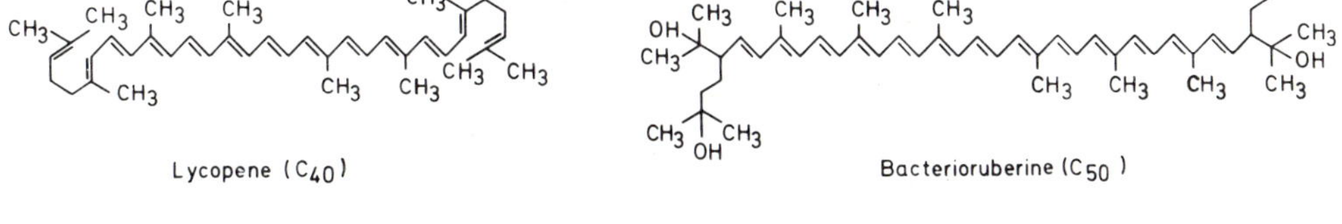

Dihydrophytol (C_{20})

Retinal (C_{20})

Lycopene (C_{40})

Bacterioruberine (C_{50})

Fig. 1. Main isoprenoid compounds in Halobacteria

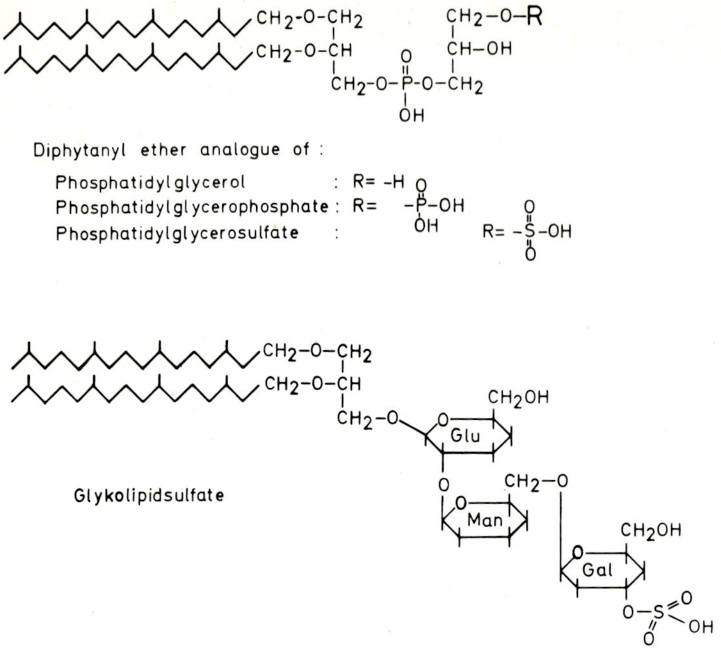

Diphytanyl ether analogue of :

Phosphatidylglycerol : R= -H

Phosphatidylglycerophosphate : R= $-P(=O)(OH)-OH$

Phosphatidylglycerosulfate : R= $-S(=O)_2-OH$

Glykolipidsulfate

Fig. 2. Lipids of known structure in Halobacteria

collaborators elucidated the structure of the 4 main lipids shown in Fig. 2 and confirmed their results by total synthesis of phosphatidylglycerophosphate, phosphatidylglycerol and phosphatidylglycerosulfate. Apparently, no differences in the chemical structure occur among Halobacteria (JOO, KATES, and SCHIER, 1968; JOO and KATES, 1969; KATES and DEROO, 1973; HANCOCK and KATES, 1973).

Three of the main lipids are highly negatively charged molecules, unable to form stable liposomal bilayer structures at higher ionic strength (0.2 M salt) except when glycolipidsulfate is added (CHEN et al., 1974). This lipid species in combination with the naturally occurring apolar lipids (squalenes and carotenoids) may therefore be responsible in part for the structural integrity of the cell membrane of Halobacteria. Squalene and other apolar lipids are apparently intercalated between the negatively charged phospholipids, making them readily accessible to cations. On this basis, membrane disintegration upon removal of salt by repulsion of the negative groups may be understood (LANYI, 1974a; PLACHY, LANYI, and KATES, 1974; LANYI, PLACHY, and KATES, 1974).

The red color of Halobacteria is due to carotenoids, the main component being bacterioruberin (Fig. 1). Its structure was proposed recently to be that of a C_{50}-tetrahydroxy-carotenoid (KELLY, NORGARD, and LIAAEN-JENSEN, 1970). Also C_{40}-carotenoids are present as expected from the occurrence of retinal. Squalenes are another group of isoprenoid compounds synthesized by these cells, occurring in much lower concentrations than the carotenoids (KUSHWAHA et al., 1972). The biosynthetic capacity therefore comprises di, tri and tetra-terpenes, in addition Vitamin K compounds containing a longer isoprenoid side chain. The assumed biosynthetic relationship of isoprenoids in Halobacteria will be discussed unter Section III.

Only few halophilic enzymes have been isolated in pure form (HUBBARD and MILLER, 1969; DUNDAS, 1972; LOUIS and FITT, 1972a; LOUIS and FITT, 1972b; HOCHSTEIN and DALTON, 1973; HOLMES and HALVORSON, 1965). Studies on pure and partially purified enzymes concentrate on the salt dependent enzyme activities and stabilities (see review of LANYI, 1974b). Two possible interpretations, although not mutually exclusive but involving different physico-chemical principles have been suggested. First, the salt can provide counter ions for the negative charge on the protein surface and reduces therefore electrostatic repulsion from like charges. That way keeps the enzyme in its active conformation (BAXTER, 1959; KUSHNER and ONISHI, 1966). This charge-shielding effect as explanation for the salt-dependence of enzyme activity is supported by their high content in acid amino acids (BROWN, 1963; KUSHNER and ONISHI, 1966). It does not explain, however, why maximal activity of some enzymes needs the presence of more than 0.2 M salt which should saturate charge shielding processes (INMAN and JORDAN, 1960). The second interpretation assumes a salt out effect of NaCl and other salts on the hydrophobic parts of the enzyme molecule. Certain amino acid residues are forced to be inside the molecule and to maintain that way the active conformation (LANYI and STEVENSON, 1970). Probably both interpretations are correct and active conformation of halophilic enzymes is guaranteed by both effects in varying proportions depending on the individual enzymes (LIEBERMAN and LANYI, 1972).

In the case of isocitrate dehydrogenase instability of the enzyme in absence of salt is caused by exposure of oxidizable SH-groups after the active conformation has changed to the inactive form (HUBBARD and MILLER, 1969; WULFF, HUBBARD, and MILLER, 1972). This irreversible inactivation following the conformational change could be a general phenomenon explaining the instability of many halophilic enzymes at low salt concentrations.

Different from halophilic enzymes, the ribosomes of Halobacteria require specifically saturating amounts of KCl and smaller amounts of magnesium for structural integrity. This property reflects accurately the internal salt content of the cells (see above).

III. Biosynthesis of Isoprenoid Compounds

By using C^{14}-labeled precursors such as acetate, malonate, mevalonate and glycerol, it was shown that mevalonate is the most efficient precursor for the phytanyl chains of the cellular lipids. Malonate is incorporated only to a very low extent, but exclusively in the phytanyl chains, and only traces of fatty acids have been detected (KATES, WASSEF, and KUSHNER, 1968).

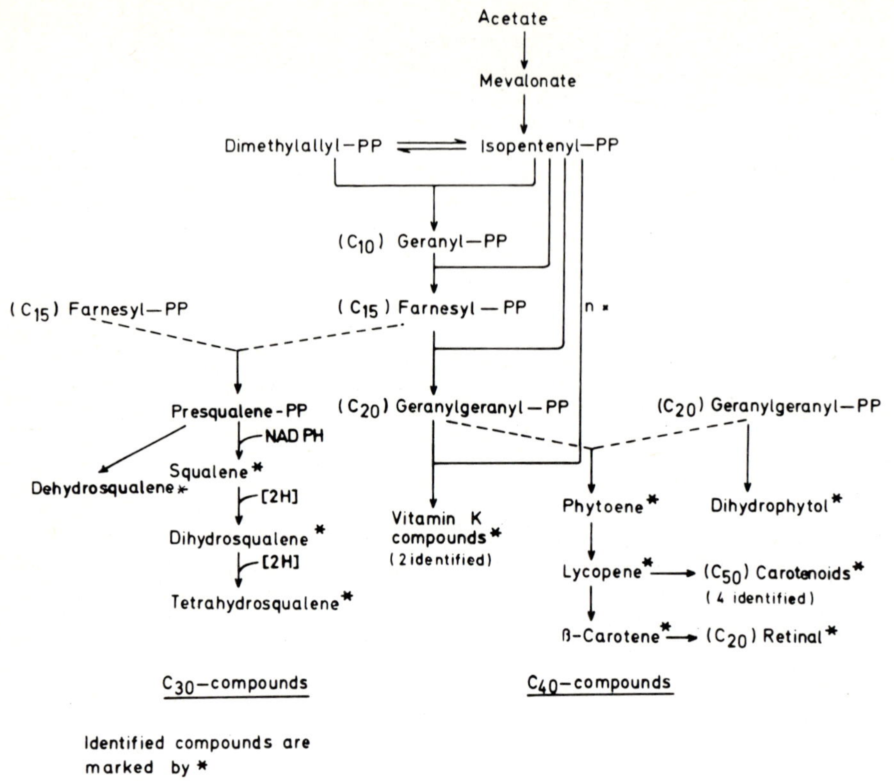

Fig. 3. Assumed biosynthetic pathways of isoprenoid compounds in Halobacteria

This suggests that under normal conditions fatty acid biosynthesis does not happen in Halobacteria. However, cell lysates show fatty acid synthesizing activity at low ionic strength (PUGH, WASSEF, and KATES, 1971). This indicates that the fatty acid synthesizing enzyme system is present in halobacterial cells but is inhibited by high salt concentrations. The same inhibition is also observed in the case of the multienzyme complex from yeast (OESTERHELT, unpublished observation). Dihydrophytol synthesis allow Halobacteria to escape the problem of synthesizing fatty acids at high salt concentrations. The methyl groups of dihydrophytol simulate a mixture of saturated and unsaturated fatty acids by lowering the melting point compared to an unbranched hydrocarbon chain. This may be important to keep the transition point of the bulk lipids in the cell membrane low enough to permit still optimal growth at medium temperatures (CHEN et al., 1974).

It should be mentioned that halotolerant organisms have been isolated which grow from 0.2 to 4 M salt concentration in the medium and synthesize the same fatty acid pattern under all growth conditions. Such organisms are especially interesting for discussions of evolution of halophilism (STERN and TIETZ, 1973).

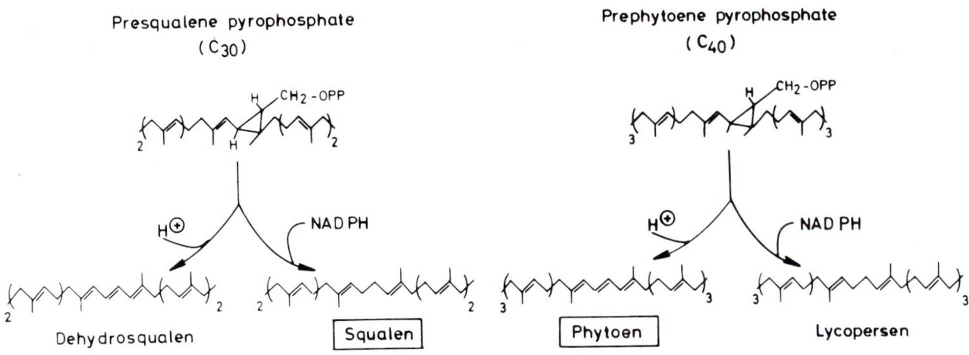

Fig. 4. Products of the head-to head condensation reactions at the C_{30} and C_{40} level

So far, no *in vitro* studies on isoprenoid biosynthesis in Halobacteria have been reported. The pathways outlined in Fig. 3 are therefore mainly based on known biosynthetic reactions in other organisms and identification of precursors as well as products in Halobacteria themselves. The main types of reactions expected to occur are:

1. Polymerization of C_5-units
2. Head-to-head-condensation of polyprenyl pyrophosphate
3. Hydrogenation of C=C double bonds
4. Dehydrogenation of C-C single bonds
5. Oxidative cleavage of C=C double bonds
6. Chain-branching reaction by addition of C_5-units.

Head-to-head condensation occurs at the C_{15}- and C_{20}-level producing the groups of squalenes and carotenoids respectively, the latter group being predominant. The squalene synthetase reaction in rat liver microsomes and yeast has been studied in some detail by several laboratories. Squalene is synthesized *via* an intermediate pyrophosphate (presqualene pyrophosphate) containing a cyclopropane ring which is subsequently cleaved by NADPH to yield squalene[2] (EPSTEIN and RILLING, 1970; EDMOND et al., 1971). Dehydrosqualene (bacterial phytoene see SUZUE, TSUKUDA, and TANAKA, 1968) is not an intermediate of the synthetase reaction (Fig. 4). At the C_{40}-level an analogous intermediate (prephytoene pyrophosphate) exists (ALTMAN et al., 1972). By the loss of a hydrogen ion it is converted directly into phytoene rather than hydrogenated to yield the C_{40}-analog of squalene called lycopersene (GREGONIS and RILLING, 1974). The two enzyme systems catalyzing the condensation reactions at the two chain lengths C_{30} and C_{40} apparently have different product specificities (i.e. reduction of the intermediate at C_{30} and proton loss at C_{40}). Halobacteria must possess both enzyme systems because squalenes and C_{40}-carotenoids have been identified (TORNABENE et al., 1969;

[2] The structure of presqualene pyrophosphate proposed by WASNER and LYNEN (1970) has been withdrawn (F. LYNEN, personal communication).

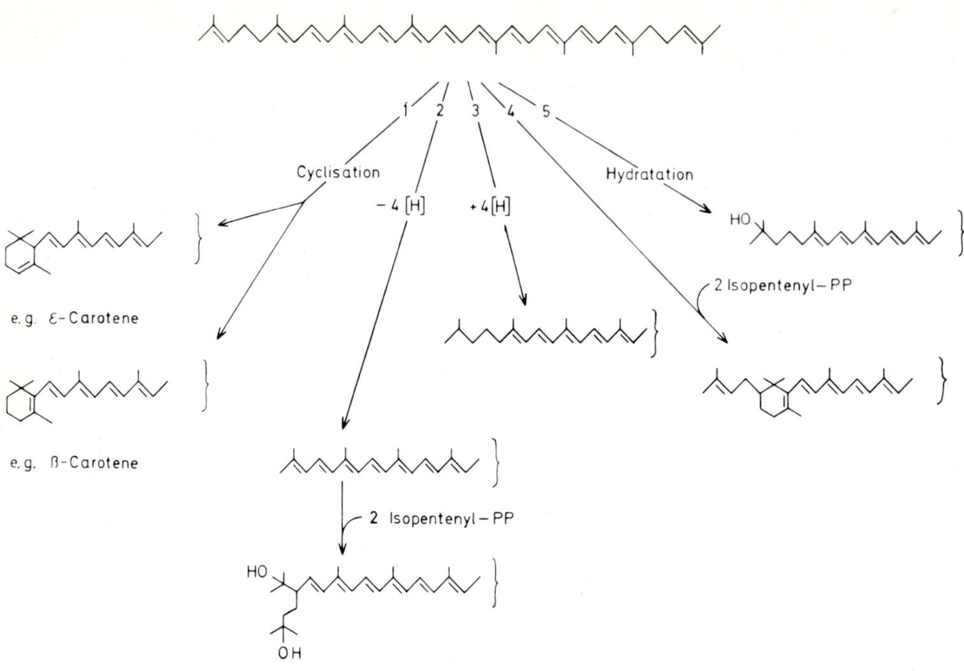

Fig. 5. Formation of carotenoid pigments starting from lycopene. Reactions 1 and 2 may occur in Halobacteria

KRAMER, KUSHWAHA, and KATES, 1972; KUSHWAHA et al., 1972). Interestingly enough, dehydrosqualene and lycopersene have also been detected so that one is tempted to speculate that the unperfect chain length specificity of the condensing enzymes produces small amounts of these two compounds.

It should be mentioned that fatty acid synthetase always produce mixtures of fatty acids of different chain length. This can be interpreted in terms of unspecific (chain-length independent) catalysis of one of the component enzymes (SUMPER et al., 1969).

Beside squalene, more saturated derivatives such as dihydro- and tetrahydrosqualene have been found. Here again unspecificity of the enyzme system catalyzing saturation of the 4 double bonds at the C_{20}-level yielding dihydrophytol (Fig. 3) could saturate some of the double bonds in squalene by a side reaction.

Stepwise desaturation of phytoene yields lycopene as a branching point for the synthesis of β-carotene and bacterioruberine. From a mutant strain of *H. halobium*, unable to synthesize bacterioruberine (MILANYTCH, 1973) all-trans lycopene can be isolated in appreciable amounts under certain growth conditions (OESTERHELT, 1972a; KUSHWAHA and KATES, 1973). The occurrence of retinal in bacteria is surprising since this biosynthetic capacity seemed to be reserved to animals where retinal is formed in the intestine by oxidative cleavage of β-carotene. No experiments have been carried out yet to prove that the same reaction is responsible also for retinal formation in Halobacteria. There are, however, indications of C_{40}-carotenoids being the precursors of retinal in *H. halobium*. Nicotine, an inhibitor of cyclization reactions of lycopene (HOWES and BATRA, 1970) stops β-carotene and retinal synthesis and accumulates lycopene (REITMEIER, H. and OESTERHELT, D., unpublished). Retinal occurs in several Halobacteria strains, e.g. *H. halobium*, *H. cutirubrum* and *H. trapanicum* and may be found still in others (GOCHNAUER et al., 1972; OESTERHELT, 1974c; KUSHWAHA and KATES, 1973). Phytoene, phytofluene and α- and β-carotene (1) were identified in *H.cutirubrum* extracts, suggesting again the biosynthetic scheme outlined in Fig. 3 (KUSHWAHA et al., 1972). Reactions starting from lycopene are summarized in Fig. 5. The formation of bacterioruberins (2) proposed in this scheme is still purely speculative, and it is unknown if dehydrogenation of lycopene has to preceed the addition of dimethylallyl pyrophosphate in order to prevent a cyclization reaction analogous to cyclic carotene (4) formation (see REES and GOODWIN, 1972). 4 bacterioruberins with varying degree of oxygenation and number of double bonds have been identified so far (KELLY, NORGARD, and LIAAEN-JENSEN, 1970; ISLER, 1971).

Control of pigment formation is found by nutritional factors, for instance glycerol (GOCHNAUER et al., 1972) and growth conditions e.g. oxygen and light (see Section VII).

A recent review on microbial carotenoids has been given (LIAAEN-JENSEN and ANDREWS, 1972) and carotenoid functions has been extensively treated by KRINSKY in "Carotenoids" (ISLER, 1971). Carotenoids in Halobacteria function most likely as protective reagents against photodamage of other cell constituents. In connection with the function of bacteriorhodopsin as part of a light-energy converting system, however, carotenoids also may have an accessory pigment function, i.e. absorb light and transfer its energy to bacteriorhodopsin (see Section VI).

IV. The Purple Membrane and Bacteriorhodopsin

A. Structure and Chemical Composition

As mentioned before, exposure of cells or cell envelopes to pure water disintegrates them. Cell membrane fragments of gas vacuole-free *H. halobium* cells were separated on sucrose-density gradients into two main fractions, the so-called red, bacterioruberin containing membrane and a purple-colored fraction (STOECKENIUS and KUNAU, 1968; OESTERHELT and STOECKENIUS, 1974). Electron microscopy of the purple fraction revealed a surprising homogenous population of round to oval sheets with an approximate diameter of 0.5 μ. It was termed the purple membrane. The identification of retinal as its pigment increased the interest in this membrane now proved to be a photoreceptor system (OESTERHELT and STOECKENIUS, 1971; OESTERHELT and STOECKENIUS, 1973).

The purple membrane shows some structural and chemical peculiarities which distinguish it from an ordinary cell membrane

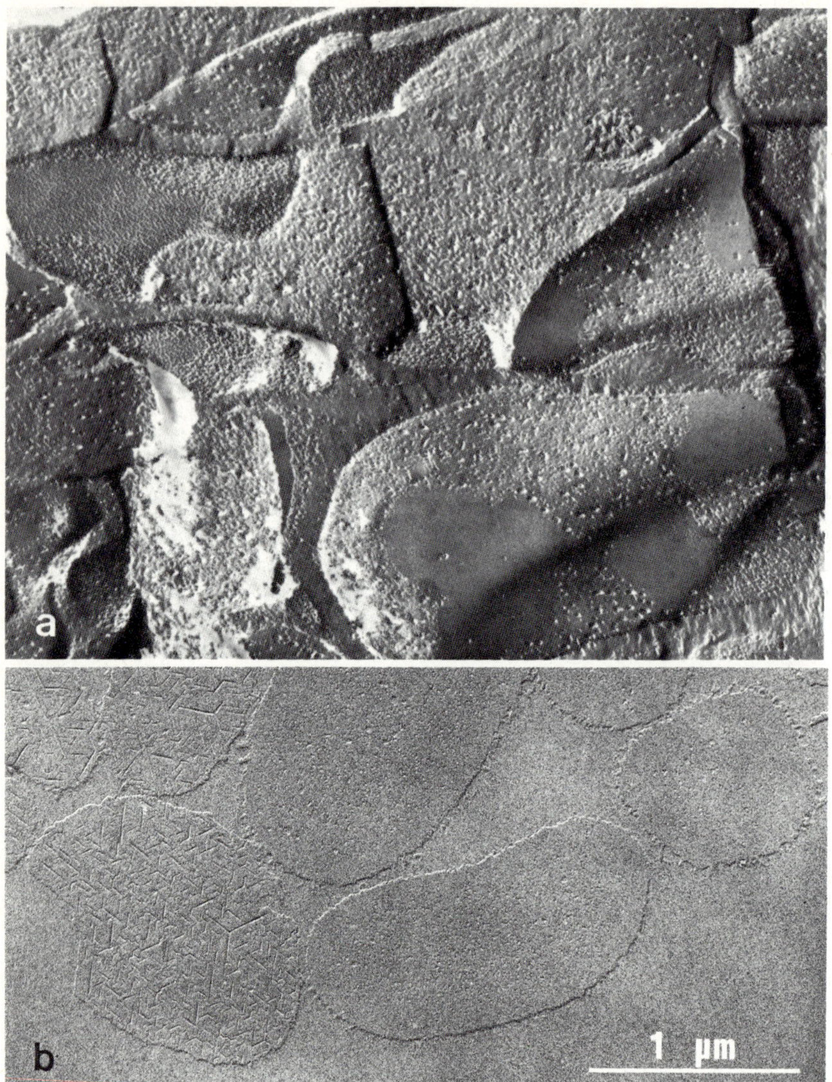

Fig. 6a and b. Electron micrographs of the purple membrane. (a) Purple membrane patches in whole cells. The smooth or granulated appearance is due to the structural assymmetry of the membrane patches discussed in BLAUROCK and STOECKENIUS, 1971. (Taken from BLAUROCK and STOECKENIUS, 1971). (b) Isolated purple membranes (picture taken by M. CLAVIER)

fragment. Information about the structure and the localization of this membrane fragment within the cell membrane was obtained by electron microscopy and X-ray diffraction. The membrane has a thickness of 48 Å, much less than that of common biological membranes. A hexagonal lattice structure extends over the plane of the membrane with a center-to-center distance of adjacent hexagons of 63 Å (BLAUROCK and STOECKENIUS, 1971). The observed hexagonal lattice structure is due to a crystalline array of bacteriorhodopsin molecules in the plane of the membrane. This certainly gives the purple membrane an extremely rigid structure and immobilizes the lipids to a large extents as shown by spin label studies of the purple membrane (HUBBELL, OESTERHELT, and STOECKENIUS, unpublished results; CHIGNELL and CHIGNELL, 1975). In addition, no rotation of bacteriorhodopsin in the purple membrane, comparable to the rotation of rhodopsin in the plane of disc membrane has been found (RAZI NAGRI et al., 1973). Bacteriorhodopsin is the only protein species in the purple membrane and amounts to 75 % of its dry weight. Correspondingly, a high buoyant density of 1.18 g/cm^3 of the membrane is found, which facilitates the separation of the purple membrane from other cell membrane fragments by isopycnic centrifugation.

The structural asymmetry of this membrane can be visualized by electron microscopy. Purple membranes which are sprayed on freshly cleaved mica, dried and shadowed with platinum/carbon show cracks which parallel each other or intersect at 60° and 120° angles confirming a hexagonal lattice structure of the membrane (Fig. 6b). Since half of the sheets appear with a smooth surface the conclusion of two different sides of the membrane (face A and face B, see BLAUROCK and STOECKENIUS, 1971) can be drawn. This is also demonstrated in freeze-etched cell preparations, where patches of smooth or particulate appearance are seen (Fig. 6a). Such electron micrographs can be analyzed by optical diffraction and the same hexagonal lattice structure as analyzed by X-ray diffraction is found (BLAUROCK and STOECKENIUS, 1971). More recent results indicate that bacteriorhodopsin is not only inserted asymmetrically into the purple membrane but may actually span as an α helix over the entire cross section of the membrane (BLAUROCK, 1975; KING and STOECKENIUS, 1974; HENDERSON, 1975). This feature would be of most important significance for the *in vivo* function of the bacteriorhodopsin molecule as an hydrogen ion pump.

As mentioned above, bacteriorhodopsin is the only protein found in the purple membrane (Fig. 7). It has a molecular weight of approximately 26,000 and contains retinal bound in a stoichiometry of 1:1 (OESTERHELT and STOECKENIUS, 1971). Under suitable growth conditions of *H. halobium* it may comprise a substantial part of the cellular protein (Fig. 7). The purple membrane then covers 50 % or more of the total cell membrane area (OESTERHELT and STOECKENIUS, 1973). Out of approximately 10 lipid molecules per protein molecule 5 - 6 molecules are phosphatidylglycerophosphate as analyzed by phosphorous determination after chromatographic separation of the membrane lipids (MILANYTCH, 1973). Another major component is glycolipidsulfate (see Fig. 2). Further quantitative analysis of lipids will possibly allow to account in stoichiometric numbers for all membrane components including salt cations and then facilitate the improvement of a membrane model based on the X-ray diffraction data.

B. The Membrane Chromophore

The absorption spectrum of the purple membrane shows the typical tryphtophan absorption band at 280 nm and a broad absorption

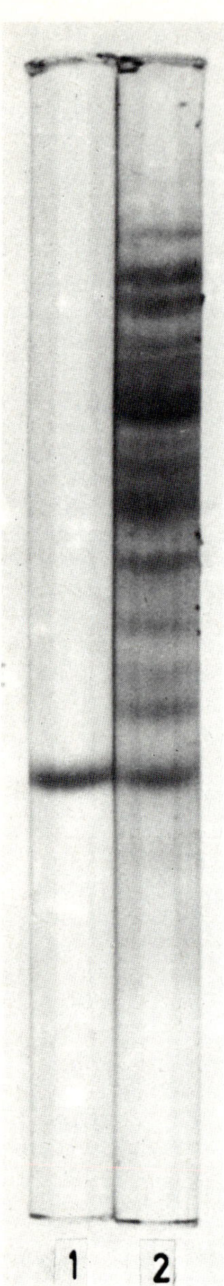

Fig. 7. SDS-acrylamide gel electrophoresis of the purple membrane (1) and a cell lysate (2)

band (half width 120 nm) centering around 560 nm which causes the purple color of the membrane (Fig. 8). During extraction of the pigment with chloroform/methanol, a color change to light yellow occurs and the extract exhibits an absorption around 380 nm typical for retinal. Identity of the pigment with retinal was proven by formation of derivatives such as retinol and retinaloxime, by the specific color reaction with antimony trichloride, thin layer chromatography and mass spectrometry

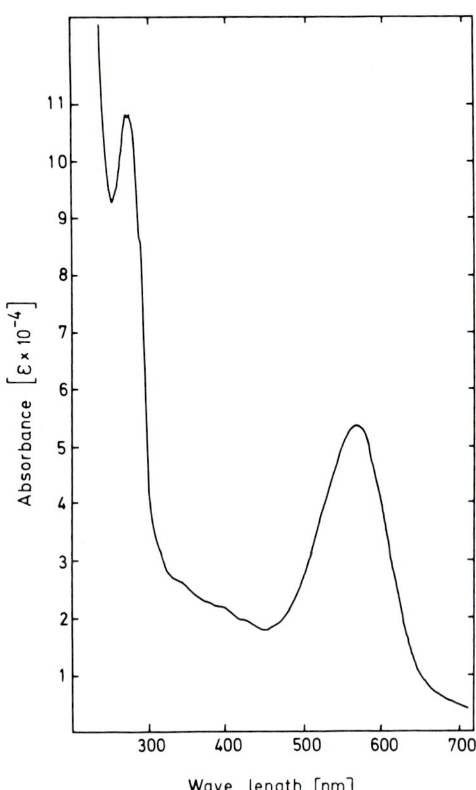

Fig. 8. Absorption spectrum of the purple membrane

(OESTERHELT and STOECKENIUS, 1971). The dramatic red shift of the retinal absorption upon association with a specific protein is common to most retinal-protein complexes. This is of essential importance for their function as photoreceptive systems of visible light.

None of the chromophores of retinal-protein complexes including bacteriorhodopsin is known in its chemical structure, although much experimental work has been accumulated (for review see HONIG and EBREY, 1974). Chemical information about the chromophore of bacteriorhodopsin was obtained by experiments using detergents, organic solvents and light-induced reaction of the chromophore with hydroxylamine and borohydride (OESTERHELT and STOECKENIUS, 1971; OESTERHELT, 1971b; OESTERHELT, MEENTZEN, and SCHUHMANN, 1973). Detergents like cetyl trimethylammonium bromide (CTAB) at neutral pH "bleach" the purple membrane, i.e. shift the absorption maximum into the UV-region of the spectrum. Subsequent treatment with ethanol precipitates retinylidenprotein which contains retinal bound *via* a C=N double bound to the protein. Borohydride reduces this Schiff base to the fluorescent retinylprotein, acid liberates retinal and hydroxylamine produces retinaloxime. In contrast to this, the native membrane reacts neither with borohydride nor hydroxylamine. Binding of retinal to bacterioopsin therefore not only induces a drastic spectroscopic change but also changes the chemical reactivity that retinal and its Schiff base exhibit. In order to describe this unique property of the molecular arrangement around retinal at its binding site in the membrane by a chemical term, the expression purple complex was chosen (OESTERHELT and STOECKENIUS, 1971). Most likely, amino acid side chains interact with the carbon skeleton of retinal under formation of the purple complex, as an intramolecular complex. The situation can be seen in analogy to the heme proteins where physicochemical properties, for instance redox potential or ligand-binding properties are modified by specific interaction with amino acid side chains of the protein part, e.g. histidine as the fifth ligand of the central iron atom. However, no detailed studies on the interaction of lipids with bacteriorhodopsin regarding the stability and photochemical behavior of the chromophor have been reported (for rhodopsin see: KONG and HUBBEL, 1973; ZORN and FUTTERMANN, 1971; APPLEBURY et al., 1974).

The action of the detergent CTAB unfolding the membrane protein reveals for the rhodopsins as well as for bacteriorhodopsin the underlying covalent structure of a Schiff base. For some time the idea was favored that the amino group of a phosphatidylethanolamine molecule and not of a lysine residue in rhodopsin binds the retinal moiety in the disc membrane of the retina. Later it was shown that a fast transimination between protein and lipid produced misleading results (BONTING, ROTMANS, and DAEMEN, 1973).

Bacteriorhodopsin changes its absorption and reactivity characteristics reversibly when it is exposed to some organic solvents described in more detail below. Using tritiated borohydride and diethylether it was tried to reduce retinal at its original binding site in "statu nascendi" before transimination reaction can occur. Alkaline hydrolysis of the product and thin layer chromatographic analysis showed lysine to be the most likely amino acid responsible for retinal binding (OESTERHELT, 1971a)

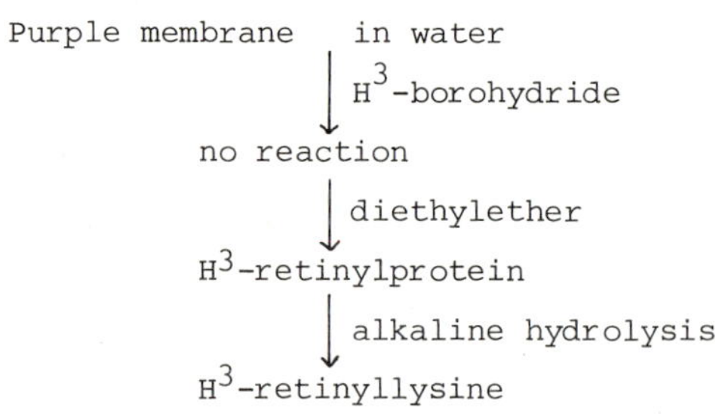

This experiment, however, gives no definite proof for covalent binding of retinal to the opsin moiety within the native purple complex, because a transimination or condensation reaction of retinal faster than borohydride reaction cannot be excluded. Support for the covalent structure at the binding site, however, has been obtained from raman spectroscopic experiments with the native membrane suspended in water. The C=N double bond frequency was identified in the spectrum of the chromophore (MENDELSOHN, 1973). More recent experiments, using the same method have shown that a protonated C=N double bond occurs in the native chromophore (LEWIS et al., 1975; MENDELSOHN et al., 1974).

During the photochemical cycle described in more detail below, it is possible to trap the retinal moiety by reaction with hydroxylamine as retinal oxime, or to reduce it by borohydride so that it becomes irreversibly linked to its original binding site. After reaction with hydroxylamine, a white apomembrane fraction is obtained which regenerates the chromophore upon addition of 13-cis retinal and all-trans-retinal but not with 9-cis or 11-cis retinal (OESTERHELT and SCHUHMANN, 1974; OESTERHELT, 1974a). In the case of the reduced chromophore, no regeneration is possible, indicating that the retinal moiety is not free to diffuse out of the binding site and therefore blocks the entry of a new retinal molecule.

C. Interaction with Organic Solvents

Most organic solvents mixed with purple membrane suspensions in water destroy the specific environment of the chromophore in the membrane as indicated by the irreversible discoloration of the membrane. Exceptions are diethylether, dimethylformamide or dimethylsulfoxide (DMSO) causing a blue shift of the purple complex absorption with is reversed upon removal of the solvent by evaporation or dilution with water. The DMSO/water system has been studied in some detail and has given useful information on the chemical nature of the chromophore system (OESTERHELT, MEENTZEN, and SCHUHMANN, 1973).

The purple complex equilibrates between 30 and 60 % DMSO in the solvent mixture with a second complex form absorbing maximally between 460 and 490 nm being called the 460 nm complex (Fig. 9). The equilibrium is influenced by pH and ions. Hydroxylamine splits retinal off the 460 nm complex as retinaloxime and borohydride reduces it to retinylprotein. For this changed reactivity the 460 nm complex is considered to be partially altered purple complex, exposing its functional group to an attack by hydroxylamine and borohydride. Its spectroscopic properties on the other hand distinguishes it from a retinylidenprotein or free retinal. One could speak as well of the two complex forms in terms of different conformational states of the bacteriorhodopsin molecule. However, the term complex involves specific interactions of molecules or parts of a molecule in addition to different atomar arrangements described by the term conformation and is preferred therefore in the present discussion.

It is important that the 460 nm complex and the purple complex are in equilibrium over a wide range of solvent concentration and thermodynamic parameters of the reaction can be measured. The results exclude for bacteriorhodopsin the possibility that the purple complex absorption band can be interpreted as a solvent shifted absorption band of retinylidenprotein (OESTERHELT, MEENTZEN, and SCHUHMANN, 1973). The system has further been used to determine the retinal isomers present in the membrane, and as a result 13-cis and all-trans retinal have been found

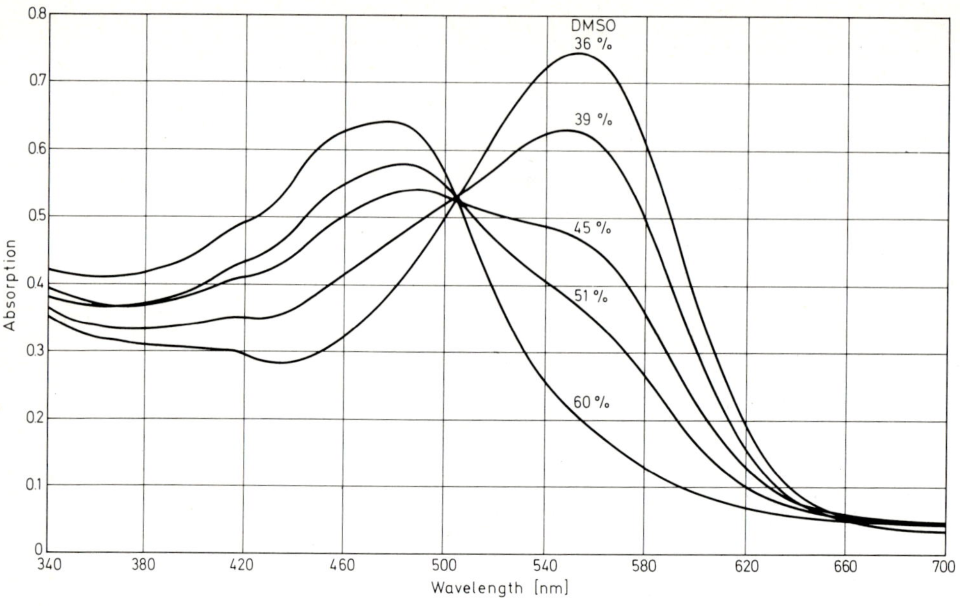

Fig. 9. Absorption spectra of the purple membrane in DMSO-water mixtures. Only the visible part of the spectrum is shown. (Taken from OESTERHELT, MEENTZEN, and SCHUHMANN, 1973)

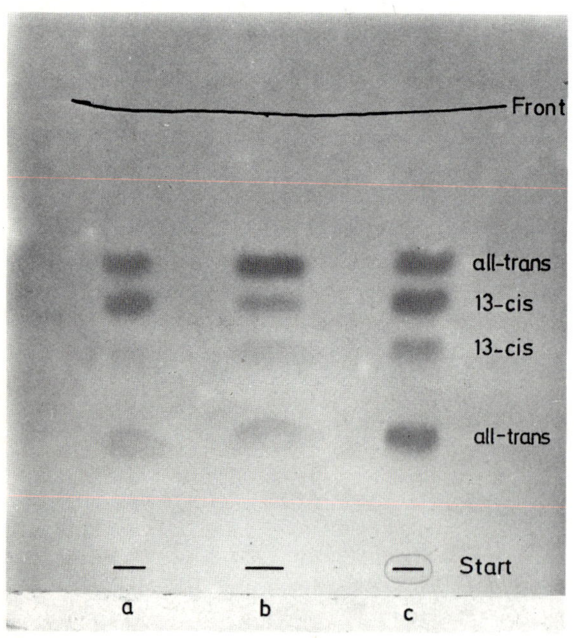

Fig. 10a-c. Thin layer chromatogramm of retinaloximes formed during the reaction between purple membrane and hydroxylamine in 50 % DMSO. Products of the initial phase (a) are compared with end products (b) and the products of CTAB-treated purple membrane (c). (Taken from OESTERHELT, MEENTZEN, and SCHUHMANN, 1973)

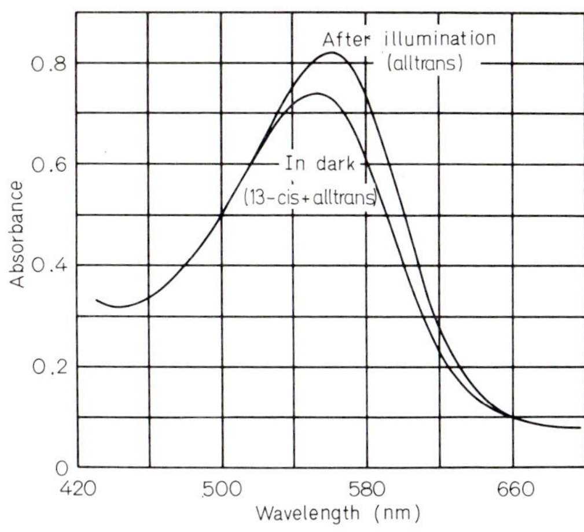

Fig. 11. Light-dependent change in absorption of the purple complex. (Taken from OESTERHELT, MEENTZEN, and SCHUHMANN, 1973)

(Fig. 10). Thin-layer chromatography of retinal-oximes for identification of retinal isomers provides a considerable advantage over the analysis of free retinals because for each C=C double bond isomer two oximes are expected to exist, the syn and the anti form:

H, R >C=N–OH (syn) and H, R >C=N–OH (anti)

syn   anti

A surprising result was obtained, when a purple complex 460 nm complex mixture was allowed to react with hydroxylamine and the product was analyzed by TLC. It contains both the 13-cis and the all-trans isomer, but only one C=N isomer, i.e. the syn form (identified by NMR Spectroscopy).

More detailed studies led to the conclusion that the chromophore occurs in two conformational states independent of the C=C isomerism in its carbon skeleton (OESTERHELT, MEENZTEN, and SCHUHMANN, 1973).

D. Interaction with Light

Two light reactions of bacteriorhodopsin are known and consist of a reversible chromophore isomerization and a cyclic sequence of reactions respectively (see under Section V: photochemical cycle). The first reaction is observed as a small red shift of the λ_{max} value of the purple complex absorption band, accompanied by a slight increase in the extinction coefficient (Fig. 11). Light absorbed by the purple complex is most efficient and the light-induced change is reversed in the dark with a half time in water at 35° C of 21 min. Thin-layer chromatography of the extracted retinaloximes after incubation of the membrane with CTAB and hydroxylamine shows a mixture of 13-cis and all-trans retinaloxime from the dark adapted sample but only the all-trans

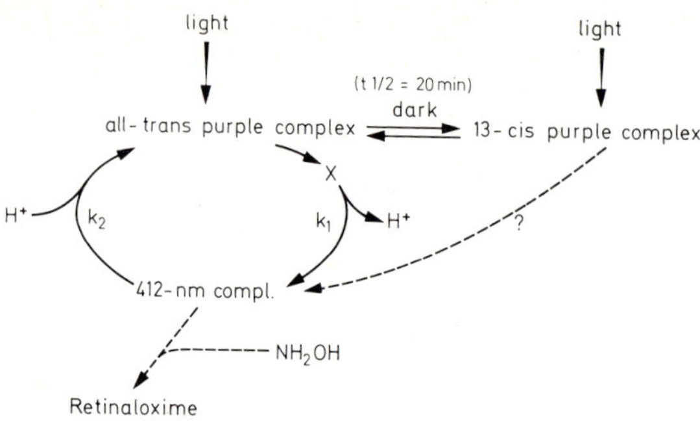

Fig. 12. The photochemical cycle of bacteriorhodopsin

isomer after illumination (OESTERHELT, MEENTZEN, and SCHUHMANN, 1973). The reaction therefore can be described qualitatively by:

$$\text{13-cis} \xrightarrow{\text{light}} \text{all-trans} \quad \text{and} \quad \text{all-trans} \xrightarrow{\text{dark}} \text{13-cis}$$

The conversion of all-trans to 13-cis in the dark may go to completion at lower temperatures (Jan 1974). However, the 13-cis form seems unlikely to be an intermediate in the photochemical cycle described below which proceeds in the cell in the millisecond range.

V. The Photochemical Cycle in Bacteriorhodopsin

Light initiates a cyclic sequence of absorption and conformation changes in bacteriorhodopsin which recall the cascade of reactions following light absorption in visual pigments. In contrast to the visual pigments bacteriorhodopsin regenerates its original state within milliseconds thermally so that light drives a photochemical cycle (Fig. 12). This points towards a photocoupling rather than a photosensing function of bacteriorhodopsin. The general term photoreceptor comprises both photocouplers and photosensors. A photosensing function implies that light absorption triggers a process that derives its required energy from other sources (e.g. visual pigments). A photocoupling function implies that the light energy absorbed is used as a driving force for such a process, for instance ion translocation or ATP synthesis (e.g. bacteriorhodopsin).

Fig. 12 illustrates that our knowledge about the intermediates of the photochemical cycle except the 412 nm complex is still very small. However, it has recently been shown by various laboratories that such intermediates (X in Fig. 12) preceeding the formation of the 412 nm complex exist (STOECKENIUS and LOZIER, 1975; DE VAULT et al., 1975; CHANCE et al., 1975; DENCHER and WILMS, 1975). Since the 412 nm complex is the longest living intermediate in the cycle it accumulates during continuous illumination to a constant steady-state concentration, which depends on light intensity and temperature (OESTERHELT and HESS, 1973; OESTERHELT, 1974b). The formation of this intermediate is accelerated and the reformation of the purple complex is slowed down

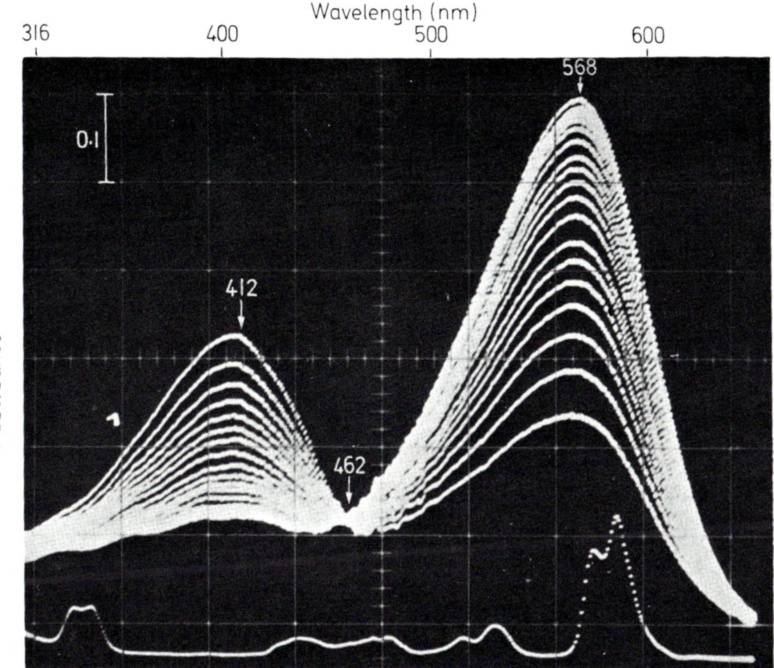

Fig. 13. Time series of spectra during regeneration of the photolyzed purple complex in the dark. The first spectrum shows maximal absorption at 412 nm and minimal absorption at 568 nm, then the absorption rises in time at 568 nm and falls at 412 nm. (Taken from OESTERHELT and HESS, 1973)

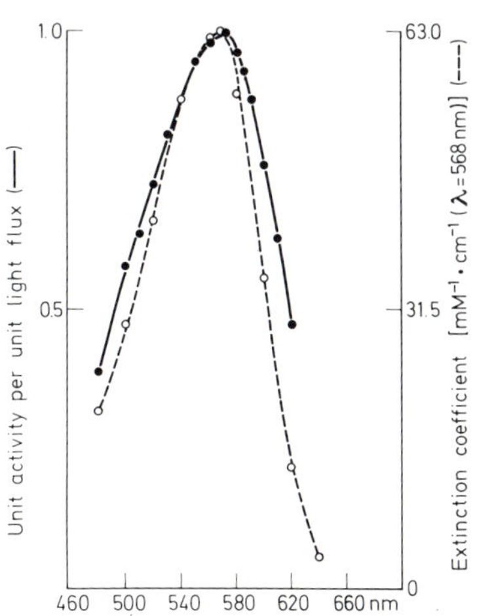

Fig. 14. Action spectrum of photolysis (●—●) and absorption band of the purple complex (o—o). Activity is defined as the initial rate of formation of the 412 nm complex. (Taken from OESTERHELT and HESS, 1973)

by diethylether in the presence of 4 M salt, therefore increasing the steady-state concentration of the 412 nm complex at a given light intensity. This system has been used for quantitative measurements. Fig. 13 shows a photosteady state mixture of the 412 nm complex (412) and the purple complex (568), as well as the complete regeneration of the purple complex after the light has been turned off. A detailed study of this system established the following facts (OESTERHELT and HESS, 1973):

1) The 412-nm complex has a broad absorption band with a λ_{max} value around 415 nm and approximately half of the molar extinction of the purple complex.

2) Only light absorbed by the purple complex is efficient in this reaction and the quantum yield amounts to 0.79 (Fig. 14).

3) The reformation of the purple complex is temperature-dependent, an activation energy of 11.4 Kcal x mol^{-1} was found.

4) Formation of the 412-nm complex changes the tryptophan fluorescence of bacteriorhodopsin and its pK-value, i.e. protons are released upon illumination and taken up during regeneration of the purple complex.

This last mentioned cyclic release and uptake of protons is the basis of bacteriorhodopsin as a light-driven proton pump (OESTERHELT, 1974b). Since it has been shown that bacteriorhodopsin is incorporated asymmetrically in the cell membrane (see Section IV) one can easily imagine that the photochemical cycle occurs as a vectorial process in the intact cell.

Considering the rate constants in Fig. 12 as overall rate constants and taking into account that the regeneration of the purple complex is the rate-limiting step, the following relationship between the speed of the photochemical cycle (i.e. proton pumping process) and the light intensity (I) is obtained (OESTERHELT, 1974d):

$$v = \frac{k_2 \cdot B}{1 + \frac{k_2}{k_1} \cdot \frac{1}{I}} \qquad v_{max} = k_2 \cdot B$$

where B is the bacteriorhodopsin concentration. k_2 was measured to be approximately 125 sec^{-1} at physiological temperatures using cell suspensions of *H. halobium*. k_2 represents the turn-over number of bacteriorhodopsin under saturating illumination conditions (OESTERHELT and HESS, 1975). This figure can be used for an estimation of the efficiency of bacteriorhodopsin function in the living cell.

Since it is possible to incorporate the purple membrane into liposomes and black films (RACKER, 1973; RACKER and STOECKENIUS, 1974; KAYUSHIN and SKULACHEV, 1974; RACKER and HINKLE, 1974) the property of bacteriorhodopsin as a light-driven electrogenic pump has been strongly suggested (DRACHEV et al., 1974a; DRACHEV et al., 1974b). Electrogenicity means that the pump transports protons without a cotransport of anions or an antiport of other cations strictly coupled to it. If, in addition, the permeability of the cell for other ions is low compared to the rate of proton extrusion by bacteriorhodopsin, a membrane potential, i.e. an electrostatic field across the liposomal membrane is created.

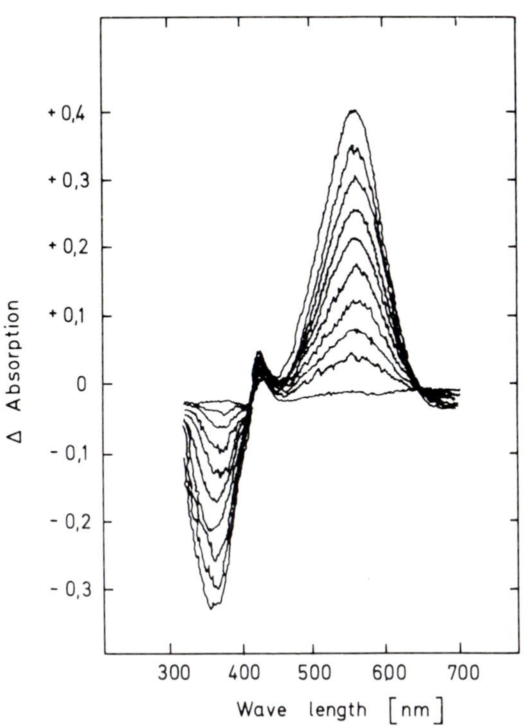

Fig. 15. Reaction of bacteriorhodopsin with hydroxylamine in illuminated cell suspensions. Difference spectra between cells in the dark (sample) and cells in the light (reference). Cells were illuminated with light from a 900 Watt Xenon high pressure lamp filtered through 10 % $CuSO_4$ and an orange glass filter. The spectra are taken at 0, 11, 23, 35, 50, 66, 87, 104, 124, and 149 min. (Taken from OESTERHELT et al., 1974)

VI. Function of Bacteriorhodopsin in the Halobacterial Cell

The photochemical cycle of bacteriorhodopsin also occurs in the intact cell as indicated by the formation of the 412 nm complex upon illumination and its decay in the dark following the rate law given above (OESTERHELT, 1974b). Furthermore retinal can be trapped during the photochemical cylce by the reaction with hydroxylamine as described for aequous purple membrane suspensions on p. 147 (OESTERHELT, SCHUHMANN, and GRUBER, 1974). Fig. 15 shows the difference spectra of the cell suspensions in the light and in the dark, both samples containing hydroxylamine. The disappearance of the bacteriorhodopsin chromophore and the formation of retinaloxime (λ_{max} 370 nm) in the illuminated sample is clearly demonstrated. No reaction occurs in the dark controls.

The vectorial nature of the proton release and uptake process accompanying the photochemical cycle of bacteriorhodopsin is the most important feature for its *in vivo* function. The asymmetric orientation of the protein within the purple membrane patches in the cell membrane guarantees its property of a light-driven proton pump. As expected from such an ion pump, light-induced changes of the pH in cell suspensions were observed (OESTERHELT, 1972b) and from more detailed studies about the effect of light on *H. halobium* cells, light-energy conversion mediated by bacteriorhodopsin was postulated (OESTERHELT and STOECKENIUS, 1973).

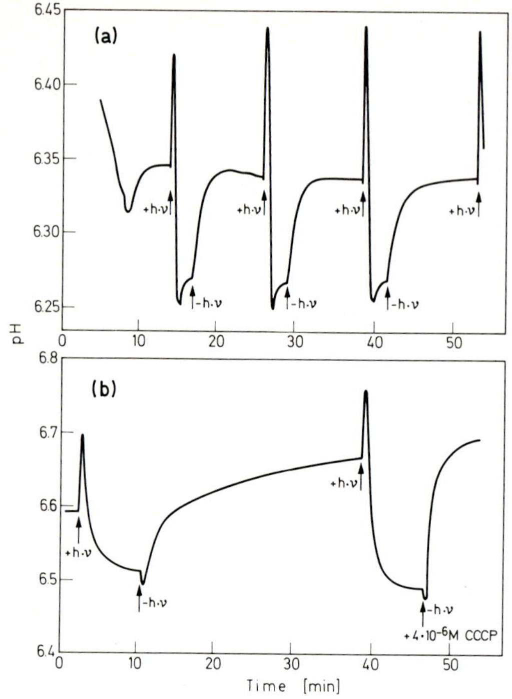

Fig. 16a and b. Light-induced pH changes in anaerobic cell suspensions. (Taken from OESTERHELT and STOECKENIUS, 1973). (a) After all oxygen has been consumed the pH reaches a nearly stable level. Successive periods of illumination and darkness cause reversible pH-changes of the medium. (b) Carbonyclanide m-chlorophenylhydrazone causes a large increase in the rate of the rise in pH in the dark

Fig. 16 shows a typical experiment of these light-induced pH-changes. Cell suspensions are brought into a closed chamber and then become rapidly anaerobic. During consumption of the residual dissolved oxygen they acidify the medium as known from other procaryotic cells (HAROLD, 1972). The pH then reaches a nearly stable level. Addition of the uncoupler carbonyl cyanide m-chlorophenylhydrazone (CCCP) which like other uncouplers increases the proton permeability of membranes (MITCHELL and MOYLE, 1967) causes an increase in pH indicating that an electrochemical gradient existed before its addition. If the cells are illuminated instead of the uncoupler being added, the pH drops after a transient increase below the starting level. This clearly indicates that the preexisting electrochemical gradient created by respiration is enhanced by illumination and light energy is converted into a proton motive force as defined by MITCHELL (MITCHELL, 1966). In the dark, the pH returns to the original level. If the uncoupler is added at the time of return to the dark condition the rise of pH is accelerated considerably and the final pH lies somewhat above the level at the beginning of the light experiment. Only light absorbed by the purple complex of bacteriorhodopsin generates this electrochemical gradient across the cell membrane. Other halobacteria able to snythesize retinal (and therefore bacteriorhodopsin) show the same pH shift when illuminated. Cells grown in the presence of diphenylamine have little bacteriorhodopsin due to a reduced retinal synthesis and show a proportionally decreased pH effect.

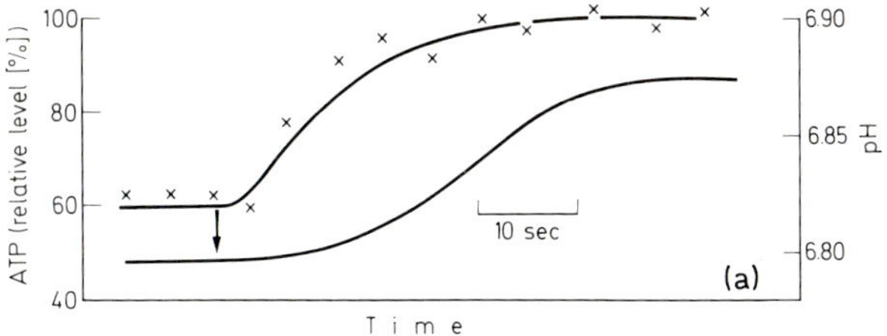

Fig. 17a and b. Photophosphorylation and pH-changes. (a) Light-induced ATP synthesis during the transient increase of pH. (b) Overall changes upon illumination and darkness

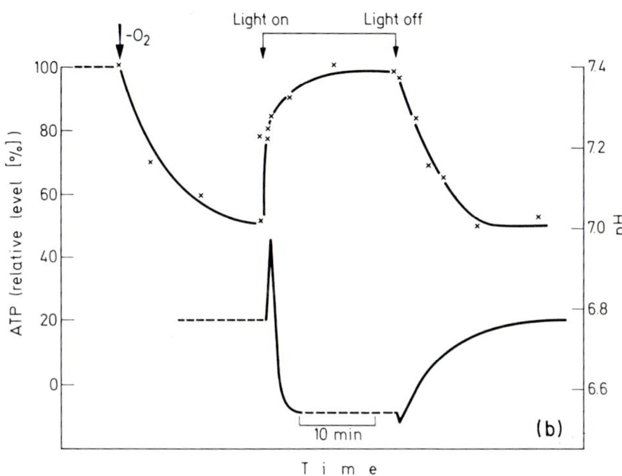

The creation of an electrochemical gradient between inside and outside of the cell leads to a consideration of MITCHELL's chemiosmotic hypothesis (MITCHELL, 1966; MITCHELL, 1972) which states that in respiration and photosynthesis a series of redox reactions is used to generate an electrochemical gradient (proton motive force). The energy stored in the gradient is then used to drive ATP-synthesis and/or transport processes. The same must hold true for a more simple proton pump driven by light but not coupled to electron flow. Indeed, light-dependent phosphorylation processes have been found (DANNON and STOECKENIUS, 1974; OESTERHELT, 1974d; OESTERHELT, 1975). This new type of photophosphorylation was shown to be insensitive to cyanide, inhibitors of the respiratory chain and inhibitors of cyclic or noncyclic electron flow in photosynthesizing organisms. It can be abolished, however, by the inhibitor of the reversible proton-translocating ATPase dicyclohexylcarbodiimide (DCCD) and by uncouplers. The first type of inhibitor no longer allows the cell to use the proton motive force created by bacteriorhodopsin for the synthesis of ATP, the uncoupler already destroys the proton motive force itself.

Fig. 17 correlated changes of pH and ATP in illuminated cell suspensions and clearly shows that photophosphorylation preceeds the measurable pH changes rather than follows it. Therefore it is important to know whether a proton motive force large enough

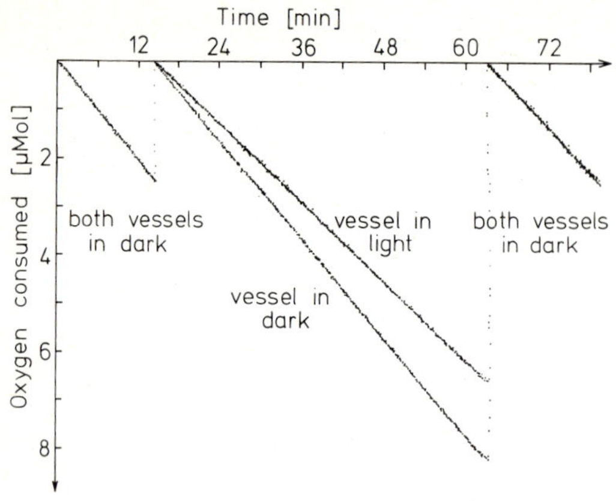

Fig. 18. Light inhibition of respiration in *H. halobium*. (Taken from OESTERHELT and KRIPPAHL, 1973)

to drive ATP synthesis can be created by bacteriorhodopsin within the time span where the change of ATP level is observed experimentally. The proton motive force as defined by MITCHELL (1966) consists of two parts, a difference in pH and a difference in charge, creating an electrostatic field across the membrane (membrane potential). With the measured values of cell volume, bacteriorhodopsin content of the cells and the turnover number of bacteriorhodopsin, it is possible to calculate that a membrane potential of 300 mV (large enough to drive ATP synthesis, see MALONEY et al., 1974) can be created in less than a millisecond. This is due to the electrogenicity of bacteriorhodopsin as a proton pump (see Section V). In contrast to this, all measured differences in pH are caused by electroneutral processes limited in their rates by the permeabilities of counterions and occur in the second to minute range.

No data on the quantum yield of photophosphorylation are available so far. In order to quantitate light-energy conversion in Halobacteria advantage was taken from the fact that respiring Halobacteria have two free energy-yielding systems: a light-driven proton pump and the respiratory chain. Both processes are expected to complete for either ADP or protons inside the cells or *via* the membrane potential created by both processes. This competition should lead to an experimentally observable light inhibition of respiration independent of the common link. By using experimental equipment originally devised by Warburg, this light inhibition was measured quantitatively (OESTERHELT and KRIPPAHL, 1973). Fig. 18 compares the oxygen consumption of cells in light and dark. Up to 30 % less oxygen is consumed by illuminated cells. This reduction in oxygen uptake is not a light-induced general inhibition of cell metabolism, for the reason that cells will grow in the light even faster than in the dark but with a reduced requirement for oxygen.

Inhibition of respiration depends, like the pH effects, on light absorbed by bacteriorhodopsin and the action spectrum fits very well the absorption band of the purple complex in the cells. If retinal synthesis is inhibited by nicotin, no light inhibition in these cells is observed anymore. Increasing light intensities saturate the system and extrapolation to zero light intensity allows the calculation of maximum efficiency. By measuring absolute light intensity with a bolometer and by determination of the absorption of the cell suspensions used for the respiration experiments it is possible to obtain a quantitative relationship between light quanta absorbed and molecules oxygen molecule not consumed. The figures vary between 25 and 40 photons per oxygen and allow an evaluation of light-energy conversion in the cell. If one assumes that no more than one proton participates in the primary reaction 25 - 40 absorbed photons would correspond to 12 protons because they can prevent consumption of one molecule oxygen. The ratio 12 is obtained from experimental data measured in mitochondria and bacterial cells which eject twelve protons into the medium during the reduction of one molecule oxygen (MITCHELL and MOYLE, 1967). The maximum quantum yield of the purple membrane system in whole cells would then be 0.3 - 0.5 with regard to the primary photochemical reaction. Even considerable variation for this figure would still permit postulation of light-energy conversion by bacteriorhodopsin (OESTERHELT and KRIPPAHL, 1973). Its photoreceptor function has consequently to be taken as a photocoupling function, clearly different from the photosensing function of the visual pigments. However, Halobacteria show phototactic behavior (STOECKENIUS and BERG, unpublished observation). By means of two photosystems the cells are able actively to find light of the wavelength absorbed by bacteriorhodopsin (HILDEBRAND and DENCHER, 1974). One of the photosystems shows an action spectrum identical to the purple complex absorption band, so that bacteriorhodopsin functions as photosensor as well as phototransducer in these cells.

VII. Biosynthesis of Bacteriorhodopsin and the Purple Membrane

The purple membrane is only synthesized in cells growing under anoxia. Two extreme conditions of aeration and their influence on purple membrane biosynthesis are shown in Fig. 19. If oxygen

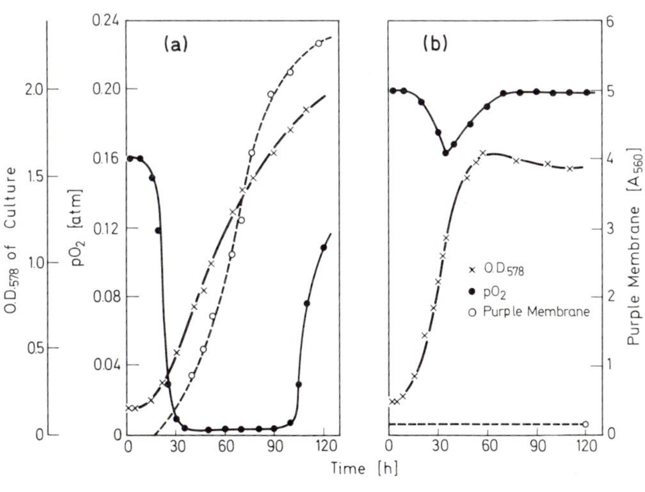

Fig. 19a and b. Synthesis of the purple membrane. (Taken from OESTERHELT and STOECKENIUS, 1973). (a) 120 liters of air per h, stirring at 300 rpm; (b) 180 liters of air per h, rpm

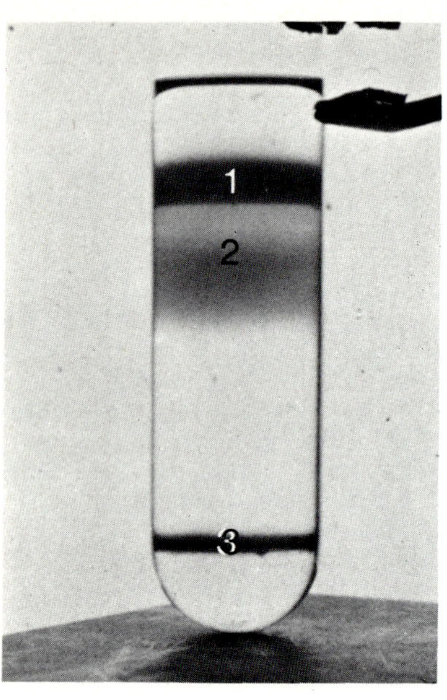

Fig. 20. Sucrose density gradient with separated cell membrane fractions from H. halobium NRL R_1M_1. (1) RM-340 (red); (2) bulk membrane fragments (yellow); (3) purple membrane (purple)

supply limits growth as indicated by a lowered cell density, increase and an oxygen steady-state concentration of virtually zero purple membrane synthesis follows closely the growth curve. If aeration keeps the oxygen concentration near saturation throughout the growth phase no measurable synthesis occurs. If the conditions of aeration are changed during growth, increasing oxygen concentrations stop synthesis, decreasing concentrations stimulate it. All these results fit very well the hypothesis that the purple membrane in Halobacteria is used as an auxiliary device for energy (ATP) production in the cell under conditions where not enough ATP can be synthesized *via* the respiratory chain because the final electron acceptor oxygen now limits the chain.

As already mentioned a mutant (*H. halobium* NRL R_1M_1) unable to synthesize bacterioruberin was used in many of the experiments discussed. Fig. 20 shows a sucrose density gradient of a lysate from R_1M_1 cells grown at high rate of aeration at which purple membrane synthesis is low. Under these conditions a light, red-colored band (1) is found on top of the yellow band (2) (color is due to β-carotene) containing the bulk cell membrane fragments. The red band has a very low buoyant density (1.06 g/cm^3) and was termed membrane fraction RM-340. This name refers to its reddish color and its main absorption peak at 343 nm with a molar extinction coefficient of 120,000. In the visible region of the spectrum, three minor absorption peaks at 566, 523 and 492 are observed, which have molar extinction coefficients around 10,000 and cause the red color of the membrane. A fifth absorption peak is found at 270 nm. It is not due to tryptophan absorption of the membrane protein but disappears upon extraction of the membrane chromophore with organic solvent. Onyl then a very weak absorption at 280 nm can be detected. If concentrated membrane samples are diluted with acetone instead of water, the

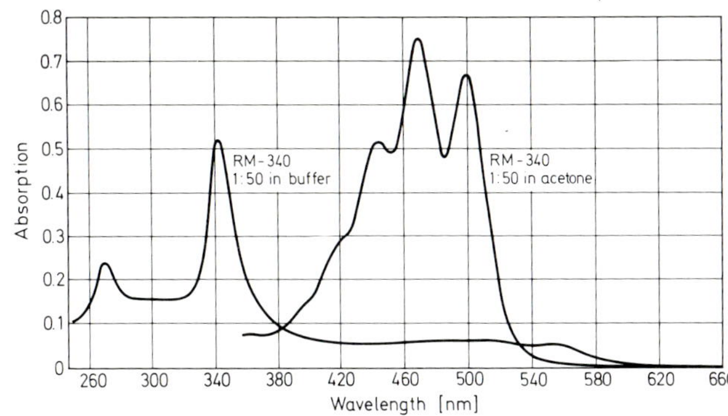

Fig. 21. Absorption spectrum of membrane fraction RM-340 in water and acetone

membrane components do not precipitate but the absorption spectrum is drastically changed to that of lycopene (Fig. 21). Identity of the membrane chromophore with all-trans lycopene has been shown by spectroscopic experiments, thin-layer chromatography and mass spectrometry (OESTERHELT, 1972a).

Lysates of R_1M_1 cells do not show any absorption attributable to lycopene and nearly all material absorbing at 343 nm sediments with membrane fraction RM-340. Therefore we assume that lycopene occurs in *H. halobium* in differentiated areas of the cell membrane as does retinal in the purple membrane. The very peculiar spectroscopic properties of lycopene in membrane fraction RM-340 suggest an unusual conformational state of the chromophore. A lycopene molecule twisted 90° around its 14 - 15 single bonds and fixed in that position would explain such high and almost exclusive absorption in its so-called cis-peak region.

The protein content of membrane fraction RM-340 is very low as suggested already by its low buoyant density and reaches only values below 10 %. Like the purple membrane RM-340 shows only one band in SDS gel electrophoresis and its position on the gels is identical to that of bacteriorhodopsin[3] (OESTERHELT, 1972a). The biosynthesis of the two membrane fragments is correlated. Low rates of aeration stimulate purple membrane biosynthesis and inhibit RM-340 formation, high rates of aeration inhibit purple membrane synthesis and accumulate RM-340. It is an attractive idea to consider RM-340 as a precursor membrane of the purple membrane with the implication that the cells have a strict economy in the sense that they synthesize bacterio-opsin to an almost negligible extent if no light-energy conversion is required. Furthermore bacterio-opsin synthesis seems to be coupled strictly to pigment synthesis. Nictoin which blocks retinal synthesis leads to accumulation of RM-340 as high oxygen supply

---

[3]bacteriorhodopsin and bacterio-opsin behave identically on SDS gels (see OESTERHELT, SCHUHMANN, and GRUBER, 1974).

does but can be expected not to interfere directly with protein synthesis. Another example of membrane synthesis regulation *via* their pigments is the accumulation of β-carotene in company with low purple membrane content when cells grow at low oxygen supply but without illumination (REITMEIER and OESTERHELT, unpublished).

It make sense that the two conditions for purple membrane synthèsis are: (1) shortage of the electron acceptor oxygen and (2) light as source of energy. Only then do the cells need to synthesize ATP by a metabolic route different from oxidative phosphorylation: light-energy conversion by the purple membrane system.

References

ALTMAN, L.J., ASH, L., KOWERSKI, R.C., EPSTEIN, W.W., LARSEN, R.B., RILLING, H.C., MUSCIO, F., GREGONIS, D.E.: Prephytoene Pyrophosphate. A new intermediate in the biosynthesis of Carotenoids. J. Am. Chem. Soc. 94, 3257 (1972).

APPLEBURY, M.L., ZUCKERMAN, D.M., LAMOLA, A.A., JOVIN, T.M.: Rhodopsin, Purification and recombination with phospholipids assayed by the metarhodopsin I → metarhodopsin II transition.

BAXTER, R.M.: An interpretation of the effects of salts on the lactic dehydrogenase of Halobacterium salinarium. Can. J. Microbiol. 5, 47-57 (1959).

Bergey's Manual of Determinative Bacteriology, 7. ed. (Eds. R.S. BREED, E.G.D. MURRAY, N.R. SMITH), p. 208, 1957; 8. ed. (Eds. R.E. BUCHANAN, N.E. GIBBONS), p. 269, 1974. Baltimore: Williams & Wilkins.

BLAUROCK, A.E.: Bacteriorhodopsin: a trans-membrane pump containing α-helix. J. Mol. Biol. 93, 139-158 (1975).

BONTING, S.L., ROTMANS, J.P., DAEMEN, F.J.M.: Chromophore migration after illumination of rhodopsin. In: Biochemistry and Physiology of Visual Pigments (Ed. H. LANGER), p. 39-44. Berlin-Heidelberg-New York: Springer 1973.

BROWN, A.D.: The peripheral structures of Gram-negative bacteria. The cationsensitive dissolution of the cell membrane of the halophilic bacterium, Halobacterium halobium. Biochim. Biophys. Acta 75, 425-435 (1963).

CHANCE, B., PORTE, M., HESS, B., OESTERHELT, D.: Low temperature kinetics of H^+-changes of bacterial rhodopsin. Biophys. Meeting 1975, Abstr.

CHEN, J.S., BARTON, P.G., BROWN, D., KATES, M.: Osmometric and microscopic studies on bilayers of polar lipids from the extreme halophile, Halobacterium cutirubrum. Biochim. Biophys. Acta 352, 202-217 (1974).

CHIGNELL, C.F., CHIGNELL, D.A.: A spin label study of purple membranes from Halobacterium halobium. Biochem. Biophys. Res. Commun. 62, 136-143 (1975).

CHRISTIAN, J.H.B., WALTHO, J.A.: Solute concentrations within cells of halophilic and nonhalophilic bacteria. Biochim. Biophys. Acta 65, 506-508 (1962).

DANON, A., STOECKENIUS, W.: Photophosphorylation in Halobacterium halobium. Proc. Nat. Acad. Sci. (Wash.) 71, 1234-1238 (1974).

DENCHER, N., WILMS, M.: Flash photometric experiments on the photochemical cycle of bacteriorhodopsin. Biophysics of structure and mechanism 1, 259-271 (1975).

DE VAULT, D., CHU KUNG, M., HESS, B., OESTERHELT, D.: Photolysis of Bacterial Rhodopsin. Biophys. meeting, Jan. 1975, Abstr.

DRACHEV, L.A., JASAITIS, A.A., KAULEN, A.A., KONDRASHIN, A.A., LIBERMAN, E.A., NEMECEK, I.B., OSTROUMOV, S.A., SEMENOV, A.Yu., SKULACHEV, V.: Direct measurement of electric current generation by cytochomoxidase, H^+-ATPase and bacteriorhodopsin. Nature 249, 321-324 (1974).

DRACHEV, L.A., KAULEN, A.D., OSTROUMOV, S.A., SKULACHEV, V.P.: Electrogenesis by bacteriorhodopsin incorporated in a planar phospholipid membrane. FEBS Letters 39, 43-45 (1974).

DUNDAS, J.E.D.: Ornithin carbomoyltransferase from Halobacterium salinarium. Europ. J. Biochem. 27, 376-380 (1972).

EDMOND, J., POPJÂK, G., WONG, S.M., WILLIAMS, V.P.: Presqualene Alcohol. Further evidence on the structure of a C_{30}-precursor of squalene. J. Biol. Chem. 246, 6254-6271 (1971).

EPSTEIN, W.W., RILLING, H.C.: Studies on the mechanism of squalene biosynthesis. The structure of presqualene pyrophosphate. J. Biol. Chem. 245, 4597-4605 (1970).

GIBBONS, N.E.: The effect of salt concentrations on the biochemical reactions of some halophilic bacteria. Can. J. Microbiol. 3, 249 (1957).

GIBBONS, N.E.: Isolation, growth and requirements of halophilic bacteria. Methods in Microbiol. III B, 169-183 (1969).

GINZBURG, M., GINZBURG, B.Z., TOSTESON, D.C.: The effects of anions on K^+ binding in a Halobacterium species. J. Membr. Biol. 6, 259-268 (1971).

GOCHNAUER, M.B., KUSHWAHA, S.C., KATES, M.: KUSHNER, D.J.: Nutritional control of pigment and isoprenoid compound formation in extremely halophilic bacteria. Arch. Mikrobiol. 84, 339-349 (1972).

GREGONIS, D.E., RILLING, H.C.: The steriochemistry of trans-phytoene synthesis. Some observations of lycopersene as a carotene precursor and a mechanism for the synthesis of *cis*- and *trans*-phytoene. Biochemistry 13, 1538-1542 (1974).

HANCOCK, A.J., KATES, M.: Structure determination of the phosphatidylglycerophosphate (diether analog) from Halobacterium cutirubrum. J. Lipid Res. 14, 422 (1973).

HAROLD, F.M.: Conservation and transformation of energy by bacterial membranes. Bacteriol. Rev. 36, 172-230 (1972).

HENDERSON, R.: The structure of the purple membrane from Halobacterium halobium: analysis of the X-ray diffraction pattern. J. Mol. Biol. 93, 123-138 (1975).

HILDEBRAND, E., DENCHER, N.: Photophobic responses in Halobacterium halobium and their relation to the retinal protein complex of the cell membrane. Ber. Dtsch. Botan. Ges. 87, 93-99 (1974).

HOCHSTEIN, L.J., DALTON, B.:.: Studies of a halophilic NADH-dehydrogenase. I. Purification and properties of the enzyme. Biochim. Biophys. Acta 302, 216-228 (1973).

HOLMES, P.K., HALVORSON, H.O.: Properties of a purified halophilic malic dehydrogenase. J. Bact. 90, 312-326 (1965).

HONIG, G., EBREY, T.G.: The structure and spectra of the chromophore of the visual pigments. Ann. Rev. Biophys. Bioeng. 3, 151-177 (1974).

HOWES, C.D., BATRA, P.P.: Accumulation of lycopene and inhibition of cyclic carotenoids in Mycobacterium in the presence of nicotine. Biochim. Biophys. Acta 222, 174-179 (1970).
HUBBARD, J.S., MILLER, A.B.: Purification and reversible inactivation of the isocitrate dehydrogenase from an obligate halophile. J. Bacteriol. 99, 161-168 (1969).
INMAN, R.B., JORDAN, D.O.: Deoxypentose nucleic acids. XI. The denaturation of deoxyribonucleic acid in aqueous solution: conductivity and mobility measurements. and XII. The denaturation of deoxyribonucleic acid in aqueous solution: changes produced by environment. Biochim. Biophys. Acta 42, 421-434 (1960).
ISLER, O., Ed.: Carotenoids. Basel: Birkhäuser 1971.
JAN, L.: Thesis 1974. California Institute of Technology.
JOO, C.N., KATES, M., SCHIER, T.: Characterization and synthesis of mono and diphytanylethers of glycerol. J. Lipid Res. 9, 782 (1968).
JOO, C.N., KATES, M.: Synthesis of the naturally occurring phytanyl diether analogs of phosphatidyl glycerophosphate and phosphatidylglycerol. Biochem. Biophys. Acta 176, 278-297 (1969).
KABACK, H.R.: Bacterial Membranes. Meth. in Enzymol. 22, 99-120 (1971).
KATES, M., DEROO, P.W.: Structure determination of the glycolipid sulfate from the extreme halophile Halobacterium cutirubrum. J. Lipid Res. 14, 438 (1973).
KATES, M., WASSEF, M.K., KUSHNER, D.J.: Radioisotopic studies on the biosynthesis of the glyceryl diether lipids of Halobacterium cutirubrum. Can. J. Biochem. 46, 971-977 (1968).
KAYUSHIN, P.L., SKULACHEV, V.P.: Bacteriorhodopsin as an elctrogenic proton pump: reconstitution of bacteriorhodopsin proteoliposomes generating $\Delta\psi$ and ΔpH. FEBS Letters 39, 39-42 (1974).
KELLY, M., NORGÅRD, S., LIAAEN-JENSEN, S.: Bacterial carotenoids XXXI. C_{50}-carotenoids. 5. Carotenoids of Halobacterium salinarium especially bacterioruberin. Acta Chim. Scand. 24, 2169-2182 (1970).
KING, G.J., STOECKENIUS, W.: Organization of Bacteriorhodopsin in the purple membrane of Halobacterium halobium. Fed. Proc. 33, 1408 (1974).
KONG, K., HUBBELL, W.L.: Lipid requirements for rhodopsin regenerability. Biochemistry 12, 4517-4523 (1973).
KRAMER, J.K.G., KUSHWAHA, S.C., KATES, M.: Structure determination of squalene, dihydrosqualene and tetrahydrosqualene in Halobacterium cutirubrum. Biochim. Biophys. Acta 270, 103-110 (1972).
KUSHNER, D.J.: Adv. Appl. Microbiol. 10, 73 (1968).
KUSHNER, D.J., BAYLEY, S.T., BORING, J., KATES, M., GIBBONS, N.E.: Morphological and chemical properties of cell envelopes of the extreme halophile, Halobacterium cutirubrum. Can. J. Microbiol. 10, 483-497 (1964).
KUSHNER, D.J., ONISHI, H.: Contribution of protein and lipid components to the salt response of envelopes of an extremely halophilic bacterium. J. Bacteriol. 91, 653-660 (1966).
KUSHWAHA, S.C., PUGH, E.L., KRAMER, J.K.G., KATES, M.: Isolation and identification of dehydrosqualene and C_{40}-carotenoid pigments in Halobacterium cutirubrum. Biochim. Biophys. Acta 260, 492-506 (1972).

KUSHWAHA, S.C., KATES, M.: Isolation and identification of "Bacteriorhodopsin" and minor C_{40}-carotenoids in Halobacterium cutirubrum. Biochim. Biophys. Acta 316, 235-243 (1973).

LANYI, J.K.: Irregular bilayer structure in vesicles prepared from Halobacterium cutirubrum lipids. Biochim. Biophys. Acta 356, 245-256 (1974a).

LANYI, J.K.: Salt-Dependent Properties of Proteins from Extremely Halophilic Bacteria. Bacteriol. Rev. 38, 272-290 (1974b).

LANYI, J.K., PLACHY, Z.W., KATES, M.: Lipid interactions in membranes of extremely halophilic bacteria. II. Modification of the bilayer structure by squalene. Biochemistry 13, 4914-4920 (1974).

LANYI, J.K., STEVENSON, J.: Studies of the electron transport chain of extremely halophilic bacteria. IV. Role of hydrophobic forces in the structure of menadione reductase. J. Biol. Chem. 245, 4074-4080 (1970).

LARSEN, H.: In: The Bacteria (Eds. I.C. GUNSALUS, R.Y. STANIER), Vol. 4, p. 297-342. New York: Academic Press 1962.

LARSEN, H.: Biochemical aspects of extreme halophilism. Advan. Microbial. Phys. 1, 97-132 (1967).

LARSEN, H.: The halobacteria's confusion to Biology. Antonie van Leeuwenhoek J. Microbiol. Serol. 39, 383-396 (1973).

LEWIS, A., SPOONHOWER, J., BOGOMOLNI, R., LOZIER, R., STOECKENIUS, W.: Tunable Laser Resonance Raman Spectroscopy of Bacteriorhodopsin. Proc. Natl. Acad. Sci. (Wash.) 71, 4462-4466 (1974).

LIAAEN-JENSEN, S., ANDREWES, A.G.: Microbial carotenoids. Ann. Rev. Microbiol. 26, 225-248 (1972).

LIEBERMAN, M.M., LANYI, J.K.: Threonine deaminase from extremely halophilic bacteria. Cooperative substrate kinetics and salt dependence. Biochemistry 11, 211-216 (1972).

LOMBARDI, F.J., REEVES, J.P., KABACK, H.R.: Mechanisms of active transport in isolated bacterial membrane vesicles XIII. Valinomycin induced rubidium transport. J. Biol. Chem. 248, 3551-3565 (1973).

LOUIS, G.B., FITT, P.S.: Isolation and properties of highly purified Halobacterium cutirubrum deoxyribonucleic acid-dependent ribonucleic acid polymerase. Biochem. J. 127, 69-80 (1972a).

LOUIS, G.B., FITT, P.S.: Purification and properties of the ribonucleic acid-dependent ribonucleic acid polymerase subunits in initiation and polymerization. Biochem. J. 128, 755-762 (1972b).

MALONEY, P.C., KAHSKET, E.R., WILSON, H.T.: A proton-motive force drives ATP-synthesis in bacteria. Proc. Nat. Ac. Sci. (Wash.) 71, 3896 (1974).

MENDELSOHN, R.: Resonance raman spectroscopy of the photoreceptor-like pigment of Halobacterium halobium. Nature 243, 22-24 (1973).

MENDELSOHN, R., VERMA, A.L., BERNSTEIN, H.I., KATES, M.: Structural studies of Bacteriorhodopsin from Halobacterium cutirubrum by resonance raman spectroscopy. Can. J. Biochem. 52, 774-781 (1974).

MESCHER, M.F., STROMINGER, I.L., WATSON, S.W.: Protein and carbohydrate composition of the cell envelope of Halobacterium salinarium. J. Bacteriol. 120, 945-954 (1974).

MILANYTCH, M.: Diploma, University of Munich 1973.

MITCHELL, P.: Chemiosmotic coupling in oxidative and photosynthetic phosphorylation. Biol. Rev. 41, 445-502 (1966).

MITCHELL, P.: Chemiosmotic coupling in energy transduction: a logical development of biochemical knowledge. J. Bioenergetics 3, 5 (1972).

MITCHELL, P., MOYLE, J.: Respiration-driven proton translocation in rat liver mitochondria. Biochem. J. 105, 1147-1162 (1967).

OESTERHELT, D.: The binding site of retinal in the purple membrane of Halobacterium halobium. Abstr. Commun. 7th Meet. Eur. Bioch. Soc., p. 205 (1971a).

OESTERHELT, D.: Effect of organic solvents on the retinal-opsin complex from Halobacterium halobium. Fed. Proc. 30, 1188 Abstr. (1971b).

OESTERHELT, D.: Isolation of a lycopene containing membrane with unusual spectroscopic properties from Halobacterium halobium. Abstr. Commun. 8th Meet. Eur. Biochem. Soc. Abstr. Nr. 125 (1972a).

OESTERHELT, D.: Die Purpurmembran aus Halobacterium halobium. Hoppe-Seyler's Z. Physiol. Chem. 353, 1554-1555 (1972b).

OESTERHELT, D.: Vergleichende Aspekte der Photorezeption von Retinal-Proteinkomplexen. In: Biochemistry of Sensory Functions, 25. Coll. Ges. Biol. Chemie, Mosbach, p. 55-77. Berlin-Heidelberg-New York: Springer 1974a.

OESTERHELT, D.: Bacteriorhodopsin as a light driven proton pump. In: Membrane Proteins in Transport and Phosphorylation, p. 79-84. Amsterdam: Elsevier 1974b.

OESTERHELT, D.: In: Comparative Biochemistry and Physiology of Transport (Ed. L. BOLIS), p. 162-174. Amsterdam: North Holland Publ. Comp. 1974c.

OESTERHELT, D.: Bacteriorhodopsin als Lichtenergiewandler. Nachr. Chem. Techn. 22, 475-476 (1974d).

OESTERHELT, D.: The purple membrane of Halobacterium halobium: a new system for light energy conversion. Ciba Foundation Symp. on Energy Transformation in Biological Systems, p. 147-167 (1975).

OESTERHELT, D., HESS, B.: Reversible photolysis of the purple complex in the purple membrane of Halobacterium halobium. Europ. J. Biochem. 37, 316-326 (1973).

OESTERHELT, D., HESS, B.: manuscript in preparation (1975).

OESTERHELT, D., KRIPPAHL, G.: Light inhibition of respiration in Halobacterium halobium. FEBS Letters 36, 72-76 (1973).

OESTERHELT, D., MEENTZEN, M., SCHUHMANN, L.: Reversible dissociation of the purple complex in bacteriorhodopsin and identification of 13-cis and all-trans-retinal as its chromophores. Europ. J. Biochem. 40, 453-463 (1973).

OESTERHELT, D., SCHUHMANN, L.: Reconstitution of bacteriorhodopsin. FEBS Letters 44, 262-265 (1974).

OESTERHELT, D., SCHUHMANN, L., GRUBER, H.: Light-dependent reaction of bacteriorhodopsin with hydroxylamine in cell suspensions of Halobacterium halobium: Demonstration of an apomembrane. FEBS Letters 44, 357-261 (1974).

OESTERHELT, D., STOECKENIUS, W.: Rhodopsin-like protein from the purple membrane of Halobacterium halobium. Nature New Biology 233, 149-152 (1971).

OESTERHELT, D., STOECKENIUS, W.: Functions of a new photoreceptor membrane. Proc. Nat. Acad. Sci. (Wash.) 70, 2853-2857 (1973).

OESTERHELT, D., STOECKENIUS, W.: Isolation of the cell membrane of Halobacterium halobium and its fractionation into red and purple membrane. Meth. in Enzymol. 31, 667-678 (1974).

PLACHY, Z.W., LANYI, J.K., KATES, M.: Lipid interactions in membranes of extremely halophilic bacteria. I. Electron spin resonance and dilatometric studies of bilayer structure. Biochemistry 13, 4906-4913 (1974).
PUGH, E.L., WASSEF, M.K., KATES, M.: Inhibition of fatty acid synthetase in Halobacterium cutirubrum and Escherichia coli by high salt concentration. Can. J. Biochem. 49, 953-958 (1971).
RACKER, E.: A new procedure for the reconstitution of biologically active phospholipid vesicles. Biochem. Biophys. Res. Commun. 55, 224-230 (1973).
RACKER, E., HINKLE, P.C.: Effect of temperature on the function of a proton pump. J. Membr. Biol. 17, 181-188 (1974).
RACKER, E., STOECKENIUS, W.: Reconstitution of purple membrane vesicles catalyzing light-driven proton uptake and adenosin triphosphate formation. J. Biol. Chem. 249, 662-663 (1974).
RAZI NAGRI, K., GONZALES-RODRIGUEZ, J., CHERRY, R.J., CHAPMAN, D.: Spectroscopic technique for studying protein rotation in membranes. Nature New Biology 245, 249-251 (1973).
REES, H.H., GOODWIN, T.: Biosynthesis of triterpenes, steroids and carotenoids. In: Biosynthesis, Vol. 1, 108-118 (1972). Special Periodical Report Chem., London.
STERN, N., TIETZ, A.: Glycolipids of a halotolerant, moderately halophilic bacterium 1. Effect of growth medium and age of culture on lipid composition. Biochim. Biophys. Acta 296, 130-135 (1973).
STOECKENIUS, W., KUNAU, W.H.: Further characterization of particulate fractions from lysed cell envelopes of Halobacterium halobium and isolation of gas vacuole membranes. J. Cell. Biol. 38, 337-357 (1968).
STOECKENIUS, W., LOZIER, R.H.: Light energy conversion in Halobacterium halobium. J. Supramol. Struct. 2, 769-774 (1974).
STOECKENIUS, W., ROWEN, R.: A morphological study of Halobacterium halobium and its lysis in media of low salt concentration. J. Cell. Biol. 34, 365-393 (1967).
SUMPER, M., OESTERHELT, D., RIEPERTINGER, G., LYNEN, F.: Die Synthese verschiedener Carbonsäuren durch den Multienzymkomplex der Fettsäuresynthese aus Hefe und die Erklärung ihrer Bildung. Europ. J. Biochem. 10, 377-387 (1969).
SUZUE, G., TSUKADA, K., TANAKA, S.: Occurence of dehydrosqualene (C_{30}-Phytoene) in staphylococcus aureus. Biochim. Biophys. Acta 164, 88-93 (1968).
TOMLINSON, G.A., KOCH, T.K., HOCHSTEIN, L.I.: The metabolism of carbohydrates by extremely halophilic bacteria: glucose metabolism via a modified Entner-Doudoroff pathway. Can. J. Microbiol. 20, 1085-1091 (1974).
TORNABENE, T.G., KATES, M., GELPI, E., ORO, J.: Occurence of squalene, di- and tetrahydrosqualenes, and vitamin MK_8 in an extremely halophilic bacterium, Halobacterium cutirubrum. J. Lipid. Res. 10, 294-303 (1969).
WASNER, H., LYNEN, F.: Die chemische Struktur der biosynthetischen Vorstufe des Squalens. FEBS Letters 12, 54-56 (1970).
WULFF, K., HUBBARD, J.S., MILLER, A.B.: Reversible inactivation of the isocitrate dehydrogenase from an obligate halophile: changes in the secondary structure. Arch. Biochem. Biophys. 148, 318-319 (1972).

WULFF, K., KERADJOPOULOS, D.: Thermophiler Charakter extrem halophiler Bakterien. Hoppe Seyler's Z. Physiol. Chem. 354, 1261 (1973).

ZORN, M., FUTTERMAN, S.: Properties of rhodopsin dependent on associated phospholipid. J. Biol. Chem. 246, 881 (1971).

Inhibitors of DNA Synthesis in RNA Tumor Viruses: Biological Implications and Their Mode of Action

P. Chandra, Linda K. Steel, U. Ebener, M. Woltersdorf, H. Laube and G. Will

I. Introduction

More than four years have passed since University of Wisconsin Virologist HOWARD M. TEMIN disclosed at the 10th International Congress of Cancer that he and his coworkers had found preliminary evidence for an enzyme that permits ribonucleic acid (RNA) to dictate the synthesis of deoxyribonucleic acid (DNA). Independently the same phenomenon was observed by DAVID BALTIMORE (1970b) of the Massachussetts Institute of Technology. Although perhaps not as fundamental a concept as the α-helical structure of DNA, the RNA-dependent DNA polymerase (reverse transcriptase) is a concept that is shedding light on several unexplained biological phenomena and has already inspired biochemists, molecular biologists, virologists and even biophysicists to look into the mystic of this process. Thus, TEMIN's discovery of "backward" transcription has set off an explosive and still unforseeable burst of activity in numerous laboratories.

The discovery of TEMIN's reverse-transcriptase in human leukemic cells by ROBERT C. GALLO (GALLO et al., 1970) of the National Cancer Institute, was a major breakthrough in the understanding of human leukemia as a molecular disease, and moreover, forming a possible link between RNA tumor viruses and human cancer. The pursuit of this discovery disclosed the fact that many other human cancers, such as Burkitt lymphoma and breast cancer exhibit enzymatic activities capable of synthesizing RNA-templated DNA. The development of very sensitive methods, such as the simultaneous detection technique by SOL SPIEGELMAN (SPIEGELMAN et al., 1972; SCHLOM and SPIEGELMAN, 1971) which allows us to screen this enzyme in very small amounts of a biological specimen, has added a great deal to advance our present day knowledge on the role of this enzyme in human cancer.

In view of the importance of reverse-transcription in cancer, and as a biologcial process, attempts have been made in our laboratory in the last years to develop inhibitors of DNA synthesis in RNA tumor viruses. This contribution will be concerned mainly with two aspects of DNA synthesis in RNA tumor viruses: 1. Inhibitors as tools in studying the biological implication of reverse transcription, and 2. The mode of action of such inhibitors. Since a number of excellent reviews on different aspects of DNA synthesis in RNA tumor viruses have appeared in the last years (TEMIN, 1971b; GALLO, 1972; TEMIN and BALTIMORE, 1972; SARIN and GALLO, 1973, 1974), in this contribution we have not referred to all the literature concerned with the topic discussed. Hence the supporting evidence for the basic, better established facts has been summarily cited, while that for the most recent developments has been discussed more extensively.

A. Reverse Transcription

Oncogenic RNA viruses have been shown to contain an enzyme – reverse transcriptase – which reverses the usual direction of information flow in a cell (DNA → RNA) so that DNA is produced on an RNA template (RNA → DNA). The discovery of reverse transcriptase (TEMIN and MIZUTANI, 1970; BALTIMORE, 1970) was, of course, prompted by TEMIN's protovirus hypothesis (TEMIN, 1964a), which was proposed to explain how an RNA tumor virus, such as Rous Sarcoma virus (RSV) can induce neoplastic transformations in the cellular gene. The belief, that RNA tumor viruses replicate *via* a DNA intermediate, not through an RNA intermediate (as do non-oncogenic RNA viruses), is the basis of Provirus theory. The support for this concept at this time, when the hypothesis was put forward, was based on the inhibitor studies, carried out by TEMIN (1964b) and BADER (1964). They had shown that actinomycin D was a potent inhibitor of RSV replication. Since actinomycin D inhibits DNA-templated reaction, and RSV is an RNA virus, it was believed that DNA is involved at some stage in the viral replication. At present the best evidence for the existence of proviral DNA comes from studies of infection of stationary chicken embryo fibroblasts by RSV in the presence of 5-bromodeoxyuridine (BOETTIGER and TEMIN, 1970; BALDUZZI and MORGAN, 1970). The fact that RNA does not incorporate bromodeoxyuridine (BDU), and BDU-containing DNA is very sensitive to light, led them to conclude from their studies that a BDU-containing DNA possessing information required for the infection was synthesized in the infected cells.

The reasons which led TEMIN and MIZUTANI (1970) to look for an enzyme(s) responsible for proviral DNA synthesis in virions itself, was an important observation of BADER (1966) that puromycin, an inhibitor of protein synthesis, failed to block infection of chicken cells with RSV. This suggested that if an RNA → DNA pathway existed in the replication, as transforming activity of RSV, it was not a cellular enzyme system induced by the virus. It was a clear indication of a "preformed" protein, probably present in the virions itself. BADER's observation (1966) was later confirmed by MIZUTANI and TEMIN (1971).

Independently, prompted by his discovery of the RNA-polymerase in vesicular stomatitis virus (1970a), BALTIMORE (1970) looked for an RNA-dependent DNA polymerase in the virions of an RNA tumor virus. He found that the virions of Rauscher murine leukemia virus contained a similar activity, as discovered by TEMIN (1970) in virions of RSV.

B. Biological Implications of Reverse Transcription

The very fact that the enzyme catalyzing RNA-templated DNA synthesis is present in all oncornaviruses, suggest its role in the neoplastic transformation by such viruses. The DNA "copy"

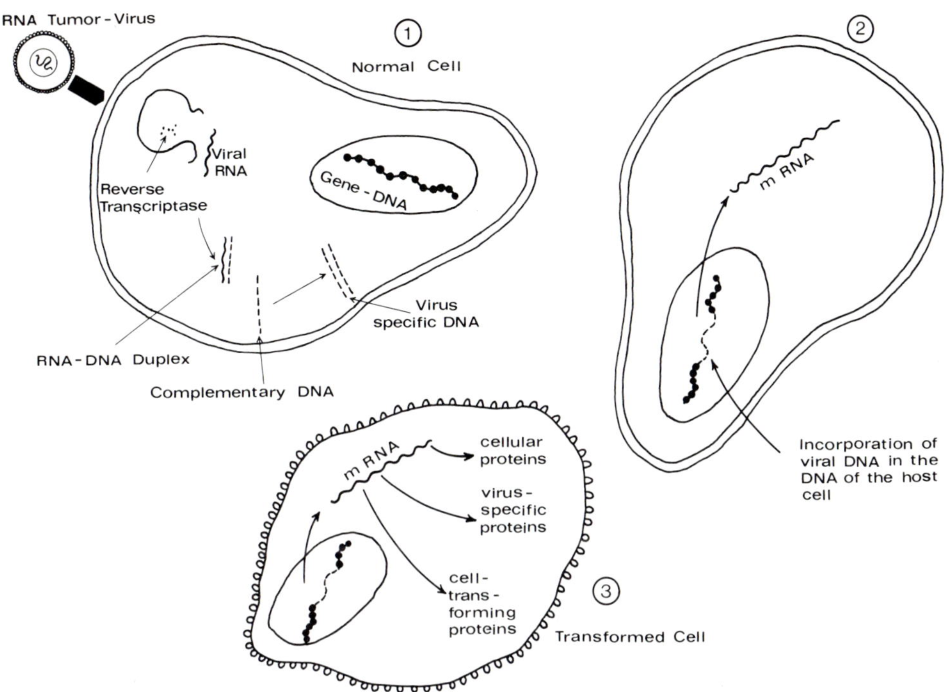

Fig. 1. Schematic presentation of oncogensis by RNA tumor viruses

(TEMIN's protovirus) of the viral genome (RNA) synthesized by this enzyme is incorporated into the host cell DNA where it carries information for viral replication and information for transformation of the normal cell to a neoplastic cell (Fig. 1).

The evidence which supports this thesis comes from three sets of experimental approaches: (1) HANAFUSA and HANAFUSA (1971) have shown that a mutant of RSV termed (RSV_{α}(O), which has no infectious capacity, is lacking in reverse-transcriptase activity. Also, no immunologically cross-reacting material could be found in RSV_{α}(O) using an antiserum against AMV reverse-transcriptase (HANAFUSA et al., 1972). (2) Studies with various types of reverse-transcriptase inhibitors (GALLO et al., 1972b; CHANDRA et al., 1972a, 1972e, 1972f, 1972g, 1973, 1974c, 1974d; CHANDRA, 1974a, 1975) have shown a definite correlation between their capacity to inhibit the enzymatic activity and their effect in blocking oncogenesis by oncornaviruses; these studies will be described in the latter part of this contribution. (3) Proviral DNA, the product of reverse-transcriptase reaction, when added exogeneously to normal cell cultures leads to cell transformation and viral production (see review by HILL and HILLOVA, 1974).

The very concept of TEMIN's protovirus theory (TEMIN, 1971a, 1971b) leads to the postulation that reverse-transcriptase should occur in normal uninfected cells. An endogenous RNase-sensitive DNA polymerase in apparently virus-free chicken embryo cells (KANG and TEMIN, 1972; TEMIN et al., 1973) and in PHA-stimulated normal human blood lymphocytes (GALLO et al., 1973) has been shown to occur. These polymerases are different to the viral reverse transcriptase, since they fail to transcribe the heteropolymeric regions of 70S RNA. The biological significance of these cellular RNA-directed DNA activities remains, at the present state of our knowledge, obscure. Nevertheless, these observations add a strong support to TEMIN's protovirus hypothesis (TEMIN, 1971a, 1971b).

Reverse-transcriptase activity in normal Rhesus monkey placenta has been recently reported by MAYER et al. (1974). This observation indicates the role of reverse-transcriptase in normal physiological processes such as information transfer for cytodifferentiation and embryogenesis. An indirect evidence for the involvement of RNA-directed DNA synthesis in cellular differentiation was obtained by TOCCHINI-VALENTINI and CRIPPA (1971), who showed that the amplification of genes for ribosomal RNA, which normally takes place in differentiating oocytes of *Xenopus*, is blocked by an inhibitor of reverse-transcriptase, namely the dimethyl benzyl derivative of rifampicin. This concept is based on the fact that the inhibitor used by these authors is specific for reverse transcriptase. However, in a later report, BROWN and TOCCHINI-VALENTINI (1972) did succeed in visualising the fundamental role of the RNA-directed reaction in gene amplification.

Among the other fascinating roles of RNA-instructed DNA snythesis one might consider antibody production. This is based on the fact that lymphoid cells of one rabbit treated *in vitro* or *in vivo* with lymphoid RNA extracted from a rabbit of different genotype, with respect to the light and heavy chains or immunoglobulin, are converted to the synthesis of specific IgM and IgG antibody of the RNA donor's light and heavy chain allotype (DRAY and BELL, 1973). There are several ways to explain the RNA mediated response in immunologic experiments, such as long time stability against RNases, and thus be able to act directly as messenger RNA; or the repetition of its information *via* the reverse-transcriptase route. More experiments are needed to explore the fact, whether reverse transcription plays any role in the transfer of immunological information.

II. Requirements for DNA Synthesis by RNA Tumor Viruses

In order to study the DNA polymerase activity found in virions of the RNA tumor viruses, an endogenous reaction can be developed and the synthesis of DNA observed. When the RNA present in disrupted virions is incubated with substrates, it acts as a template. The addition of exogenous nucleic acids to this system results in a more rapid rate of DNA synthesis, with properties similar to that of purified DNA polymerase with the same templates. This reaction has revealed valuable information as to the mode and specificity of the enzyme reactions *in vitro*, and perhaps a clue to its biological activity in cells after infection. A great deal of literature has been devoted to reporting the conditions for assaying the endogenous virion DNA polymerase activity. Some major reviews and references are: BALTIMORE (1970), SPIEGELMAN et al. (1970), GREEN et al. (1970), GALLO and TING (1972), and TEMIN and BALTIMORE (1972).

Concentration and purification of viral preparations are necessary to remove cellular contaminants which could result in irrelevant endogenous DNA synthesis. SPIEGELMAN et al. (1970) suggest ammonium sulfate precipitation in combination with differential centrifugation in glycerol and sucrose density gradients. The purified viral preparations at this point have little endogenous activity. Apparently the polymerase system is located within the viral core, and the viral envelope inhibits contact with the necessary components for endogenous DNA synthesis. Treatment with a nonionic detergent such as Nonidet P-40, Triton X-100, or Tween 40 disrupts the envelope and the viral cores sediment at a density of 1.22 - 1.24 g/ml. in a sucrose density gradient (GERWIN et al., 1970). The concentration of detergent required is dependent on the type of virus and the concentration of protein (GARAPIN et al., 1970). The DNA polymerase and the nucleic acid primer (70s RNA) within the disrupted virions generally show an increase in endogenous DNA synthesis activity of 20- to 50-fold over intact virions. However some murine leukovirus preparations do not require detergent activation (BALTIMORE, 1970; SPIEGELMAN et al., 1970), and the storage of virus, in particular, the avian myeloblastosis virus (BAUER and SCHÄFER, 1966) disrupts labile virions.

Full activation of the enzyme system requires the presence of all four deoxyribonucleoside triphosphates. The rate of DNA synthesis decreases or is negligible if one or more is withheld from the reaction mixture. Ribonucleoside triphosphates are not incorporated into the endogenous reaction and hence do not replace the deoxyribonucleoside triphosphates (BALTIMORE, 1970; TEMIN and MIZUTANI, 1970; SPIEGELMAN et al., 1970; SCOLNICK et al., 1970). GARAPIN et al. (1971) have demonstrated a direct proportionality between the concentration of tritiated TTP and the rate of DNA synthesis. Varying the concentration of TTP from 8×10^{7}M to 5×10^{-4}M in the presence of excess dATP, dGTP, and dCTP resulted in a 100-fold increase in the rate of DNA synthesis.

The addition of ATP to the system has been shown to increase RNA directed DNA synthesis (MIZUTANI and TEMIN, 1970; GARAPIN et al., 1971). This stimulation may result from kinase and phosphatase contaminants using ATP as a substrate and thus maintaining the deoxyribonucleotide triphosphate levels available.

Viral endogenous DNA polymerase activity requires the presence of a divalent cation (Mg^{++} or Mn^{++}). For Mg^{++}, an optimum concentration of 5 - 10 mM and for Mn^{++} an optimum of 1 - 2 mM will yield optimum activity (TEMIN and MIZUTANI, 1970; BALTIMORE, 1970; SPIEGELMAN et al., 1970; ABRELL and GALLO, 1973). Low concentrations of sodium or potassium monovalent cations (20 - 30 mM) stimulate the reaction 15 - 20 %. Higher concentrations are slightly inhibitory.

The optimum temperature for the reaction mixtures is about 37 - 40°C for mammalian viruses and 40 - 45°C for avian viruses, and the optimum pH centers around pH 8.0. The system also requires a reducing agent such as mercaptoethanol or dithiothreitol for full activity (TEMIN and MIZUTANI, 1970; BALTIMORE, 1970; SPIEGELMAN et al., 1970).

Preincubation of disrupted virions with pancreatic ribonuclease inhibits the endogenous DNA polymerase activity. Concentrations of 3 - 50 μg/ml results in 75 % inhibition of the endogenous reaction (MIZUTANI and TEMIN, 1970). Susceptibility to ribonuclease treatment indicates that viral 70S RNA plays a role in DNA synthesis.

In summary, the virion polymerase system will achieve optimal endogenous activity when the virus is disrupted and incubated at about 40°C, pH 8.0, in a reaction mixture containing all four deoxyribonucleoside triphosphates, a divalent cation (Mg^{++} or Mn^{++}) and a reducing agent. The reaction products are precipitated with trichloracetic acid and collected on millipore filters; the amount of DNA synthesized is measured.

A. Primer-template Requirements

In addition to the 60 - 70S RNA molecules acting as a template-primer for viral DNA polymerase in the synthesis of DNA, a number of synthetic polynucleotides have been employed. Because DNA polymerase can copy only single-stranded regions of templates or those which have been "primed" with the addition of a complimentary polynucleotide, double-stranded molecules or single-stranded ones missing the complimentary nucleotide are ineffective as templates (BALTIMORE and SMOLER, 1971a). The primer provides a free 3'-OH end for producing a double-stranded molecule from a single-stranded template. However, several exceptions exist such as the formation of "hairpins" resulting in self-priming of single-stranded DNA (GOULIAN et al., 1968; ENGLUND, 1971), the copolymer poly (dA-dT) and strand-displacement ability exhibited by *E. coli* DNA polymerase I (MASAMUNE and RICHARDSON, 1971). Without the addition of a primer, homopolymers such as poly A, poly G, poly C and poly U cannot serve as templates. When primed with polymers, oligodeoxyribonucleosides or in some cases oligoribonucleotides, they will function as templates (BALTIMORE and SMOLER, 1971b; HURWITZ and LEIS, 1972; WELLS et al., 1972).

The efficiency with which DNA polymerase utilize templates and template primers can be used to characterize these enzymes and distinguish between those of viral and cellular origin (GOODMAN and SPIEGELMAN, 1971; ROBERT et al., 1972; SARIN and GALLO, 1973; GALLO et al., 1972a; BALTIMORE and SMOLER, 1971; WELLS et al., 1972). The template poly rA · poly dT has been shown to stimulate viral reverse transcriptases very effectively (BALTIMORE and SMOLER, 1971b); poly rA · dT_{12-18} having a greater stimulatory effect than poly dA · dT_{12-18}. Poly rC when primed with oligo dG is also effectively transcribed by all tumor virus DNA polymerase. This duplex has exhibited specificity for DNA polymerase of viral origin and hence can be useful in detecting their presence and activity form that of DNA polymerases of normal cells.

B. Analysis of the Endogenous Product

Analysis of the endogenous DNA product of short-time reactions (less than 20 min) shows the formation of DNA-RNA complexes which sediment at 60 - 70S RNA region in sucrose gradients (GARAPIN et al., 1970; SPIEGELMAN et al., 1970; ROKUTANDA et al., 1970; MANLY et al., 1971). Treatment with alkali or ribonuclease results in pieces of DNA sedimenting at 8 - 10S in sucrose gradients, indicating a detachment of DNA from its template. Centrifugation in Cs_2SO_4 gradients further supports the concept of DNA-RNA hybridization (SPIEGELMAN et al., 1970); sedimenting in the viral RNA region. When subjected to *Neurospora* endonuclease, an enzyme which degrades only single-stranded DNA (LINN and LEHMAN, 1965) only part of the DNA shows sensitivity. Thus, short-time reactions consist of RNA-DNA hybrids, single-stranded DNA and double-stranded DNA.

Studies of short-time reactions (10 min) in Rous Sarcoma Virus (RSV) using detergent-activated virions and limiting precursor concentration indicates the DNA synthesized is comprised of mainly dA, and alkaline hydrolysis reveals that 90 % or more of the pdA residues are attached to the 3'-terminal nucleoside (adenosine) of the primer molecule by a 5'-terminal in the DNA. Electrophoresis of this material on polyacrylamide gel shows migration with 4S RNA. Further incubation of the reaction (up to 90 min) results in the accumulation of longer DNA molecules, but the DNA at 4S continues to be comprised of mostly dA deoxynucleoside, indicating further initiation of DNA synthesis throughout the reaction (BISHOP et al., 1973).

After several hours of incubation, the DNA product is found to be small molecules of single- and double-stranded DNA which sediment at 10S in sucrose gradients (MIZUTANI and TEMIN, 1970; FUJINAGA et al., 1970; FANSHIER et al., 1971). Nearest neighbor analysis indicates the product is a heteropolymer (SPIEGELMAN et al., 1970). The small size of the DNA molecules, at present, has not been explained. Perhaps they are the result of degradation by a virus-associated nuclease (MIZUTANI et al., 1970) or are intermediates in provirus synthesis (SUGINO and OKAZAKI, 1973).

It is probable that all of the viral RNA, except the poly rA stretches (REITZ et al., 1972), is transcribed into DNA sequences, although for unknown reasons, some portions appear to be selectively copied to a greater extent. To determine what portion of the viral RNA template is transcribed into the DNA product, hybridization experiments involving the annealing of viral RNA with endogenous DNA products treated with alkali have been performed in several laboratories (SPIEGELMAN et al., 1970; DUESBERG and CANAANI, 1970; HATANAKA et al., 1970; ROKUTANDA et al., 1970; FUJINAGA et al., 1970; COFFIN and TEMIN, 1971, 1972). Analysis by Cs_2SO_4 gradient centrifugation showed one DNA product with a density characteristic similar to DNA and RNA-DNA hybrid product with a viral RNA density characteristic. Short-time reactions consisted of more RNA-DNA hybrid product and single-stranded DNA, whereas further incubation resulted in a larger proportion of double-stranded DNA (GARAPIN et al., 1971). DNA isolated from the DNA density region has been shown to rehybridize to viral RNA to a lesser extent than the DNA isolated from the RNA-DNA hybrid product (COFFIN and TEMIN, 1972) possibly due to complementary sequences with the viral RNA. In contrast to homologous viral RNA, no cross-hybridization has been observed with foreign RNAs from tobacco mosaic virus, Qβ virus, polio virus, chicken RNA yeast RNA, or other heterologous RNAs, indicating hybridization specificity. Ribonuclease resistance of viral RNA before and after hybridization indicates the DNA synthesized is composed of sequences complementary to 75 % or more of the 60 - 70S RNA (DUESBERG and CANAANI, 1970).

The kinetics of reassociation of annealed double-stranded DNA product has indicated that the DNA synthesized in virions is of two classes with respect to size. The majority of the DNA product is rapidly reassociating DNA, representing 5 - 25 % of a viral genome weighing 5 x 10^5 to 3 x 10^6 daltons. Slower reassociating DNA represents 30 - 100 % of a second RNA molecule weighing 2 - 10 x 10^6 daltons (DUESBERG and CANAANI, 1970; GELB et al., 1971; VARMUS et al., 1971).

C. Purification and Properties of Viral Reverse-Transcriptases

Purification of reverse transcriptase from a number of RNA tumor viruses has been reported (HURWITZ and LEIS, 1972; KACIAN et al., 1971; GRANDGENETT et al., 1973; LEIS and HURWITZ, 1972; ABRELL and GALLO, 1973). The lipoprotein envelope of the virion is removed with nonionic detergent (7 % NP-40) and salt (0.8M KCL). Part of the enzyme is found in the virion core and the rest is solubilized; possibly due to disruption of the core itself.

Purification of the soluble enzyme has been accomplished using anion and cation exchange, gel filtration, glycerol gradient centrifugation and ammonium sulfate precipitation; with the addition of a nonionic detergent, glycerol and dithiothreitol to stabilze the enzyme. The degree of purification and yield varies, depending on the type of virus, presence of inhibitors, and methods employed. Both RNA and DNA directed DNA polymerases are found in leukoviruses, and purification has failed to separate these activities (GALLO and TING, 1972; GALLO, 1971; TEMIN and BALTIMORE, 1972).

The size of the virion DNA polymerase varies with its origin. The avian enzyme AMV is composed of 2 polypeptide chains (KACIAN et al., 1971) one of molecular weight 65 - 70,000 and one 110 - 120,000 (GRANDGENETT et al., 1973). Rous sarcoma virus (RSV) enzyme has also 2 chains, molecular weight 65,000 and 105,000 (FARAS et al., 1972). Rauscher leukemia virus has a molecular weight of 70 - 90,000 (HURWITZ and LEIS, 1972; ABRELL and GALLO, 1973; ROSS et al., 1971; GERWIN and MILSTEIN, 1972; TRONICK et al., 1972). Mason-Pfizer monkey virus at 70 - 110,000 (ABRELL and GALLO, 1973; TRONICK et al., 1972), wooly monkey (SSV-1) virus at 70,000 (TRONICK et al., 1972; ABRELL and GALLO, 1973), and human leukemic cells at 130,000 (SARNGADHARAN et al., 1972). These findings indicate that avian and murine reverse transcriptases may differ with respect to their structure and polypeptide composition.

Avian RNA-dependent DNA polymerases have been shown to be associated with a nuclease, Ribonuclease H (RNase-H) (MÖLLING et al., 1971; BALTIMORE and SMOLER, 1972; KELLER and CROUCH, 1972; WATSON et al., 1973; GRANDGENETT et al., 1972, 1973; LEIS et al., 1973). Recent work of WEIMANN and SCHMIDT (1974) has reported the presence of RNase-H in murine Friend virus complex. MÖLLING and coworkers first reported the copurification of this enzyme with DNA polymerase of avian myeloblastosis virus (AMV). Because RNase-H did not degrade single or double stranded DNA or the DNA strand of RNA-DNA hybrids, it is believed that it specifically degrades the RNA strand of the RNA-DNA hybrid (HAUSEN and STEIN, 1970). Previous attempts to separate these two enzymatic activities have proven unsuccessful (MÖLLING et al., 1971; GRANDGENETT et al., 1973; BALTIMORE and SMOLER, 1972; WATSON et al., 1973). Using a variety of methods including sucrose and glycerol gradient centrifugation, cation and anion exchange chromatography and nondissociating gel electrophoresis, it was concluded that reverse transcriptase and RNase-H activities of AMV are located on a single polypeptide (GRANDGENETT et al., 1973). Recently, however, WU et al. (1974) have succeeded in separating RNase-H from reverse transcriptase in a mouse murine type-C RNA virus. Using sucrose density gradient centrifugation, the core fractions of the virus were isolated and found to contain reverse transcriptase activity but no detectable RNase-H activity. This would indicate that these two enzymes are not parts of the same polypeptide, but rather two distinct molecules.

Zinc as a structural entity of DNA polymerase in Avian Myeloblastosis Virus (AMV) has recently been established (AULD et al., 1974). Using a highly sensitive system, employing microwave-induced emission spectrometry, metal concentrations as low as 10^{-13} to 10^{-14} g-atoms were detected. Gel-exclusion chromatography was employed to remove metal quenching agents and protein contaminants of low molecular weight. Analysis showed that in purified DNA polymerase, zinc is a stoichiometric component. About 1.8 to 2.0 g-atoms of zinc was detected per molecular weights of protein ranging from 1.6 to 1.8 x 10^5 (KACIAN et al., 1971; HURWITZ and LEIS, 1972; GRANDGENETT et al., 1973). Agents which chelate zinc inhibit DNA polymerase, further implicating its role in nucleic acid metabolism.

The immunological characteristics of reverse transcriptase shows antigenic distinction from other viral antigens. Antibodies have been produced by immunization with enzyme (AARONSON et al., 1971; OROSZLAN et al., 1971). PARKS and coworkers (PARKS et al., 1972) have prepared a rabbit antiserum against an avian C-type virus (SR-RSV) which has been partially purified. Their results showed antigenic cross-reactivity among mammalian C-type viral polymerase, suggesting immunological relatedness; and immunological distinction between those of avial and murine origin. Cross-reaction studies of DNA polymerase (SCOLNICK et al., 1972a, 1972b;

TODARO and GALLO, 1973; SARIN and GALLO, 1973b) have lent further support to the immunological specificity of these enzymes. They have found that antiserum to the avian virus enyzems (AMV and RSV) cross-react only with avian DNA polymerases; feline antiserum cross-reacts only with murine and feline DNA polymerases; murine antiserum cross-reacts with feline, gibbon, wooly, murine and human leukemic cells; gibbon antiserum cross-reacts with gibbon, wooly and leukemic human cells. The immunological specificity of reverse transcriptase appears to be group specific. Normal cellular DNA polymerases are not affected by antibodies to viral DNA polymerases (ROSS et al., 1971).

Immunological studies of various avian viruses and normal chicken cells by MIZUTANI (MIZUTANI et al., 1974) has shown some relationships in the DNA polymerases and RNAs. This suggests that the avian viruses may have arisen from normal cellular RNA-directed DNA polymerase activity, perhaps by some genetic variation (TEMIN, 1970).

Studies of Hodgkins disease and Burkitts lymphoma in humans, RD114 feline virus in cats and Rauscher leukemia virus in mice have indicated these cancers are associated with the addition of new sequences to the normal DNA complement (KUFE et al., 1973; RUPRECHT et al., 1973; SWEET et al., 1974). The biochemical properties of these sequences are similar to those of RNA tumor viruses.

Examination of the ultrastructure of 60 - 70S RNA by means of electron microscopy has been attempted by several laboratories (GRANBOULAN et al., 1966; KAKEFUDA and BADER, 1969; KLEINSCHMIDT et al., 1972; WHALLEY, 1973; SARKAR and MOORE, 1970). Using the Kleinschmidt technique of spreading AMV, RNA, WEBER (WEBER et al., 1974) has observed molecules ranging in length from 0.7 to 1.9 μm, with corresponding molecular weights ranging from 0.7 - 2.2 x 10^6. Examination of RNA from Newcastle disease virus (KOLAKOFSKY et al., 1974) and vesicular stomatities virus (KILEY and WAGNER, 1972) showed a correspondence between length and molecular weight. Molecules of AMV RNA observed may be subunits of disrupted molecules, and the actual length of the intact viral genome cannot be determined by present methods.

III. Inhibition of Oncogensis by RNA Tumor Viruses

There are several approaches to inhibiting the molecular process leading to a manifestation of the oncogenic information. Of these, two approaches are of interest from the experimental standpoint: (1) Inhibition of the expression of proviral DNA, and (2) Inhibition of the proviral DNA synthesis.

A. Inhibition of the Expression of Proviral DNA

The rationale for the investigation of compounds which may selectively block the expression of proviral DNA is justified by

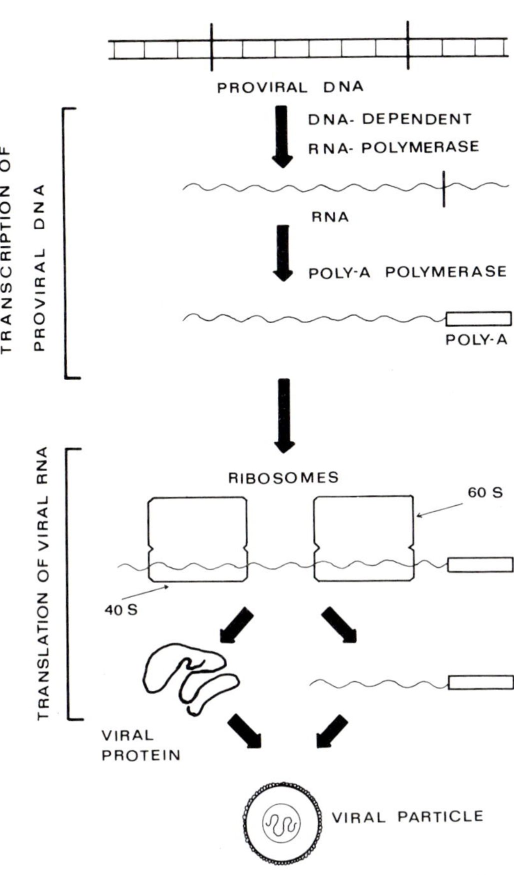

Fig. 2. Schematic presentation of molecular processes involved in the expression of proviral DNA

several facts. (1) The current thoughts on viral oncogenesis based on the protoviral theory of TEMIN (1971a, 1971b) and the oncogene theory of TODARO and HUEBENER (1972) emphasize the presence of viral specific DNA integrated into the genome of normal cells. This integrated viral-specific DNA can be transcribed to produce viral RNA (70s RNA) which may serve as messenger to code for the viral specific structure of the proteins, and for proteins required for the transformation process. The molecular reactions which lead to the formation of 70S viral RNA and its subsequent translation are shown in Fig. 2. (2) The virus specific sequences of DNA integrated into the DNA of a normal cell can be selectively induced by halogenated pyrimidines (LOWY et al., 1971) to produce C-type particles. This seems to be a novel approach to develop a test system for compounds blocking the proviral DNA expression.

Making use of the above system WU et al. (1972) succeeded in showing that cordycepin (3'-deoxy-adenosine) blocks the induction of virus production by 5-iododeoxyuridine (IDU); other adenine-containing nucleosides did not block this induction. The effect of cordycepin was found to occur only at a critical time interval

(12 - 18 h after IDU treatment). The cordycepin inhibition was shown to be independent of the state of growth of cell cultures. The authors (WU et al., 1972) believe that cordycepin acts by inhibiting the formation of polyadenylic acid (Poly A) by blocking chain extension due to the absence of 3'-OH-group. Since Poly A sequences are a characteristic of RNA tumor viruses (GILLESPIE et al., 1972; LAI and DUESBERG, 1972; GREEN and CARTES, 1972), though their exact role is unknown, such inhibitors are of future interest. However, one may argue the selectivity in blocking the proviral DNA expression since Poly A sequences are known to occur in the normal cellular messenger RNA (DARNELL, 1968). It has been suggested that Poly A stretches are required for transport of the messenger RNA from nucleus to cytoplasm (DARNELL, 1968). It the Poly A sequences for the viral 70S RNA and for the cellular messenger RNA are synthesized by the same Poly A polymerase, the selectivity of this adenosine analog in blocking the formation of 70S RNA is questionable.

The most selective approach to inhibit the proviral DNA expression could be to search for compounds which may specifically act at a very early stage of induction of the integrated proviral DNA. However, such an approach requires an understanding of the mechanism involved in the induction of the proviral genome. Unfortunately, this aspect has drawn very little attention; how the halogenated pyrimidines induce the proviral DNA in a selective manner is yet to be elucidated.

B. Inhibition of Proviral DNA Formation

The inhibition of proviral DNA formation might be possible at several stages: (1) blocking the penetration of the membrane by the virus; (2) blocking the release of viral components inside the host cell; (3) inhibiting the RNA-dependent DNA-polymerase; and (4) damaging the viral 70S RNA. Of these, the last two are of immediate interest since the reverse-transcriptase and the 70S RNA are novel features of oncornaviruses.

Compounds which inhibit DNA synthesis in RNA tumor viruses may be divided, according to their chemical structure into several classes; the important ones will be described here.

1. *Ansamacrolides*

The most important member of this class of compounds are the rifamycins; they consist of a chromophore naphthoquinone or naphthohydroquinone ring which is spanned by a long aliphatic bridge. One of the substances originally produced by *S. mediterranei* is rifamycin B, which can be oxidized and hydrolyzed to rifamycin S; reduction yields the naphthohydroquinone derivative rifamycin SV (see review by WEHRLI and STAEHELIN, 1971). The fact that rifamycin S and rifamycin SV are very potent inhibitors of bacterial growth (0.0025 μg/ml; SENSI et al., 1967) has led to the synthesis of a vast number of structural derivatives to obtain analogs with greater antibacterial activity. The main feature of their biochemical action is their selectivity in blocking the

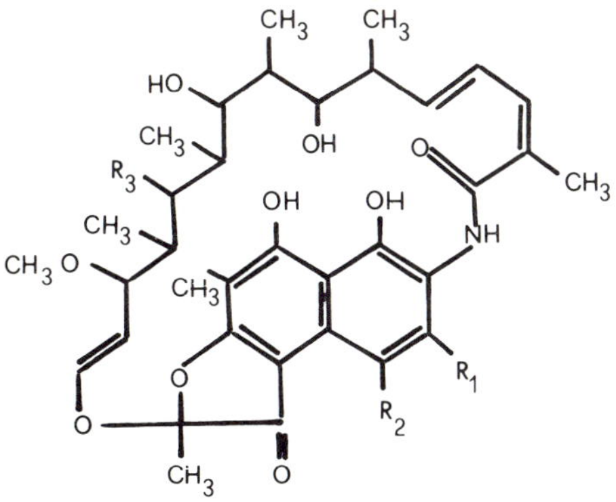

DERIVATIVE	R_1	R_2	R_3
RIFAMPICIN	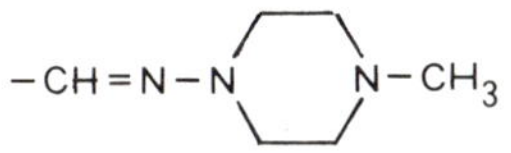 $-CH=N-N\langle\ \rangle N-CH_3$	$-OH$	$-OOCCH_3$
M/14 DA	$-H$	$-OCH_2CON(C_2H_5)_2$	$-OH$
M/14	$-H$	$-OCH_2CON(C_2H_5)_2$	$-OOCCH_3$
RIFAMYCIN SV	$-H$	$-OH$	$-OOCCH_3$
AF/AP	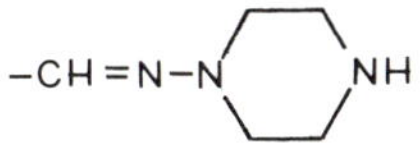 $-CH=N-N\langle\ \rangle NH$	$-OH$	$-OOCCH_3$
AF/013	$-CH=NO(CH_2)_7CH_3$	$-OH$	$-OOCCH_3$
AF/DNFI	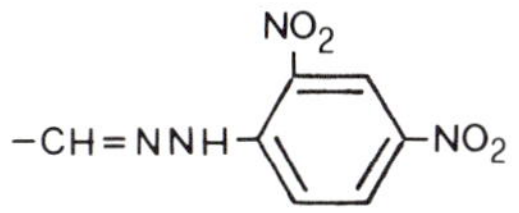 $-CH=NNH-C_6H_3(NO_2)_2$	$-OH$	$-OOCCH_3$
AF/ABDP	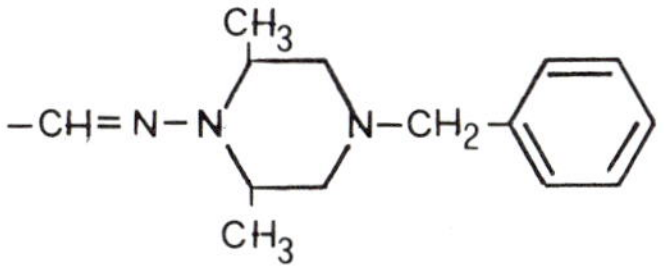 $-CH=N-N\langle(CH_3)(CH_3)\rangle N-CH_2-C_6H_5$	$-OH$	$-OOCCH_3$

Fig. 3. Structures of some rifamycin derivatives

RNA polymerase activity of bacterial cells, whereas the RNA polymerase of eukaryotic cells is insensitive to their action. Surprisingly, some of these derivatives were reported to inhibit the RNA-dependent DNA activities in leukemic cells (GALLO et al., 1970), and later in RNA tumor viruses (YANG et al., 1972; GURGO et al., 1972; TING et al., 1972). Some of the structural derivatives tested in these laboratories are shown in Fig. 3.

Of particular interest is the structure-activity relationship of these derivatives in viral DNA-polymerase. From these studies (YANG et al., 1972; GURGO et al., 1972), it is apparent that for the inhibition of viral DNA-polymerases by rifamycin derivatives, some structural entities are of major importance; (1) the ansa ring structures should be intact, since free aminopiperazines are inactive as DNA polymerase inhibitors (see also CHANDRA et al., 1972a), (2) a large side chain at the 3-position of rifamycin SV molecule (see Fig. 3, R_1) enhances the inhibitory activity. Thus rifamycin derivatives $AF/_{O13}$, $AF/_{DNF1}$ and $AF/_{ABDP}$ (Fig. 3) were found to be the most active inhibitors of viral reverse-transcriptase.

The structure-activity relationship of rifamycin derivatives in the viral DNA-polymerase system could be demonstrated also in biological experiments (TING et al., 1972; GALLO et al., 1972a). They found that if virus is pre-incubated with different (but structurally similar) rifamycin derivatives, the derivative then removed and the virus tested for ifectivity, compounds which have no enzyme inhibitory effect also have no effect of the biological activity (Fig. 4).

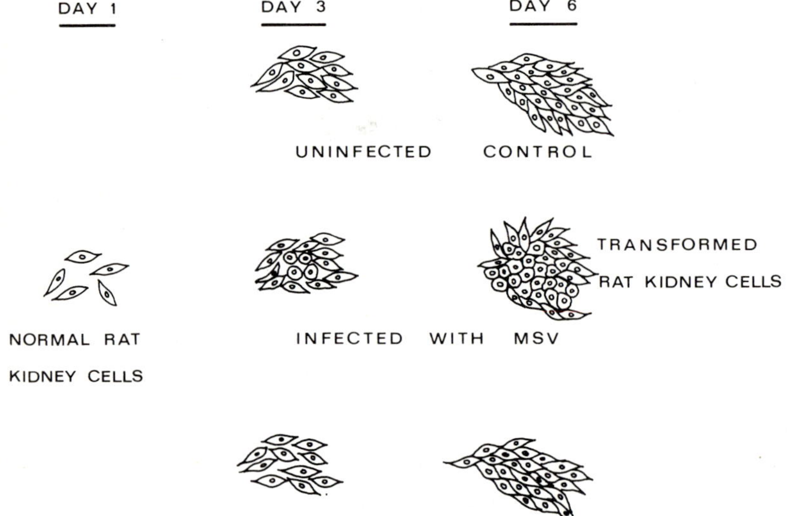

Fig. 4. Transformation of normal rat kidney (NRK) cells by Moloney Sarcoma virus (MSV) and prevention by selective rifamycin derivatives (GALLO et al., 1972b, reproduced with the permission of Dr. R.C. GALLO)

The mechanism of viral DNA-polymerase inhibition by rifamycin derivatives has recently been studied by WU and GALLO (1974). The inhibitory effect is independent of the choice of template-primer but is sensitive to the presence of non-ionic detergent. Potent inhibitors were found to bind to the viral enzyme as shown by co-sedimentation of the inhibitor with purified enzyme during glycerol-gradient centrifugation. This binding is reversible with a relatively low concentration of Triton X-100 (0.05 %). The inhibition takes place probably at a step before initiation. The authors therefore suggested that conditions which facilitate initiation may interfere with the inhibitory response of these compounds.

Three groups of antibiotics that are chemically very similar to the rifamycins have been described: the streptovaricins (RINEHART et al., 1968), the tolypomycins (KISHI et al., 1969) and geldanamycin (DEBOER et al., 1970). Like the rifamycins, they all contain an aromatic ring system spanned by an aliphatic bridge. These antibiotics have biochemical properties closely resembling those of the rifamycins, as one would expect from their chemical similarity. They inhibit the initiation step of bacterial RNA-polymerase (MIZUNO et al., 1968), although to a considerable lesser degree than the most active rifamycins.

Streptovaricins have been shown to inhibit the reverse-transcriptase activity of murine leukemia virus (BROCKMAN et al., 1971), focus formation by Moloney Sarcoma Virus and splenomegaly induced by Rauscher leukemia virus in mice (CARTER et al., 1971). These studies were carried out with a streptovaricin complex composed of seven macrolides; A, B, C, D, E, F, and G. Streptovaricins A, E, and F have been reported to be inactive, B and C weakly active and D and G most active in inhibiting MLV reverse-transcriptase activity (CARTER et al., 1972), while GURGO et al. (1972) reported that A, C, and D are all inactive against MLV. Since these compounds are labile, particularly in solution (GALLO, 1972), it should be of interest to reexamine this issue on the most potent component of Streptovaricin complex. Only recently geldanamycins and tolypomycins have been reported to inhibit the reverse-transcriptase activity in oncornaviruses (RINEHART et al., 1973).

2. *Pyrrole-amidine Antibiotics*

In the past years several compounds such as distamycin A, netropsin, congocidin, anthelvencin A and the kikumycins A and B have been grouped as members of pyrrole-amidine class of antibiotics. Of these, distamycin A, netropsin and congocidin (Fig. 5) have been of great interest in virological and biochemical investigations. In the foreign pages only studies with distamycin A and its structural analogs will be described; for details on other compounds one should refer to a very extensive review by HAHN (1974).

Chemical structure and total synthesis of the antibiotic distamycin A, a metabolite of *Streptomyces disthallicus*, were reported in 1964 (ARCAMONE et al., 1964). This compound represented

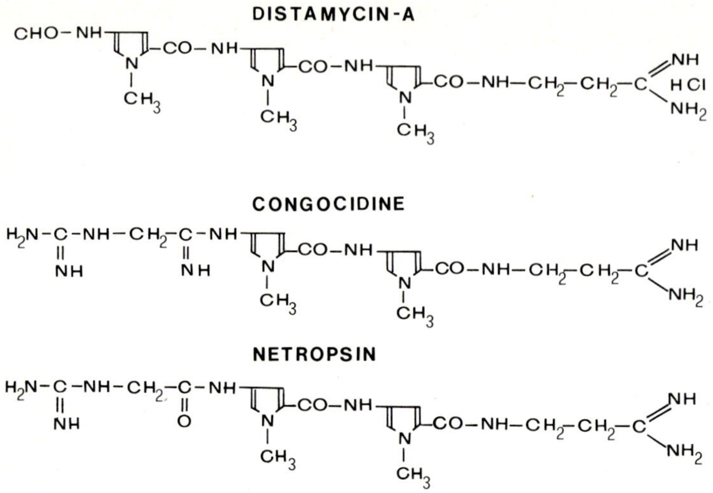

DISTAMYCIN-A

CONGOCIDINE

NETROPSIN

Fig. 5. Chemical structures of some pyrrole-amidine antibiotics

a new molecule of microbial origin endowed with many interesting biological properties: (1) it has a cytostatic property evidenced by its activity on some experimental tumors of the mouse and the rat (DI MARCO et al., 1962); (2) it suppresses the multiplication of T_1 and T_2 phages in *E. coli* K_{12} (DI MARCO et al., 1963a and b); (3) it interferes with the multiplication of some DNA-viruses such as vaccinia, herpes simplex and adenoviruses (WERNER et al., 1964); and it prevents the induction of bacterial adaptive enzymes of *E. coli* (SANFILIPPO et al., 1966; HOLLDORF et al., 1970).

The interaction of distamycin A with nucleic acids, especially with DNAs, has been investigated in several laboratories (CHANDRA et al., 1970, 1972a-c; KREY and HAHN, 1970; HAHN, 1974; ZIMMER et al., 1971a, 1971b; ZUNINO and DI MARCO, 1972). The details of these interactions, and their implication on DNA and RNA synthesis in cell-free systems have recently been reviewed by HAHN (1974).

The structure of this antibiotic is characterized by three residues of 1-methyl-4 amino-pyrrole-2-carboxylic acid and two side chains, the first constituted by a formyl group, and the second by a propionamidine chain (Fig. 5). The recent interest on distamycin A in experimental and clinical research has led to the chemical synthesis of a vast number of its structural analogs. Synthetic analogs of distamycin A may be divided into two major classes. To the first class belong those analogs which, unmodified in the side chain, contain a different number of residues of 4-amino-1-mehtylpyrrole-2-carboxylic acid. In the second class are considered those analogs which show modifications in side chains of the distamycin molecule. These modifications have been achieved either in the propionamidine side chain, or in the formyl-amino side chain. These analogs have been synthesized by ARCAMONE et al. (1969a, 1969b).

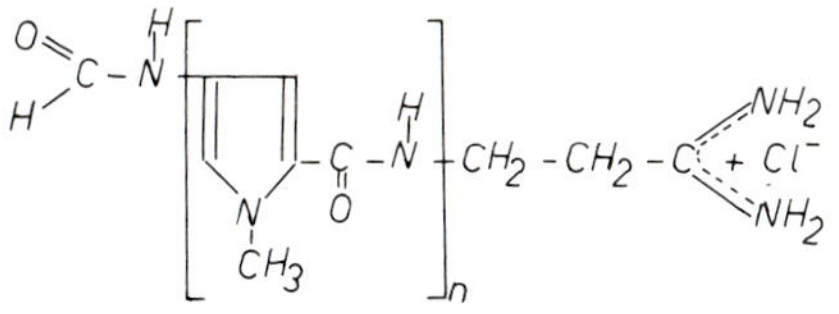

Fig. 6. Distamycin A analogs with different numbers of pyrrole residues

a) Analogs with Different Number of Pyrrole Residues

The initial compound of the series (n = 1) was obtained (ARCAMONE et al., 1969a) from β-(1-methyl-4-nitropyrrole-2-carboxamido)-propionitrile. This compound is completely devoid of biological, as well as biochemical activity. Compounds containing at least two (n = 2) or more pyrrole groups are of great interest to study structure-activity relationship in various systems. The present study describes the biochemical basis of action of distamycin analogous containing two or more pyrrole groups (Fig. 6).

Fig. 7 compares the effects of native and denatured DNA and of RNA on the ultraviolet spectra of the oligopeptide antibiotics containing 3, 4, and 5 pyrrole rings. Three general observations contained in this Figure need emphasis. First, yeast RNA and (in the case of distamycin-A) apurinic and apyrimidinic DNA at r = 0.025 did not change the characteristic absorption band of the free compounds. The absorption spectrum of distamycin-5 was enhanced in the presence of RNA. Second, heat-denatured DNA at pH 7.0 bound all the compound tested and, at corresponding values of r, the spectra of the bound antibiotics were similar to those found with native DNA. Finally, in each case, the red shift of the characteristic absorption band was accompanied by an increase in the intensity of the absorption maximum. This hyperchromic effect depended upon DNA concentration and became more pronounced as the number of pyrrole residues in the distamycin molecule increased. Under identical conditions of r and ionic strenght, the spectral changes induced by denatured DNA were somewhat weaker.

The differences in hyperchromic effect can hardly be attributed to differences in the stability of the compounds tested, as the spectra of the compound both in the absence and in the presence of DNA were read immediately after preparation. These observations provide strong support for the view that the chromophore of the antibiotic contributes to the binding mechanism.

The cytotoxicity and antiviral activity of distamycin derivatives containing 2, 3, 4, and 5 pyrrole rings was investigated. Cultures of HeLa cells (grown in HLS medium) or mouse-embryo cells

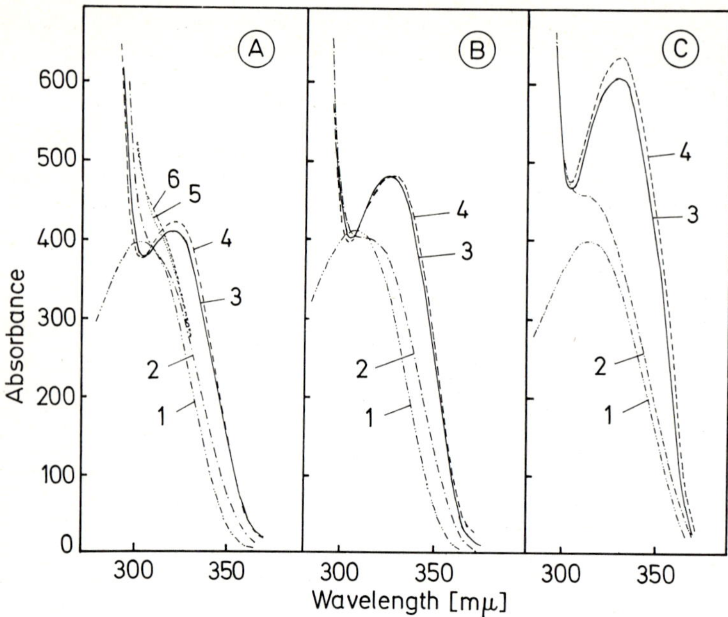

Fig. 7. Effect of native DNA, denatured DNA and yeast RNA on the ultraviolet absorption spectra of distamycin (A), distamycin/4 (B) and distamycin/5 (C), in 0.01 M Tris-HCl buffer (pH 7.0). Curve 1 is the spectrum of free antibiotics. Other curves are spectra of the antibiotics in the presence of yeast RNA (curve 2), denatured DNA (curve 3), native DNA (curve 4), apurinic acid (curve 5), and apyrimidinic DNA (curve 6) at a molar antibiotic/nucleic acid-P ratio of 0.025 (CHANDRA et al., 1972a)

Table 1. Cytotoxicity and antiviral activity of the distamycin derivatives against vaccinia virus

Compound	Number of pyrrole rings (n =)	Inhibition of vaccina virus multiplication[a] (% Inhibition)	Cytotoxicity[a]
DST-2	2	13	100
DST-A	3	100[b]	100[b]
DST-4	4	400	50
DST-5	5	1000	25

[a]Activity calculated with respect to distamycin A considered = 100.
[b]Absolute values (ID_{50} μg/ml): Cytotoxicity = 80; WR = 2. (CHANDRA et al., 1972a).

(grown in HLS medium plus 0.1 % yeastolate) infected with vaccinia virus (strain WR/ATCC) were used. The results shown in Table 1 indicate that the cytotoxicity of the derivatives with 2 and 3 pyrrole residues is the same, but that the compounds containing 4 and 5 pyrrole rings (DST-4 and DST-5) are less toxic. The cytotoxicity of DST-4 and DST-5 is 50 % and 25 % that of the natural antibiotic (DST-A), respectively. It seems therefore, that the cytotoxicity decreases as the number of pyrrole rings increases. This is true at least for DST-A, DST-4 and DST-5. Our studies on DST-6 (distamycin with 6 pyrrole rings) have, however, shown that no such relationship strictly exists; DST-6 was found to be as toxic as DST-4.

The antiviral activity of distamycin derivatives is dependent on the pyrrole ring. Taking the antiviral activity of the natural compound (DST-A) as 100, one observes a 4-fold increase for DST-4 and a 10-fold increase for DST-5. On the other hand, we found about 85 % reduction of the antiviral activity of DST-A on removing 1 pyrrole ring (DST-2).

It is known that RNA oncogenic viruses require DNA synthesis for their replication. As DST-A blocks some early steps in the growth cycle of DNA viruses, probably connected with DNA replication (CASAZZA et al., 1971), it was of interest to investigate the effect of DST, DST-4, and DST-5 on MSV (Moloney).

Experimental studies carried out on *in vitro* infected mouse embryo cell cultures have shown that treatment with DST-A, DST-4, and DST-5 reduced the number of MSV(M) foci formed *in vitro* (CHANDRA et al., 1972g). The inhibitory activity was dependent on the dose used, and increased according to the number of pyrrole residues in the molecule. The cytotoxic activity of these compounds increased at higher doses; however, it was always less than their inhibiting activity on foci formation. Since the dose-dependent curves for cytotoxicity and antiviral activity are almost parallel, it was possible to calculate the ratio cell-ID_{50}/virus ID_{50} (therapeutic index: TI), which increased from 2.5 for DST-A, to 7.0 for DST-4 and to 7.7 for DST-5.

To find out whether the reduced foci formation was due to inhibition of MSV(M) replication, the effect of these compounds on virus growth was studied. DST-A, even at cytotoxic doses, poorly inhibited the virus replication; DST-4 was very active on virus replication and DST-5 at non-cytotoxic doses was able to reduce viral yields by 92.3 %.

In mice DST-4, administered intraperitoneally 6 times/day for 6 days, starting 2 days after the infection, was able to inhibit and to reduce the growth of tumors induced by giving MSV(M) intramuscularly. Good results were also obtained by administering DST, DST-4, and DST-5 intramuscularly in the MSV(M)-infected leg. Less activity was observed *in vivo* than *in vitro*. This may be due to uneven distribution in the animal body, metabolism and/or rapid elimination (G. DI FRONZO and G. LENAZ, unpublished results). Further studies will investigate the possibility of obtaining a more significant activity against oncogenic viruses in the animal by varying schedule and route of treatment.

Table 2. Inhibition of reverse-transcriptase activity of FL virions by distamycin derivatives in the absence of exogenous template

System	Antibiotic used[a]	^{3}H-TMP incorporation into DNA (cpm./reaction mixture)	% of control
Without virions	-	14	4.1
Without Triton	-	40	11.7
Virions + RNase[b]	-	61	17.6
Complete	-	343	100
Complete	DST-2	336	98
Complete	DST-A	242	70.5
Complete	DST-4	203	59
Complete	DST-5	206	59.8

[a]Antibiotic concentration = 20 µg/reaction mixture.
[b]Virions containing Triton were preincubated at room temp. for 25 min. with 50 µg/ml of pancreatic RNase. (CHANDRA et al., 1972a).

The activity of DST-A on FLV leukemogenesis in mice has been studied. The studies were carried out on male and female albino mice weighing 20 - 30 g. FLV suspensions were prepared by filtering the homogenates from infected spleens through Seitz K-filters (Seitz Company, Bad Kreuznach, Germany). This suspension was diluted 1:20 with Hanks solution and 0.1 ml (ID_{90}) was injected intraperitoneally. In the experimental group the suspensions were preincubated with DST-A (20 µg/ml), in the control group they were not; each group contained 50 animals. At various intervals 10 animals from each group were killed and their leucocyte number and spleen weight determined. The results showed a very significant reduction in the number of leucocytes in the DST-A treated group and a slight reduction in spleen weight, as compared to controls (CHANDRA et al., 1972a).

The mechanism by which DST-A and its structural analogs exert their activity against MSV(M) and FLV could be explained by inhibition of the reverse transcriptase activity. The activity of distamycin derivatives on the DNA polymerase activities of FLV and MSV(M) was tested. As shown in Table 2, the reverse transcriptase activity (without exogenous template) of FL virions is inhibited by distamycin derivatives containing 3, 4, and 5 pyrrole rings. The compound containing 2 pyrrole rings (DST-2) is ineffective, which confirms our earlier observations on vaccinia virus. The compound with 4 pyrrole rings inhibits the reverse transcriptase reaction more than the one with 3 pyrrole rings (DST-A). Further progression could not be demonstrated with the compound containing 5 pyrrole rings (DST-5) which is very unstable; longer incubation (90 min) may cause some degradation of this compound.

Table 3. Inhibition of DNA-polymerase activity of FL virions by distamycin A and its analogs in the presence of various templates

Antibiotic used	^{3}H-TMP incorporation[b] into DNA poly (dA - dT)	^{3}H-TMP incorporation[b] into DNA poly (rA) · $(dT)_8$
None	1020 (100)	520 (100)
DST-2	357 (35.8)	450 (85.5)
DST-A	137 (14)	330 (63.5)
DST-4	117 (11.9)	290 (55.5)

[a]Antibiotic concentration = 20 μg/reaction mixture;
[b]cpm./reaction mixture. The figures in brackets are percentages.

The effect of distamycin derivatives on the DNA polymerase activity of FL virions was studied in the presence of poly (dA-dT) and poly (rA) · $(dT)_8$ (Table 3). The DNA polymerase activity was found to be most sensitive to distamycin inhibition with poly (dA-dT) as primer-template and weaker with poly (rA) · $(dT)_8$. This is in good agreement with the observations of KOTLER and BECKER (1972) who found that distamycin A preferentially inhibits the synthesis of double stranded DNA by RSV reverse transcriptase. In both cases the inhibitory response of the antibiotic increase according to the number of pyrrole rings in the molecule.

Inhibition of DNA polymerase activity of MSV(M) by distamycin A and its derivatives in the presence of various templates was also studied with similar results to those observed in FL virions. The fact that distamycin derivatives inhibit both reverse transcriptase activity and oncogenesis by RNA tumor viruses suggests that both activities derive from the same structural component(s) of the molecule. We have seen that the antiviral activity of DST analogs on MSV(M) and their inhibition of the DNA polymerase activity of RNA tumor viruses increases with the number of pyrrole rings in the molecule. The higher antiviral activity, as against cytotoxicity, of DST-4 and DST-5 suggests that more pyrrole residues specifically increase the inhibition of virus-associated enzyme activity.

b) Analogs with Side Chain Modification

The cytotoxicity and antiviral activity of distamycin derivatives (Fig. 8) obtained by substitutions of the formyl group (II) or the propionamidine chain (III and IV), has been studied (CHANDRA et al., 1972b). Substitution of the formyl group with a cyclopentyl propionyl chain does not influence its cytotoxicity but the compound loses its antiviral activity completely. The substitution of the propionamidine group with a benzamidine moiety doubles the cytotoxicity of the compound. The antiviral activity

SUBSTITUTIONS →	R_1	R_2
I.	OHC–NH–	$-NH-CH_2-CH_2-C(=NH)NH_2$
II.	$CH_2-CH_2-CO-NH-$ (cyclopentyl)	$-NH-CH_2-CH_2-C(=NH)NH_2$
III.	OHC–NH–	$-NH-C_6H_4-C(=NH)NH_2$
IV	OHC–NH–	$-NH-CH_2-CH_2-CH_2-C(=NH)NH_2$

Fig. 8. Analogs of distamycin A with side chain modifications

of this compound is, however, only 44 % of that of distamycin A. The analog containing butyramidine group in place of the propionamidine moiety (IV) also exhibits a higher cytotoxicity than the compound I. This has only 31 % of the antiviral activity of compound I.

These results allow the conclusion that the presence of the formyl group in distamycin is necessary for its antiviral activity. Studies with other derivatives, where the formyl group was substituted by a nitro, amino or acetyl group have shown that all these derivatives are completely inactive against viruses. However, substitutions at the formyl group do not influence the cytotoxicity of the compound. Compounds having substitutions at the propionamidine are active against viruses but exhibit a much higher toxicity. An interesting compound of this group is the acetamidine derivative which showed a higher antiviral activity (150 %) than distamycin A.

To investigate the role of formyl group some amino acid derivatives of des-formyldistamycin A were synthesized, in which the formyl group was linked to the amino group of glycine or alanine. Thus, the side chains of these derivatives are constituted by N-formyl-glycyl, or N-formyl-alanyl groups (Fig. 9A). Studies on various systems reported here indicate that the compound with formyl group linked to 4-amino-1-methylpyrrole-2-carboxylic acid residue (distamycin/A) is more active compared to its amino acid derivatives.

The effect of distamycin/A and its amino acid derivatives (distamycin/Gly and distamycin/Ala) on the template activity of calf thymus DNA in RNA-polymerase (*E. coli*) reaction was studied (CHANDRA et al., 1974b). In these experiments, equimolar concentrations of the antibiotics (8 x 10^{-5} M) were added to the incubation mixture, and the reaction was started immediately with the enzyme. Under these conditions distamycin/A inhibits appx. 70 % of the incorporation of ^{3}H-AMP into RNA. Distamycin/Gly and distamycin/Ala at the same molar concentrations are much

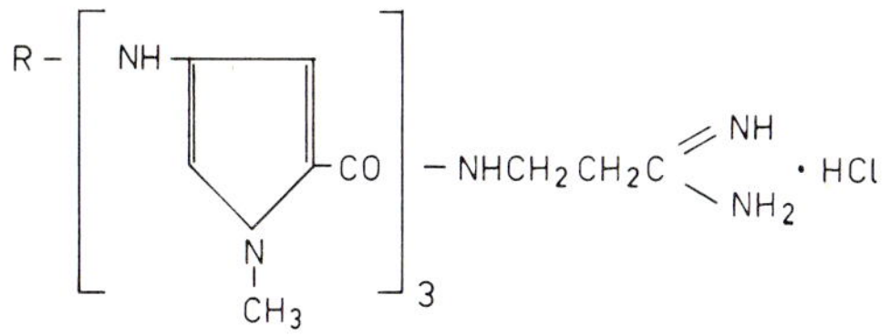

$$R-\left[NH-\text{(N-}CH_3\text{ pyrrole)}-CO\right]_3-NHCH_2CH_2C\begin{matrix}\nearrow NH\\ \searrow NH_2\end{matrix}\cdot HCl$$

Fig. 9a. Amino acid derivatives of des-formyl-distamycin A

COMPOUND	R
Dist./A	OHC-
Dist./Gly.	$CHONHCH_2CO$ -
Dist./Ala.	$CHONHCH(CH_3)CO$ -

Table 4. Inhibition of reverse-transcriptase activity of FL-virions by distamycin derivatives in the absence of exogenous template

System	Antibiotic (concentration = 2×10^{-4}M)	^{3}H-TMP incorporation into DNA (cpm/reaction mixt.)	% of control
Without virions	-	11	3
Without Nonidet P-40	-	45	12.2
Complete	-	364	100
Complete	Distamycin/A	93	25.6
Complete	Distamycin/Gly	146	40
Complete	Distamycin/Ala	258	71

less effective in inhibiting the template activity of DNA. These results indicate that the binding affinity of amino acid derivatives for DNA is much less, compared to that of the parent compound. Thus, the position of the formyl group is important in its binding mechanism to DNA. This has been further shown by studies on the DNA-polymerases of oncornaviruses and of bacterial cells.

Table 4 shows the activities of distamycin/A and its amino acid derivatives on the DNA polymerase activity of Friend leukemia virions (FLV). These studies were carried out without the exogenous template. It has been emphasized that the study of endogenic reaction has greater implications in viral cancerogenesis. The results show a strong inhibition of the endogenic reaction by distamycin/A at 2×10^{-4}M; distamycin/Gly was also very effective, distamycin/Ala showed a poor activity. The activity of these derivatives on the viral oncogenesis *in vivo* is currently being investigated.

CHEMICAL STRUCTURES OF DAUNOMYCIN DERIVATIVES

General structure

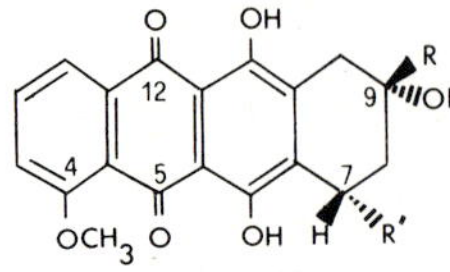

Structure of daunosamine

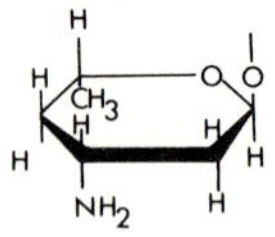

SUBSTITUTIONS

	R	R′
DAUNOMYCIN	$-CO-CH_3$	DAUNOSAMINE
ADRIAMYCIN	$-CO-CH_2OH$	DAUNOSAMINE
13-DIHYDRO-DAUNOMYCIN	$-CHOH-CH_3$	DAUNOSAMINE
N-GUANIDINE-ACETAMIDE-DAUNOMYCIN	$-CO-CH_3$	N-GUANIDINE-ACETAMIDE-DAUNOSAMINE
N-ACETYL-DAUNOMYCIN	$-CO-CH_3$	N-ACETYL DAUNOSAMINE

Fig. 9b. Chemical structures of daunomycin derivatives

3. *Anthracyclines*

Some of the major advances in the chemotherapy of cancer are attributed to the use of some natural products of this group such as daunomycin, alone or in combination with other compounds, as chemotherapeutic cocktails. The basic idea in using such chemotherapeutic cocktails is to increase the therapeutic efficy of the principal constituent. This can be achieved by (1) facilitating the favorable distribution of the drug at its site of action; or (2) by altering the pharmacodynamics of the compound in such a way, that the degradation of the compound is slowed down so that the compound can exist in its therapeutically active form for a longer time.

Another possibility which may directly lead to the potentiation of chemotherapeutic efficacy of a compound is its structural modification. This has been observed by us with respect to the biochemical and cytostatic activity of daunomycin, adriamycin and their structural analogs, which were synthesized in the research laboratories of Farmitalia, Milan, Italy. Our initial studies (CHANDRA et al., 1972a, 1972f, 1973) were carried out with the daunomycin derivatives with substitutions in the anthracycline ring (adriamycin and dihydro-daunomycin) or in aminosugar moiety (N-guanidino-acetamido-daunomycin and N-acetyl-daunomycin), see Fig. 9b.

Table 5. Effect of daunomycin and its derivatives on the viral oncogenesis in mice and chicken

Antibiotic used	Oncogenesis in mice by FLV No. of animals survived[a] / No. of animals infected	Oncogenesis in chicken by RSV[b] Mean survival time (days)
None	1/6	12.3
Daunomycin	6/6	30
Adriamycin	5/6	28.3
Dihydro daunomycin	3/6	14.2
N-guanidine-daunomycin	1/6	13.6
N-acetyl daunomycin	0/6	12.0

[a]13 days after infection; for details see text.
[b]Each experimental group contained 6 chicken. The tumor suspension (1:10) was incubated with 50 μg/ml of the antibiotic at 37° for 1 h. Control suspension was incubated with the antibiotic. 0.1 ml of this suspension was injected intraperitoneally.

The inhibitory effect of different daunomycin derivatives (Fig. 9b) on the viral oncogenesis by FLV and RSV is shown in Table 5. FLV suspensions were prepared by filtering the homogenates from infected spleens (AKR mice) through Seitz EK-filters (Seitz Company, Bad Kreuznach, Germany). This suspension was diluted 1:20 with Hank's solution. The diluted suspension was incubated with daunomycin or its derivatives (50 μg/ml) for 1 h at 37°. The control suspensions were incubated under similar conditions, without the antibiotic. 0.1 ml of this suspension (ID_{50}) was injected intraperitoneally. Each experimental group contained 6 animals. As follows from Table 5, 5 of the 6 control animals died after 13 days of infection. At this time the animals injected with daunomycin-treated viral suspensions were all alive. Adriamycin also has a very benefical effect, whereas dihydro daunomycin shows a moderate activity. The derivatives substituted in the amino sugar moiety are ineffective. The oncogenesis in chicken by RSV is also inhibited by daunomycin and its analogs. Thus, the mean survival time is prolonged from 12.3 days to 28.3 days by adriamycin. Daunomycin is even more effective, whereas the other derivatives have no significant effect.

The inhibitory activity of daunomycin and its structural analogs in viral oncogenesis suggests that the virus-associated enzymatic activities are sensitive to these antibiotics. The inhibition of reverse transcriptase activity of MSV(M), FLV and RSV by various daunomycin derivatives is shown in Table 6. Daunomycin and adriamycin inhibit the DNA-polymerase reaction (endogenous) to 60 - 70 % at 10 μg/reaction mixture (0.25 ml). The dihydro derivative is also quite effective, whereas the N-guanidine derivative has a moderate activity. However, the N-acetyl derivative is com-

Table 6. Inhibition of DNA-polymerase activity[a] (endogenous) of RNA tumor viruses by daunomycin derivatives

System	Antibiotic used[b]	^{3}H-TMP incorporation into DNA (cpm/reaction mixt.)		
		MSV (Moloney	FLV	RSV
Without virions	-	7 (3.4)	7 (3.7)	7 (2.9)
Virions + RNase[c]	-	26 (13)	25 (13.4)	40 (16.8)
Complete	None	202 (100)	187 (100)	237 (100)
	Daunomycin	65 (32.1)	68 (36.3)	80 (33.7)
	Adriamycin	66 (33.1)	83 (44.4)	86 (36.3)
	Dihydro daunomycin	86 (42.5)	87 (46.5)	113 (47.7)
	N-guanidine daunomycin	106 (52.4)	97 (51.8)	117 (49.3)
	N-acetyl daunomycin	196 (97)	192 (102.6)	136 (57.3)

[a]DNA-polymerase assay: DNA-polymerase activity was assayed essentially by the method of ROSS et al. (1971). The reaction mixture, regardless of template, was similar to that of ROSS et al. (1971), except that we used 0.04 M Tris and the end concentration of Nonidet P-40 (Shell Chemie, Hamburg) was 0.2 %. The reaction mixture contained 0.25 µg of the template used. The reaction mixture (0.25 ml) containing virions (28 µg of protein) was incubated at 37° for 90 min. Acid precipitable material was counted on "Millipore" filters (HAWP 02500) in a liquid scintillation counter. Protein was estimated by the method of LOWRY et al. (1951).

[b]Antibiotic concentration = 10 µg/reaction mixt. (0.25 ml).

[c]Virions containing Nonidet P-40 were preincubated at room temp. for 25 min. with 50 µg/ml of pancreatic RNase. Figures in parentheses are the percent of control.

pletely ineffective in MSV(M) and FLV systems. The RSV system was moderately inhibited by the N-acetyl derivative.

Table 7 shows the inhibition of poly (dA-dT)-dependent DNA-polymerase activity of FLV by various daunomycin derivatives.

Table 7. Inhibition of DNA-polymerase activity of FL-virions by daunomycin and its derivatives in the presence of poly (dA-dT)

Antibiotic used[a]	^{3}H-TMP incorporation into DNA (cpm/reaction mixt.)	% of control
None	1223	100
Daunomycin	127	10.3
Adriamycin	106	8.7
Dihydro daunomycin	151	12.3
N-guanidine daunomycin	322	26.3
N-acetyl daunomycin	1412	115.6

[a]Antibiotic concentration = 5 µg/reaction mixt. (0.25 ml). Reaction conditions are shown under Table 2.

The reactions catalyzed by poly (dA-dT) is highly sensitive to the action of daunomycin and its derivatives. Even under these conditions daunomycin and adriamycin are most effective, and the N-acetyl derivative is completely inactive. It is interesting to notethat poly (dA-dt)-dependent reaction is much more sensitive to antibiotics than the endogenous reaction (see Table 6).

The results show that the inhibiting activity of daunomycin requires specific structural parameters. Thus substitutions in the amino sugar moiety, especially N-acetylation, inhibit its activity against oncogenic viruses and influence its inhibitory action of the DNA polymerases of RNA tumor viruses.

To evaluate the therapeutic efficacy of these compounds, the activity of daunomycin and adriamycin on DNA-polymerase from various sources was measured. These studies were carried out using a constant concentration of the template poly (dA-dT) in DNA-polymerase reactions, catalyzed by preparations from MSV(M), rat liver and *M. lysodeiktícus*. The MSV-DNA-polymerase was found to be most sensitive to both the antibiotics.

The antitumor activity of some derivatives of daunomycin at the amino and methyl ketone functions has been studied by YAMAMOTO et al. (1972). Their studies were carried out mainly on leukemia 1210 in mice. At 2 mg/kg dose, the N-piperidinoimine derivative was found to have the same antitumor activity as daunomycin; other derivatives were not active at this dose level. The N-acetyl derivative was found to possess only a little antitumor activity, but displayed no acute toxicity even at very high doses. According to our experience, the N-acetyl derivative was in most of the cases ineffective against tumor growth (Table 8).

As follows from Table 8 adriamycin inhibits the growth of Ridgeway Osteo-sarcoma (ROS) in mouse to more than 80 %. Under similar experimental conditions one finds a slight inhibitory effect by daunomycin and its dihydro derivative. However, the derivatives

Table 8. Inhibition of growth of some transplanted tumors by daunomycin derivatives. Tumor suspensions were incubated with 50 μg/ml of the antibiotic at 37°C for 1 h

Antibiotic	10 animals transplanted with: Ridgeway-Osteo-Sarcoma (Mouse) Tumor weight (g)	L-1210 (Mouse) Ascites (ml)	SV 40 (Hamster) Tumor weight (g)
None	7.0 ± 2.4	0.78 ± 0.16	47 ± 12
Adriamycin	1.0 ± 0.3	0.00	0.0
Daunomycin	5.0 ± 2.3	0.00	0.0
Dihydro daunomycin	5.3 ± 2.0	0.53 ± 0.2	0.0
N-guanidine-acetamine daunomycin	7.8 ± 2.8	0.63 ± 0.1	18.0 ± 8
N-acetyl daunomycin	9.8 ± 1.4	0.64 ± 0.07	46 ± 8

Table 9. Effect of daunomycin and its derivatives on the thermal transition temperature (T_m) and viscosity of DNA

Antibiotic	T_m [a]	$\frac{(\eta)r = 0.1}{(\eta)r = 0}$ [b]
None	70.5	1.00
Adriamycin	85.5	1.75
Daunomycin	83.9	1.92
Dihydro daunomycin		1.65
N-guanidine acetamide daunomycin	77.1	1.30
N-acetyl daunomycin	71.5	1.24

[a] All experiments were carried out in 0.01 M tris-HCl buffer (pH 7.0) at a ratio of antibiotic to DNA-P (r) of 0.1. DNA concentration in all the experiments was 1 x 10^{-4}M.
[b] Ratio of intrinsic viscosity of antibiotic-DNA complex (r = 0.1) to that of DNA alone. Conditions of viscosity measurements: 20°C, 0.1M tris-HCl buffer (pH 7.0). r is the ratio of bound antibiotic to total DNA-P (CHANDRA et al., 1972a).

with substitutions in the aminosugar moiety, N-guanidino-acetamide-daunomycin and N-acetyl-daunomycin, are completely ineffective. In L-1210 and SV 40 systems tumor suspensions preincubated with adriamycin or daunomycin failed to grow in their hosts. The dihydro derivative was not effective in L-1210, however a total inhibition was achieved in case of SV 40. Unexpectedly, the N-guanidino-acetamide derivative showed a significant activity against SV 40. However, the N-acetyl derivative was ineffective against all types of tumors studied by us.

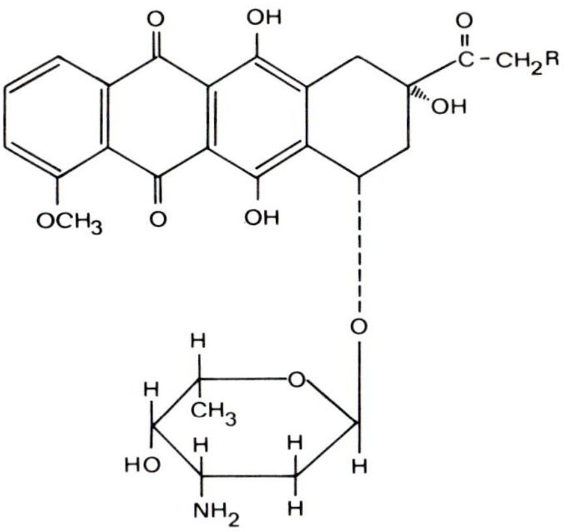

Fig. 10. Chemical structure of adriamycin derivatives

Daunomycin	R = H
Adriamycin	R = OH
Adriamycin 14 - Acetate	R = $OCOCH_3$
Adriamycin 14 - Octanoate	R = $OCO(CH_2)_6CH_3$

The structure-activity relationship exhibited in enzymatic and biological systems was also observed with respect to their physicochemical interaction with DNA (Table 9). Daunomycin causes a large increase in the thermal transition temperature (T_m) of calf thymus DNA. This effect depends on the ratio antibiotic/DNA · P (r). The effect of daunomycin derivatives on the thermal transition of calf thymus DNA at r = 0.1 is shown in Table 9. Adriamycin was found to be most effective in stabilizing the secondary structure of DNA ($\Delta T_m = 15.3^{\circ}C$), whereas very little increase in T_m was observed for N-guanidine-acetamide daunomycin and N-acetyl daunomycin, the derivatives with substitutions in the amino sugar moiety. In attempting to obtain further information on the affinity of the compounds tested for DNA, we studied their effect on the viscosity of DNA. Table 9 shows the intrinsic viscosity of antibiotic-DNA complexes (r = 0.1) relative to intrinsic viscosity of DNA alone (r = 0). Under these conditions the daunomycin-DNA complex has the highest intrinsic viscosity, followed by adriamycin and dihydro daunomycin. Again we find only a moderate increase in intrinsic viscosity in the presence of compounds substituted at the amino sugar residue.

The fact that adriamycin is one of the most powerful cytostatic compounds among the derivatives tested so far, has led to the synthesis of some adriamycin derivatives in the laboratories of Farmitalia. We have carried out some experiments with adriamycin-14-acetate and adriamycin-14-octanoate (Fig. 10).

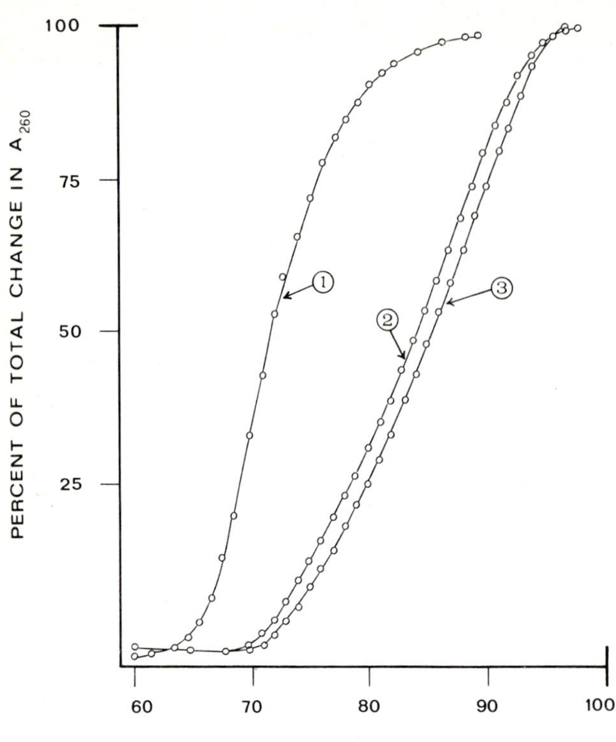

Fig. 11. Effect of adriamycin-14-acetate on the thermal transition temperature (T_m) of calf thymus DNA. Solvent ia 0.01 M Tris-HCl pH 7.0, and the DNA concentration in all experiments was 1×10^{-4}M; the ratio of the antibiotic to DNA-P (r) = 0.1. 1 = DNA alone; 2 = DNA and Adriamycin-14-acetate; 3 = DNA and Adriamycin

Effect of adriamycin and adriamycin-14-acetate on the thermal transition temperature (T_m) of calf thymus DNA at a drug to DNA-P ratio (r) of 0.1 is shown in Fig. 11. As follows from both the curves (numbered 2 and 3) the rise in T_m of calf thymus DNA in the presence of adriamycin and adriamycin-14-acetate exhibits a similar profile. However, under similar experimental conditions we observed large differences between the activities of adriamycin and adriamycin-14-octanoate (Fig. 12, curves 2 and 3).

The same structure-activity relationship could be seen in the poly (dA-dT)-catalyzed DNA-polymerase reaction in Friend leukemia virions (FLV). Adriamycin inhibits this reaction to about 70 % at 5×10^{-6}M, whereas octanoate derivative was almost ineffective; adriamycin-14-acetate, under similar conditions showed about 60 % inhibition.

It is interesting to note that the *in vitro* data on adriamycin derivatives presented above are not consistent to their cytostatic activity against FLV. The structure-activity relationship of adriamycin derivatives in the enzymatic system and with respect to their interaction with DNA is not observed in the biological system. The *in vitro* effect of adriamycin derivatives on the leukemogenic potential of cell-free spleen extracts prepared from spleen of FLV-treated mice is shown in Table 11.

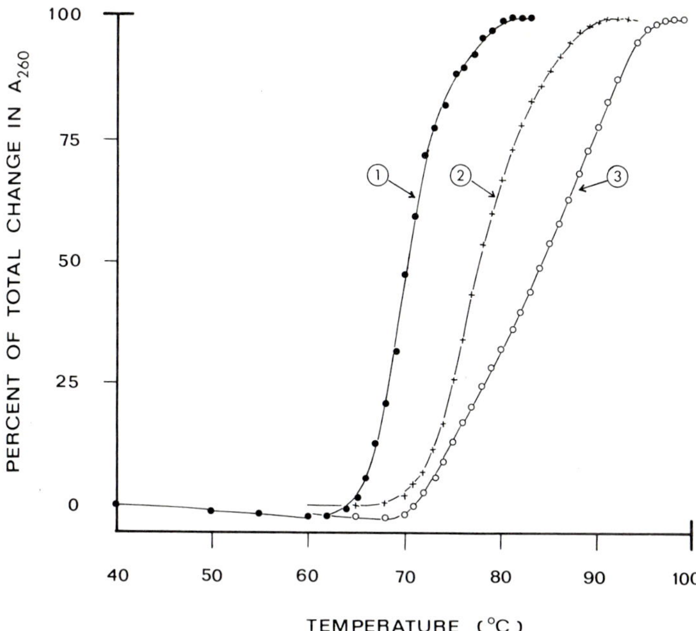

Fig. 12. Effect of adriamycin and adriamycin-14-octanoate on the thermal transition temperature (T_m) of calf thymus DNA. Experimental conditions are described under Fig. 11. 1 = DNA alone; 2 = DNA and Adriamycin-14-octanoate; 3 = DNA and Adriamycin

Table 10. Inhibition of FLV-DNA-polymerase reaction by adriamycin derivatives
Template: Poly (dA-dT)

System	Concentration of compound (1×10^{-5}M)	^{3}H-TMP incorporation into DNA (% of the control)
Control	-	100 (3789)[a]
Adriamycin	0.5	29
	1.0	6.2
Adr. acetate	0.5	39.0
	1.0	8.0
Adr. Octanoate	0.5	95.0
	4.0	47.8

[a]Designates cpm incorporated in 60 min per reaction mixture (0.25 ml). The reaction mixture contained 1 mcg of the template.

Table 11. Effect *in vitro* of adriamycin derivatives on the leukemogenic potential of cell-free spleen extracts prepared from spleens of FVL-infected mice (AKR-strain)

System	Spleen weight (g) 12th day post infection
control (infected)	3.75, 2.56, 1.78, 3.51 (2.89)[a]
Adriamycin (5 x 10^{-4}M)	2.12, 1.37, 0.93, 0.33 (1.19)
Adr. Acetate (5 x 10^{-4}M)	0.92, 1.10, 0.43, 1.98 (1.10)
Adr. Octanoate (5 x 10^{-4}M)	2.29, 1.88, 2.28, 0.45 (1.72)

Preincubation of cell-free extracts, with or without compound was carried out for 2 h at 37°C.
[a]Figures in brackets are the arithmetic mean of individual values.

As follows from the data of the Table, the octanoate derivative has a very good activity in inhibiting FLV-leukemogenesis. GOLDIN and JOHNSON (1975) have compared the activities of adriamycin and adriamycin-14-octanoate against Lewis Lung carcinoma, and found that both the compounds have a similar effect.

The differences in the biological activity of adriamycin derivatives and their *in vitro* effects on DNA structure (CHANDRA, 1975) indicate that these derivatives undergo some metabolic alteration leading to the formation of a derivative, which behaves similarly to adriamycin. It has been recently observed by LENAZ (L. LENAZ, Farmitalia, Milan, personal communication) that the octanoate derivative is converted *in vivo* to hydroxyadriamycin. This explains the significant activity of the octanoate derivative in biological systems, though its interaction with DNA under *in vitro* conditions is very moderate.

4. *Bis-basic Substituted Tricyclic Aromatic Compounds*

Bis-DEAE-fluorenone, also referred to as tilorone hydrochloride (non-proprietary name) is a preferred representative of this new class of antiviral compounds. It is a broad spectrum antiviral compound (KRUEGER and YOSHIMURA, 1970) with antitumor activity (ADAMSON, 1971; MUNSON et al., 1972; ROYE et al., 1971; RHEINS et al., 1971). MAYER and co-workers (MAYER and KRUEGER, 1970; MAYER and FINK, 1970; MAYER et al., 1972) have identified this compound as an interferon inducer and established a relationship with the antiviral activity. However, recently a lack of correlation between interferon induction and viral protection by tilorone hydrochloride has been reported (GIRON et al., 1972). The possibility that this compound may react directly with DNA was indicated by the cytogenetic studies of GREEN and WEST (1971). Tilorone was found to inhibit mitosis significantly at 3.0 µg/ml. Soon it was discovered by CHANDRA et al. (1972d, 1972e, 1972f) that tilorone does form molecular complexes with DNA and polydeoxynucleotides. Some of these studies will be described here.

Tilorone hydrochloride in water shows two absorption maxima, in the ultraviolet region around 271 nm, and in the visible region around 470 nm. Thus, the investigation of the long wavelenght band, where DNA and RNA do not absorb, should provide some evidence whether or not the chromophore of tilorone hydrochloride is involved in the binding process. There is a characteristic change in tilorone spectrum in the presence of DNA. In the presence of calf thymus DNA the visible absorption spectrum of tilorone hydrochloride is depressed and red shifted. This hypochromic effect of DNA on the absorption of tilorone chromophore is dependent on DNA concentration. The largest hypochromic effect is observed at 2 x 10^{-3}M DNA-P in a 4.25 x 10^{-4}M solution of tilorone hydrochloride. Further studies showed that DNA in its double helical state produces largest changes in the absorption spectrum of tilorone, whereas, the effect of single-stranded DNA is slightly weaker. In contrast, the yeast RNA exerts only a slight effect on the visible spectrum of tilorone hydrochloride. These data indicate a specificity of the tilorone chromophore towards DNA.

Further information on the binding of tilorone with DNA was derived by studying the thermal melting of the complex. In order to characterize the stability of DNA secondary structure in the presence of tilorone, temperature profiles were run at tilorone/DNA-P molar ratio of 1 : 5. Tilorone hydrochloride shows a large increase in the thermal transition temperature (T_m) of native DNA; the T_m of calf thymus DNA was raised from 71.6 to 85.2°C under these conditions.

Hypochromic effect of native DNA on the absorption of tilorone chromophore is very sensitive to magnesium ions. The effect of magnesium ions on tilorone binding to DNA was confirmed by density-gradient studies using ^{14}C-labeled tilorone hydrochloride.

These studies indicate that electrostatic forces contribute greatly to the binding process. The interaction between tilorone and DNA may, however, involve other kinds of forces. Tilorone forms a reversible complex with DNA, since the drug could be completely dissociated from a DNA-cellulose column. Interaction of apurinic and apyrimidinic DNA's with tilorone hydrochloride also gave spectral changes. However, only with the apyrimidinic DNA, the spectrum of the bound drug was similar to that found with native DNA. The absorption spectrum studies reflect the electronic environment of the molecule and do not give specific information about the type of interaction. The data which must be accounted for in considering a physical mode for the binding process can be derived from several different approaches. Hydrodynamic measurements on the DNA-drug complex are of interest, since LERMAN (1964, 1961) has established that an increase in the intrinsic viscosity of DNA and a decrease in the sedimentation coefficient of the polymer are two criteria for intercalation of ring systems between base pairs of a double-helical DNA.

The relationship between the intrinsic viscosity of DNA and the amount (r) of bound tilorone was studied (CHANDRA et al., 1972d). The intrinsic viscosity of the complex increases with r up to

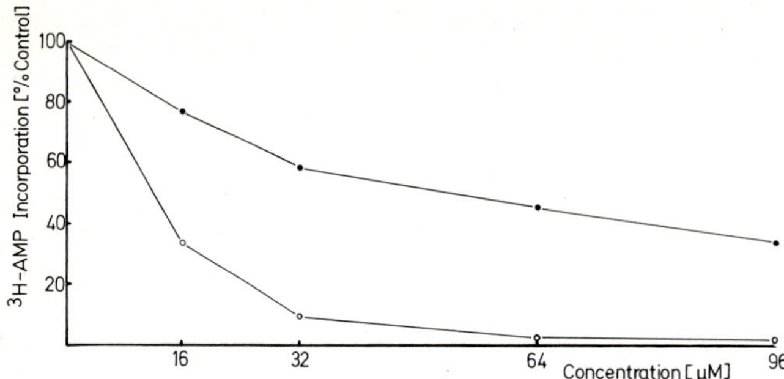

Fig. 13. Inhibition by tilorone of the RNA polymerase reaction catalyzed by the templates: Calf thymus ●—●—●; and poly (dA-dT) o—o—o. Regardless of the template, the reaction mixture (0.25 ml) contained 1 μg of the template. In the poly (dA-dT)-catalyzed reaction, GTP and CTP were omitted from the reaction mixture

a limiting value of about 0.05. The maximum relative enhancement of viscosity was about 1.7. In addition, at the same ionic strenght and at a ligand to DNA-P molar ratio of 0.1, the sedimentation rate of DNA was decreased to 78 % of the value in the absence of ligand.

These observation are consistent with an intercalative mode of binding in the interaction of tilorone hydrochloride with double-helical DNA. These results were not examined in an attempt to verify whether they agree with measurements of the length increase on sonicated DNA. For this reason, the intercalation model of the DNA complex remains tentative.

The interaction of tilorone hydrochloride with native DNA stabilizes the double helical structure of the macromolecule towards thermal denaturation. The effect of tilorone hydrochloride on the thermal denaturation of DNAs from various sources having different base composition has been studied (CHANDRA et al., 1972d). At a drug to DNA-P molar ratio of 0.21 the ΔT_m increased with increasing AT content of the DNA. This observation indicates that tilorone hydrochloride preferentially binds to the dAT portions of the DNA molecule. This hypothesis is confirmed by the strong effect of tilorone hydrochloride on the thermal transition temperature of poly d(A-T), $\Delta T_m = 29^{\circ}C$.

An intercalative mechanism for binding of a ligand to DNA is consistent with a stabilization of the double helix. Such a stabilization, however, does not constitute proof of intercalation. But, when considered with the evidence of the results reported here, showing increased viscosities and decreased sedimentation rate of DNA, one may conclude that the large increase of T_m points to an intercalative mode of binding.

The interaction of tilorone hydrochloride to DNA encouraged us to study the template activity of the complexes in DNA- and RNA-polymerase systems from *E. coli*. Both the activities were found to be inhibited by tilorone; the DNA-polymerase reaction being more sensitive towards tilorone. The inhibition of the RNA-

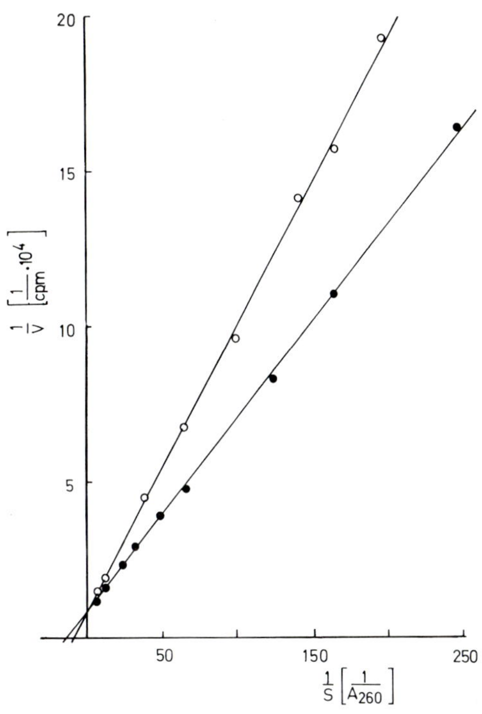

Fig. 14. Lineweaver-Burk plot of the effect of increasing concentrations of template DNA on the incorporation of ^{3}H-AMP into RNA by RNA polymerase of *E. coli* K-12, in the absence of tilorone (●–●), or in the presence of tilorone, 3,2 x 10^{-5}M (o–o). Abscissa: 1/S = (DNA template, A_{260} per reaction mixture)$^{-1}$. Ordinate: 1/V = (^{3}H-labeled AMP incorporated into RNA)$^{-1}$

polymerase reaction by tilorone was dependent on the A-T content of DNA-template (Fig. 13).

The template activity of poyl (dA-dT) (o–o) in the RNA polymerase reaction is distinctly more sensitive towards tilorone than that of DNA (●–●); particularly at low drug concentrations the poly (dA-dT)-catalyzed reaction is three times more sensitive than the DNA-catalyzed activity of RNA polymerase reaction.

The nature of the inhibitory response, as depicted in Fig. 13 indicates that these compounds compete for the binding sites on DNA. This is clearly shown in the Lineweaver Burk plot of the kinetic data obtained by measuring the RNA polymerase activity at various concentrations of template DNA (Fig. 14). Fig. 14 depicts the kinetic curves of the reactions in the absence of the inhibitor (●–●), and in the presence of 3.2 x 10^{-5}M tilorone (o–o). This shows that tilorone is a competitive inhibitor of DNA template activity in the RNA polymerase reaction.

The interaction of tilorone with DNA and synthetic polydeoxynucleotides can be influenced by modifying tilorone structure. Such studies are indeed important to elucidate the role of structural entities in their complex formation with DNA. Structure-activity relationship of tilorone derivatives (Fig. 15) has recently been studied in our laboratory (CHANDRA et al., 1974d; CHANDRA and WOLTERSDORF, 1974).

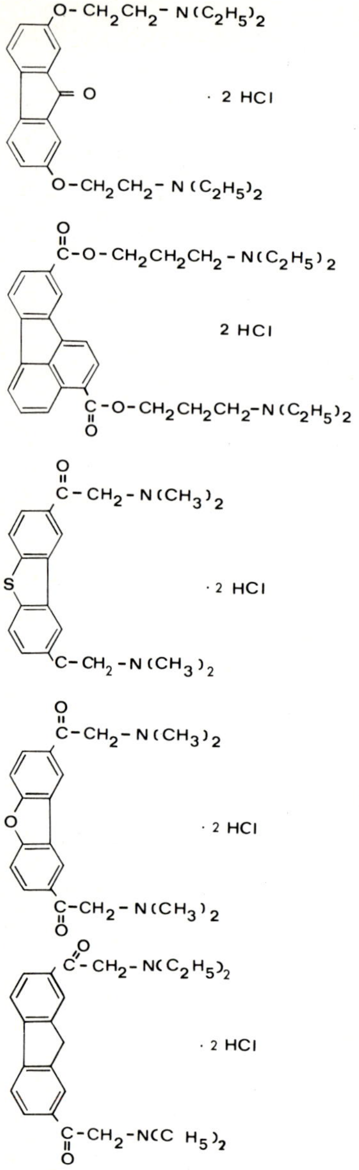

$O-CH_2CH_2-N(C_2H_5)_2$

$=O$ $\cdot$ 2 HCl

$O-CH_2CH_2-N(C_2H_5)_2$

DEAE-FLUORENONE

TILORONE

$C(=O)-O-CH_2CH_2CH_2-N(C_2H_5)_2$

2 HCl

$C(=O)-O-CH_2CH_2CH_2-N(C_2H_5)_2$

DEAP-FLUORANTHENE

RMI-9563 DA

$C(=O)-CH_2-N(CH_3)_2$

S $\cdot$ 2 HCl

$C-CH_2-N(CH_3)_2$

DMAA-DIBENZOTHIOPHEN

RMI-11877 DA

$C(=O)-CH_2-N(CH_3)_2$

O $\cdot$ 2 HCl

$C(=O)-CH_2-N(CH_3)_2$

DMAA-DIBENZOFURAN

RMI-11567 DA

$C(=O)-CH_2-N(C_2H_5)_2$

$\cdot$ 2 HCl

$C(=O)-CH_2-N(C\ H_5)_2$

DEAA-FLUORENE

RMI-11002

MEAA-FLUORENE

RMI-11829 A

monosubstituted RMI 11002

Fig. 15. Chemical structures of tilorone and cogeners

The effect of tilorone and cogeners (Fig. 15) on the DNA-dependent RNA polymerase reaction is shown in Fig. 16. The template activity of native DNA is strongly inhibited by DEAE-fluoranthene, showing an 80 % inhibition at a concentration 8 x 10^{-6}M. Other derivatives at this concentration do not show any significant inhibition of the template activity of DNA. However, at higher concentrations one observes a dose-dependent inhibition of DNA-template activity by DEAE-fluorenone, DMAA-dibenzothiophene,

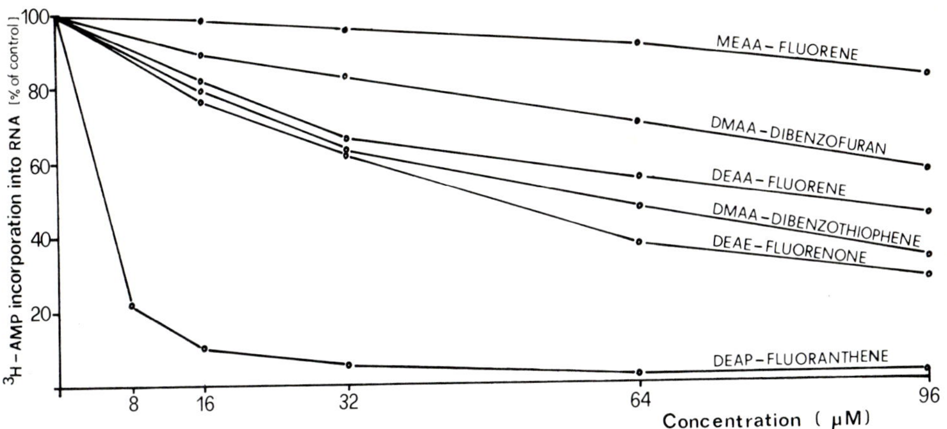

Fig. 16. Inhibition of DNA-dependent RNA polymerase reaction (*E. coli* K 12) by tilorone and cogeners

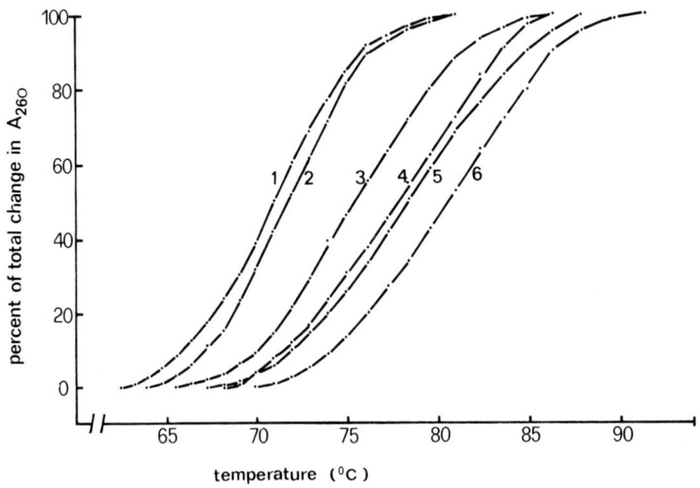

Fig. 17. Effect of tilorone and cogeners on the thermal transition temperature (T_m) of calf thymus DNA. Solvent is 0.01 M Tris-HCl pH 7.0 and the concentrations of DNA-P and cogeners are 5 x 10^{-5}M and 5 x 10^{-6}M, respectively. Curve 1 = DNA; 2 = DNA + MEAA-fluorene; 3 = DNA + DEAA-fluorene; 4 = DNA + DMAA-dibenzothiophene; 5 = DNA + DMAA-dibenzofuran and 6 = DNA + DEAE-fluorenone

DEAA-fluorene and DMAA-dibenzofuran. The monoethyl derivative, MEAA-fluorene does not show any activity even at higher concentrations.

The inhibition of DNA template activity by tilorone and its cogeners is strongly influenced by substitutions in the ring (e.g. thiophene, furan, etc.), as well as in the side chains. It was therefore, of interest to study whether, such substitutions influence their interaction to DNA. Fig. 17 depicts the melting curves of calf thymus DNA alone (curve 1), or in the presence of MEAA-fluorene (curve 2), DEAA-fluorene (curve 3), DMAA-dibenzothiophene (curve 4), DMAA-dibenzofuran (curve 5), and DEAE-fluorenone (curve 6). These studies were carried out at a drug/DNA-P ratio (*r*) of 0.1. Tilorone hydrochloride (DEAE-fluorenone) shows a large increase in the thermal transition temperature (T_m) of native DNA (curve 6). The cogener DEAE-fluoranthene, under these conditions, showed a very similar response (curve not shown). It is interesting to note that DMAA-dibenzofuran though, less active, than DMAA-dibenzothiophene and DEAA-fluorene in the RNA polymerase reaction, has a higher effect on the T_m of calf thymus DNA, than exhibited by these two derivatives.

The structure-activity relationship observed in the RNA polymerase reaction, is not strictly exhibited by the melting curves of DNA and cogener complexes. The latter studies were done in the absence of magnesium ions, whereas the RNA polymerase reaction requires magnesium ions for its enzymatic activity. The fact that compounds of tilorone type form chelates with Mg^{+2} may explain differences in the structure-activity relationships of cogeners in the RNA polymerase reaction, and their effects on the melting behavior of DNA. Thus, the dibenzofuran cogener, though less active than DEAA-fluorene in the RNA polymerase reaction, has a higher affinity for DNA than DEAA-fluorene. The dibenzofuran derivative should have a higher tendency for chelating with magnesium ions, which leads to its partial inactivation in the RNA polymerase reaction.

To measure the effect of various tilorone cogeners (Fig. 15) on the oncogenic activity of MSV(M), viral suspensions were incubated with 5 x 10^{-7} moles/ml of each compound at 37° for 1 h. In the control group, where no compound was used, virus was preincubated with the solvent. Tris-buffer, 0.01 M, pH 7.4, 0.2 ml of this mixture containing 1 x 10^{-7} moles of the drug, was injected intraperitoneally. The amount of compound introduced this way had no direct physiological effect in the host (unpublished results). The mortality and the survival period were significantly influenced by tilorone and two of its congeners, DEAP-fluoranthene and DEAA-fluorene (Table 12).

Viral suspensions, pretreated with MEAA-fluorene, were as active as the untreated viral suspensions. It is important to note that at the time of writing this report 30 percent of the animals in the tilorone and DEAP-fluoranthene groups and 20 percent of the animals in DEAA-fluorene group were still alive. Thus, the mean survival period in these groups depicts a mean of 6 - 7 animals only.

The effect of prior treatment of cell-free spleen extracts from FLV-infected mice with tilorone congeners to induce splenomegaly is shown in Table 12. The experimental conditions are described under Table 12 and in the text. Under these conditions, only tilorone and DEAP-fluoranthene showed a significant inhibition of splenomegaly induced by FLV. DEAA-fluorene showed only slight activity. It is interesting to note that MEAA-fluorene does not show any activity in this system.

Table 12. *In vitro* effect of tilorone and cogeners on the oncogenic activity of MSV(M) and leukemogenic potential[a] of cell-free spleen extracts from mice infected with FLV

Compound	MSV(M)-Oncogenesis No. of survivors per No. of infected animals	Survival time	FLV-Oncogenesis Spleen weight (g)[b]
Control	0/10	13.2 ± 1.4	1.34 ± 0.27
DEAE-fluorenone (tilorone)	3/10	24.3 ± 1.2	0.86 ± 0.15
DEAP-fluoranthene (RMI-9563 DA)[c]	3/10	24.4 ± 1.6	0.82 ± 0.17
DEAA-fluorene (RMI-11002)	2/10	20.9 ± 0.7	1.06 ± 0.13
MEAA-fluorene (RMI-11829 A)	0/10	14.2 ± 0.4	1.32 ± 0.11

[a]Determined by measurement of spleen weights, 12 days after FLV infection.
[b]Mean ± S.E. derived from ten animals in the control group, and five animals in each of the experimental groups, where *n* is the no. of survivors.
[c]Mean ± S.E. derived from ten animals in the control group, and five animals in each of the experimental groups.
Pharmaceutical code numbers of Merrell-National Laboratories, Ohio, U.S.A.
Abbreviations: MSV(M): Murine Sarcoma Virus (Moloney); FVL: Friend Leukemia Virus.

Since none of the compounds at the concentrations used showed a complete supression of splenomegaly, one would expect a residual viral activity in spleen extracts of mice, which received FLV suspensions preincubated with these compounds. Studies are now in progress to evaluate the leukemogenic activity of cell-free spleen extracts, prepared from mice inoculated with pretreated suspensions. WU et al. (1973) have carried out such studies with RLV and rifamycin derivatives. They reported that the inoculation of mice with inocula from mice infected with RLV pretreated with AF/ABDP and AF/DNF1 did not cause splenomegaly.

If the suppression of biological activity of RNA tumor viruses is due to a block of some molecular event(s) involved in oncogenesis, one would expect an inhibition of DNA polymerase by these compounds. That tilorone does inhibit the DNA-polymerase activity of oncornaviruses has been shown already by us (CHANDRA et al., 1972e). An attempt was made to correlate the biological response of various congeners with their inhibitory activity in the DNA-polymerase system on oncornaviruses. These studies were done using purified FLV, since in our hands this system showed a good endogenous activity.

The inhibition of the endogenous activity of FLV-DNA polymerase by tilorone and congeners is shown in Fig. 18. A maximum inhibition was obtained with DEAP-fluoranthene. The inhibitory responses of tilorone (DEAE-fluorenone), DMAA-dibenzothiophene and

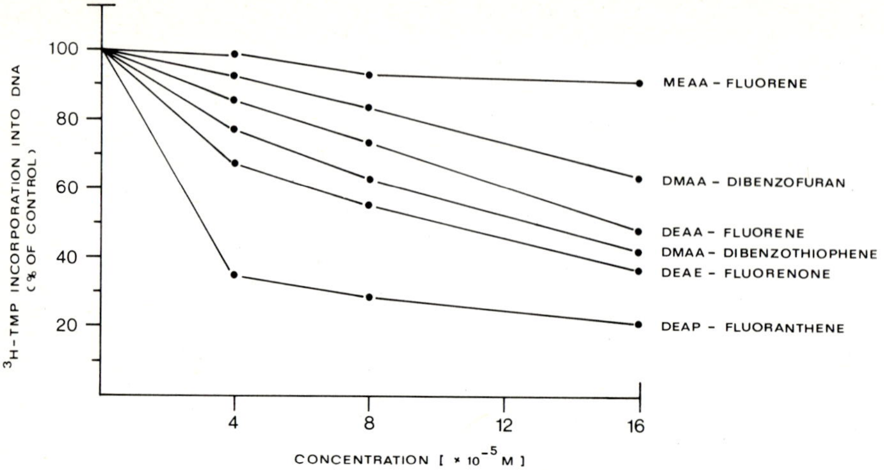

Fig. 18. Inhibition of endogenous RNase-sensitive FLV-DNA-polymerase activity by tilorone and congeners

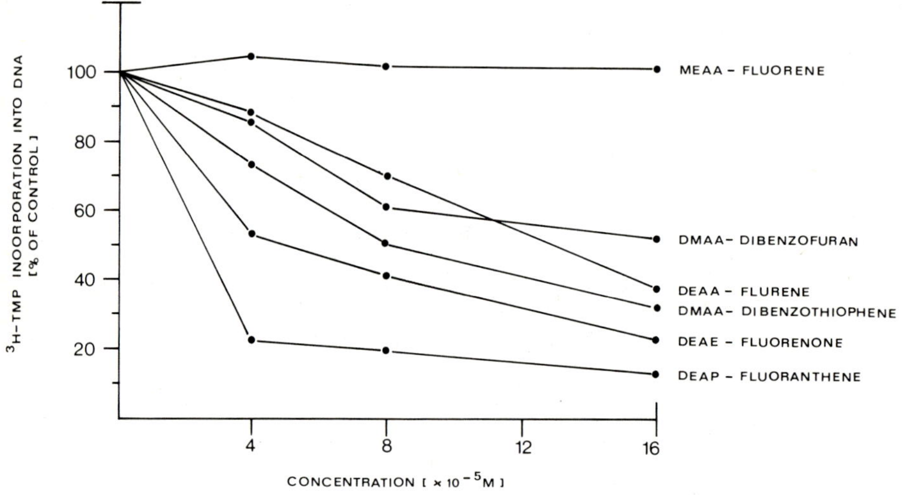

Fig. 19. Inhibition of the poly rA · $(dT)_{12}$-catalyzed FLV-DNA-polymerase activity by tilorone and congeners

DEAA-fluorene were of the same magnitude; whereas, DMAA-dibenzofuran showed a weak response. It is interesting to note that the monosubstituted congener MEAA-fluorene did not inhibit the endogenous reaction at any concentration.

A similar inhibitory response by the tilorone congeners was exhibited in the DNA-polymerase system of FLV, catalyzed by the template poly rA · $(dT)_{12}$, as shown in Fig. 19. The effect of tilorone and congeners on the FLV-DNA-polymerase reaction, catalyzed by poly (dA-dT), is shown in Fig. 20. This reaction was

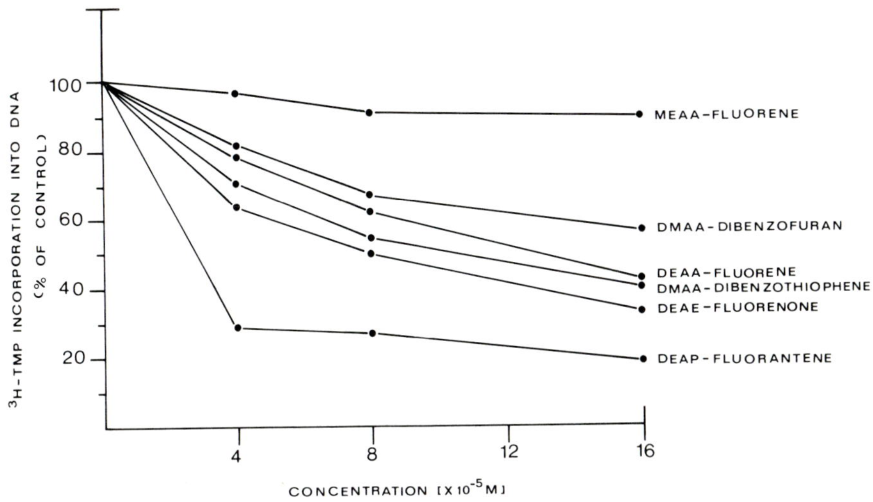

Fig. 20. Inhibition of the poly (dA-dT)-catalyzed FLV-DNA-polymerase activity by tilorone and congeners

more sensitive towards tilorone and its congeners than the endogenous, or poly rA · $(dT)_{12}$ -catalyzed reactions. This is in accordance to our findings on tilorone action, reported earlier (CHANDRA et al., 1972e).

The data reported in Table 12 and Figs. 18-20 shows that tilorone congeners which inhibit the DNA polymerases of the virus, suppress also the oncogenic activity of the virus. Since the poly (dA-dT)-catalyzed reaction of viral DNA-polymerase is most sensitive to these compounds, it is possible that the biological activity of these compounds is due to their interaction with the hybrid RNA-DNA (hy-DNA), single stranded DNA (ss-DNA), or the DNA-DNA duplex (ds-DNA). It was therefore, interesting to locate the site of action of tilorone in the viral DNA-polymerase system. It is still not clear whether a particular site or target in the DNA-polymerase system, other than the true RNA-dependent reaction, can be correlated with the biological role of the oncornaviruses. The key role of the RNA-directed reaction in *in vivo* leukemogenesis using purified enzyme and rifamycin derivatives, has been nicely demonstrated by WU et al. (1973).

We have conducted some model studies to analyze the products of the FLV-DNA-polymerase reaction under the influence of tilorone. The procedure we adopted was based on a recent report by KOTLER and BECKER (1972) on distamycin A, which has been shown to react with ss-DNA and ds-DNA (CHANDRA et al., 1970).

The product analysis of the DNA-polymerase reaction (FLV) in the absence and in the presence of tilorone (1 x 10^{-4} M) is depicted in Fig. 21. The products of the viral DNA-polymerase reaction were, under these conditions, eluted in three species. The first species to be eluted from the column contained ss-DNA, the second contained the RNA-DNA hybrid molecules (hy-DNA) and finally,

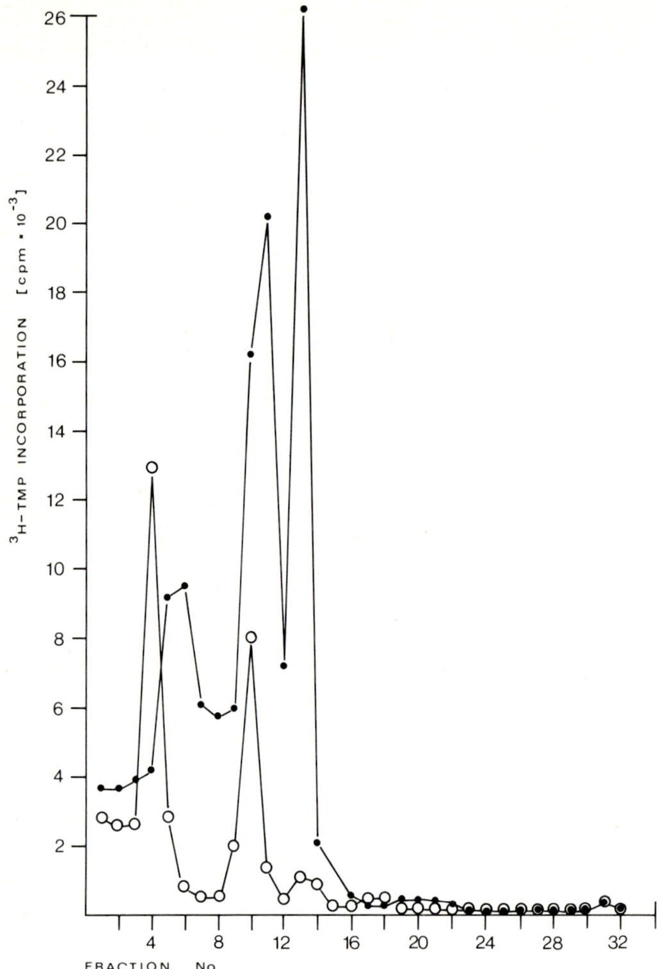

Fig. 21. Analysis of the DNA species synthesized by FLV-DNA polymerase by elution from hydroxylapatite column. Each column was filled with 1 g of hydroxylapatite and carefully washed with 0.05M sodium phosphate (approx. 50 ml). The columns were loaded with the reaction products, as described in text. The columns were washed with 0.05M sodium phosphate buffer, pH 6.8, until equilibrium was reached. Macromolecules were eluted from the columns by a linear gradient of sodium phosphate (0.05M - 0.4M). ●—●, DNA species synthesized in the absence of tilorone; o—o, DNA species synthesized in the presence of 10^{-4}M tilorone

the ds-DNA, eluted in the last species. Analysis of products synthesized in the presence of tilorone showed that the ss-DNA and the hybrid species, but not the ds-DNA species, were synthesized. This indicates that tilorone has a low affinity to viral RNA, but can block the synthesis of ds-DNA by interacting with ss-DNA or hy-DNA.

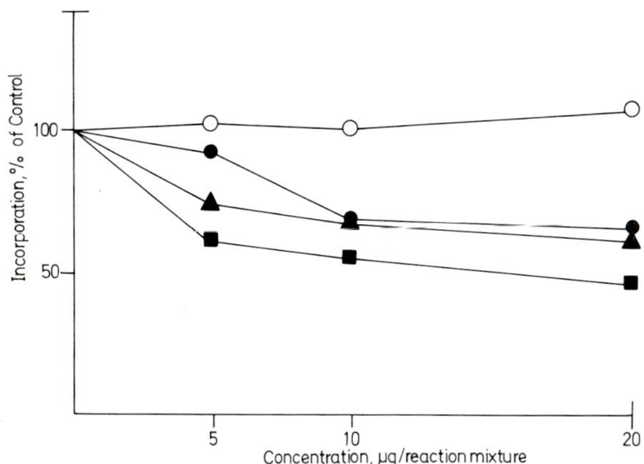

Fig. 22. Effects of polycytidylic acid Poly(C) and of three different thiolated polycytidylic acids (MPC) on the activity of the DNA polymerase of Friend leukemia virus, as measured by the incorporation of the radioactive precursor into DNA relative to the control, in endogenous reaction (i.e., with the viral RNA as template). Abscissa: Polynucleotide added, µg/reaction mixture: ○ Poly(C), ● MPC I (1.7 % - thiolated); ▲ MPC II (3.5 % - thiolated); ■ MPC III (8.7 % - thiolated). Ordinate: Incorporation of radioactive nucleotide into acid insoluble material, expresses as percent of the control (= 100 %; no polynucleotide added)

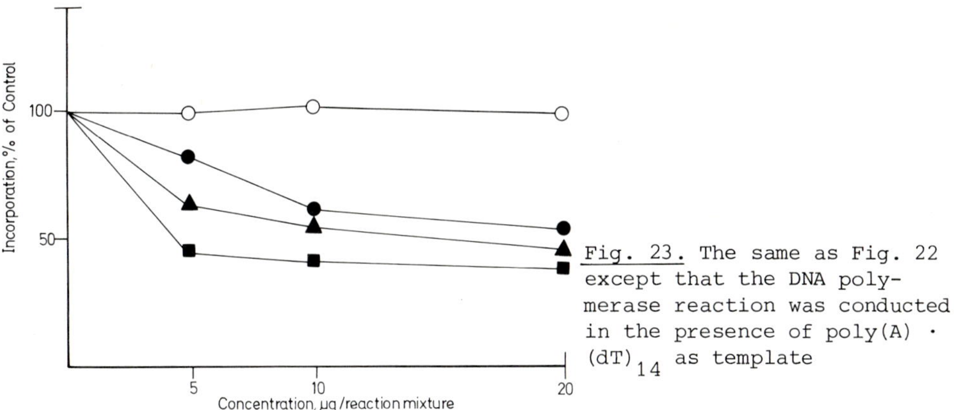

Fig. 23. The same as Fig. 22 except that the DNA polymerase reaction was conducted in the presence of poly(A) · $(dT)_{14}$ as template

5. Modified Nucleic Acid

Single-stranded polyribonucleotides are known to act as efficient templates for the viral DNA polymerases in the presence of the complimentary oligo-deoxyribonucleotide primer (Section 2 of this Chapter). Chemical modification of such templates would be expected to alter the interaction between the template and the viral enzyme (CHANDRA and BARDOS, 1972; CHANDRA et al., 1974a, CHANDRA, 1974b; PITHA et al., 1973; ERICKSON and GROSCH, 1974). This appears to be a very useful approach for designing specific

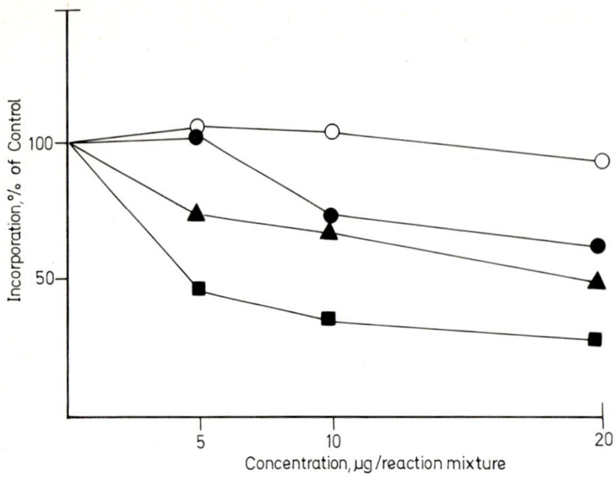

Fig. 24. The same as Fig. 22 except that the DNA polymerase reaction was conducted in the presence of poly[d(A-T)] as template

inhibitors of viral DNA polymerases, which might find application in the chemotherapy of cancer (CHANDRA et al., 1974c).

In a preliminary communication (CHANDRA and BARDOS, 1972), we reported the inhibitory effect of a partially thiolated polycytidylic acid (MPC) on the DNA-polymerases from RNA tumor viruses. The initial results indicated that chemically modified polynucleotide template analogues may serve as effective and potentially selective inhibitors of DNA synthesis in the presence of various templates. It was suggested that the inhibitory effects of modified polynucleotides may depend upon their structural similarities to the functional templates with which they presumably compete for their binding site(s) on polymerases. In the following pages the selectivity and mechanism of this inhibition is described.

Fig. 22-24 show the effects of poly(C) and three different partially thiolated poly(C) preparations (MPC I to III) on the DNA polymerase from Friend leukemia virus, using the endogenous viral nucleic acid, poly(A) · $(dT)_{14}$, or poly [d(A-T)], respectively, as the template. It is apparent that poly(C) shows no activity, while the MPC samples inhibit the enzyme in direct relation to their relative extent of thiolation. Thus, at 20 µg per reaction mixture concentration, MPC I in which 1.7 % of the total cytosine bases are thiolated, shows about 30 % inhibition of the endogenous reaction, while MPC III which contains the highest ratio (8.6 %) of 5-mercaptocytidylate untis, produces 50 % inhibition, and the activity of the 3.5 %-thiolated MPC II falls between the former two values (Fig. 22). Essentially analogous results were obtained when the "reverse transcriptase" template poly(A) · $(dT)_{14}$ was used (Fig. 23). The difference between the activities of the three MPC samples was larger in the case of the poly [d(A-T)] template (Fig. 24) where MPC III gave 55 % inhibition at 5 µg, and 70 % at 20 µg/reaction mixture concentration, while the corresponding results for MPC I were 0 and 30 %, respectively.

The greater inhibitory activity of MPC III against the poly [d(A-T)] template, as compared to the poly(A) · $(dT)_{14}$ template,

Table 13. Effects of the soluble RNA fraction (tRNA) from Ehrlich ascites cells, before and after modification on DNA polymerase of Friend leukemia virions

System	% Incorporation of Control		
	Endogenous RNA	Poly(A)·$(dT)_{14}$	Poly [d(A-T)]
Polynucleotide added, µg/reaction mixture			
Non-thiolated tRNA			
10	91	115	97
20	87	121	96
40	85	126	81
3.3 % - thiolated tRNA			
10	50	34	22
20	45	22	15
40	36	16	12

is consistent with our earlier results obtained for the FLV-DNA-polymerase (CHANDRA and BARDOS, 1972). It should be noted that in the case of MSV (Moloney) DNA polymerase, MPC showed, in contrast, significantly more activity against the poly(A) · $(dT)_{14}$ template (CHANDRA and BARDOS, 1972).

The inhibitory effects of various natural nucleic acid isolates from Ehrlich ascites cells, before as well as after partial thiolation, on the DNA polymerase of FLV in three different template-primed reactions are given in Tables 13-15. The effect of partial thiolation on the activity of a polynucleotide is most dramatically demonstrated by the results obtained for the soluble RNA fraction from Ehrlich ascites cells, before and after thiolation, respectively, using either the endogenous RNA, the poly(A) · $(dT)_{14}$, or the poly [d(A-T)] template (see Table 13). It should be pointed out that the thiolated soluble RNA contained only one mole-unit of 5-mercaptopyrimidine nucleotide per average molecular weight. In the case of the ribosomal RNA, the unmodified nucleic acid showed significant stimulation of the viral DNA polymerase in the presence of the poly(A) · $(dT)_{14}$ template, and essentially no activity when poly [d(A-T)] or, the endogenous RNA was used as template. In comparison, the 3.6 %-thiolated ribosomal RNA caused significant inhibition of this enzyme in each of the three assays (Table 14). In contrast to the RNA isolates, the unmodified DNA from Ehrlich ascites cells caused significant inhibition of the viral DNA polymerase using either of the three different templates; however, the corresponding 3.9 %-thiolated DNA (which contained one 5-mercapto-2'-deoxytidylate per approximately 100 nucleotides) showed about 8 - 10 times higher inhibitory activity than the unmodified DNA, in each case (Table 15).

Table 14. Effects of rRNA from Ehrlich ascites cells, before and after modification, on the DNA polymerase of Friend leukemia virions

System	% Incorporation of Control		
	Endogenous Reaction	Poly(A)·$(dT)_{14}$	Poly d(A-T)
Polynucleotide added, µg/reaction mixture			
Non-thiolated rRNA			
10	84	111	89
20	82	122	89
40	79	170	88
3.6 % - thiolated rRNA			
10	86	98	69
20	65	75	51
40	47	58	31

Table 15. Effects of DNA from Ehrlich ascites cells, before and after modification, on DNA polymerases of Friend leukemia virions

System	% Incorporation of Control		
	Endogenous RNA	Poly(A)·$(dT)_{14}$	Poly d(A-T)
Polynucleotides added, µg/reaction mixture			
Non-thiolated DNA			
10	80	63	44
20	76	51	27
40	53	37	18
3.9 % - thiolated DNA			
10	35	26	12
20	30	23	10
40	26	16	8

In order to determine the selectivities and mode of action of the partially thiolated polynucleotides, further studies are being conducted using DNA polymerase from different sources.

Table 16 shows the effects of the three MPC samples I - II (see above) and of the unmodified poly(C) on the DNA polymerase of

Table 16. Effect of polycytidylic acid and partially thiolated polcytidylic acid (MPC) on the DNA polymerase[a] of *E. coli* - K_{12} in the presence of denaturated DNA

Compound added 5 µg/reaction mixture	^{3}H-Labeled dAMP Incorporation into DNA (cpm/reaction mixture)	% of control
None	575	100
Poly(C)	608	106
MPC[b] I	572	99.5
MPC II	554	96
MPC III	591	103

[a]DNA-polymerase was isolated and purified by the procedure of RICHARDSON (1966).
[b]MPC-I, II and III are partially thiolated polycytidylic acids containing 1.7 %, 3,5 % and 8.6 % 5-mercaptocytidylate units, respectively.

E. coli - K_{12}, using denatured calf thymus DNA as template. All of the polynucleotides tested were virtually inactive in this assay system.

It is interesting that the DNA polymerase of *E. coli* - K_{12} is not inhibited by MPC. The *E. coli* enzyme has a single sulfhydryl group which is "located in a relatively exposed site on the surface of the molecule, remote from the active site" (JOVIN et al., 1969; KORNBERG, 1969). Thus, inhibition due to irreversible binding of MPC would not be expected to occur. The lack of reversible inhibition by MPC might be explained, possibly, by a "shielding action" of the non-essential "exposed" sulfhydryl group which could bind the MPC outside of the active site. Alternatively, the denatured DNA used as template in this assay might bind to the enzyme with a much lower dissociation constant than MCP, in which case it could not be effectively displaced by the latter in a competitive manner.

The mode of inhibition of viral DNA synthesis by MPC was further investigated by the product analysis of DNA-polymerase reaction in the absence, or in the presence of MPC, as described earlier (see Section III.B.4). The reaction mixtures were dissolved with Na-dodecyl sulfate (1 %, w/w, final concentration), loaded on a hydroxylapatite column (1 g, Bio-Rad Lab., Munich), eluted with a Na-phosphate gradient (0.05-0.4M), collected into about 40 tubes (total vol. 100 ml) and the TCA insoluble radioactivity was collected on GF/C filters from Whatman, and counted in a liquid scintillation counter.

As follows from Fig. 25, in the presence of MPC there is an over-all inhibition of ^{3}H-dTMP incorporation, indicating that the formation of all the 3 DNA species is blocked. This is to

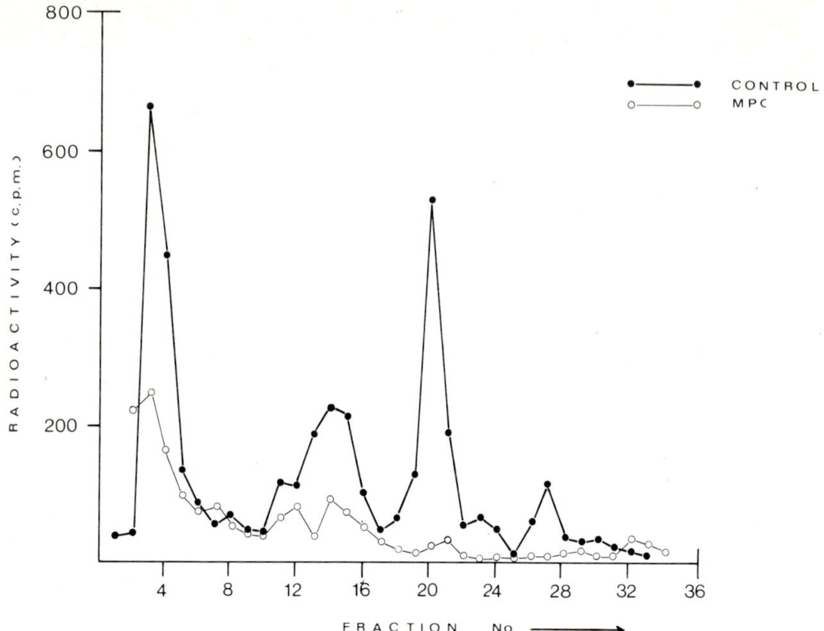

Fig. 25. Analysis of the DNA species synthesized by FLV-DNA polymerase by elution from hydroxylapatite column. Each column was filled with 0.05M sodium phosphate (approx. 50 ml). The columns were loaded with the reaction products, as described in text. The columns were washed with 0.05M sodium phosphate buffer, pH 6.8, until equilibrium reached. Macromolecules were eluted from the column by a linear gradient of sodium phosphate (0.05 - 0.4M). The first species to be eluted from the column contained ss-DNA, the second contained hy-DNA and finally, the ds-DNA eluted in the last species. o–o DNA species synthesized in the absence of MPC. o–o DNA species synthesized in the presence of MPC (20 μg/reaction mixture)

be expected since the inhibitor binds to the enzyme. This has been confirmed by the ultracentrifugation studies in which the binding of 35_S-labeled MPC to a purified FLV enzyme fraction was investigated (manuscript in preparation). However, these experiments do not rule out the possibility that some heteropolymeric regions of the functional endogenous template might undergo hydrogen-bonding with the added inhibitor. This would mean a bifunctional mode of MPC activity in the viral DNA-polymerase reaction.

In view of the fact that all the oncornaviral DNA polymerases examined so far do require a primer template-like double-stranded secondary structure for the initiation of DNA synthesis, it is no surprise that single-stranded synthetic polynucleotides (unprimed templates) can act as inhibitors of the polymerization reaction. This, presumably, is due to hydrogen bonding of the base sequences between the added polymer and the functional template. Thus, the specificity of inhibition by such polymers is not limited to the viral enzyme system only. On the other hand, minor modifications in the chemical composition of synthetic polynucleotides might be useful to develop inhibitors, which interact directly with the DNA-polymerase. Of more practical value will be such polymers which interact with the enzyme, but fail to be transcribed, i.e. they function as "dead template" for the enzyme. The recent data from our laboratory indicate that the partially thiolated polyC (MPC) is functioning as a dead template in the DNA-polymerase system of FLV. It is interesting that though MPC acts as dead template in the DNA-polymerase system, its messenger-coding properties in cell-free protein-synthesizing systems are similar to those of polycytidylic acid (VAZQUEZ and CHANDRA, unpublished results). Synthesis of such template analogs may lead to the development of more selective inhibitors of reverse transcriptase of oncornaviruses.

Preliminary experiments in our laboratory have shown that MPC III inhibits the leukemogenic potential of cell-free spleen extracts from FLV-infected mice, measured on the basis of spleen weight. Further studies are in progress to evaluate the dose-response relationship under *vitro-vivo* and *in vivo* conditions.

6. Other Inhibitors

Reverse transcriptase inhibitors other than reported here include ethidium bromide (HIRSCHMAN, 1971; MÜLLER et al., 1971; FRIDLENDER and WEISSBACH, 1971); several cations (Ag^{+}, Co^{++}, Zn^{++}, Cd^{++}, and Ni^{++}) (LEVINSON et al., 1973); thiosemicarbazones (LEVINSON et al., 1973); elenolates (HIRSCHMAN, 1972); Ara-CTP (TOUMINEN and KENNEY, 1972; MÜLLER et al., 1972); cactinomycins (APPLE, 1973); heparin (CHANDRA and GÖTZ, 1974); pyran copolymer (PAPAS et al., 1973); Streptonigrin (CHIRIGOS et al., 1973); Silicotungstate (HAAPALA et al., 1973); and several other compounds.

IV. Conclusion and Future Prospects

Soon after the discovery of DNA polymerase in virions of RNA tumor viruses, a great deal of hope was expressed that the discovery might lead to resolving the possibility of the involvement of oncorna viruses in an inapparent form in "spontaneous" or chemically induced tumors, especially in man. So far, there has been some evidence to support that RNA tumor viruses are related to human neoplasia.

The presence of enzymes which catalyze RNA-directed DNA synthesis in virus-like particles, or neoplastic cells of human origin described here, do not necessarily confirm the role of RNA viruses in human neoplasia. However, these discoveries lead to a possibility that studies with compounds, which are potent and specific inhibitors of such enzyme(s) may be useful to design antitumor compounds of future interest.

Acknowledgements

The authors wish to acknowledge the generous gifts of distamycin and daunomycin derivatives from Farmitalia Research Institute, Milan; and tilorone cogeners from Merrel-National Labs., Cincinnati, Ohio. We also acknowledge the valuable contributions to these studies made by Miss R. THORBECK, Miss S. HAUSMANN, Mrs. A. GÖTZ, Dr. F. ZUNINO, Dr. A.M. CASAZZA, and Dr. D. GERICKE. We thank Drs. F.E. HAHN (Washington), R.C. GALLO (Bethesda), and R.F. KRUEGER (Cincinnati) for sharing with us results of their studies prior to publication as well as providing valuable suggestions on several aspects of our work. This research was supported by the Deutsche Forschungsgemeinschaft (L.K.S.), Hermann-Schlosser-Stiftung (U.E., and G.W.), Berta-Heilscher-Vermächtnis, and Kind-Philipp-Stiftung.

References

AARONSON, S.A., PARKS, W.P., SCOLNICK, E.M., TODARO, G.J.: Antibody to the RNA-dependent DNA polymerase of mammalian C-type RNA tumor viruses. Proc. Nat. Acad. Sci. (Wash.) 68, 920 (1971).

ABRELL, J.W., GALLO, R.C.: Purification, characterization and comparison of the DNA polymerases from two primate RNA tumor viruses. J. Virol. 12, 431 (1973).

ADAMSON, R.H.: Antitumor activity of tilorone hydrochloride against some rodent tumors. J. Nat. Cancer Inst. 46, 431 (1971).

APPLE, M.A.: Reverse transcription and its inhibitors. Ann. Rep. Med. Chem. 8, 251 (1973).

ARCAMONE, F., PENCO, S., OREZZI, P., NICOLELLA, V., PIRELLI, A.: Structure and synthesis of distamycin A. Nature (Lond.) 203, 1064 (1964).

ARCAMONE, F., NICOLELLA, V., PENCO, S., REDAELLI, S.: Sintesi di analoghi con diverso numero di resti dell'acido 1-metil-4-amminopyrrolo-2-carbossilico. Gazetta Chim. Ital. 99, 632 (1969a).

ARCAMONE, F., PENCO, S., DELLE MONACHE, F.: Sintesi di analoghi contenenti modificazioni nelle catene laterali. Gazetta Chim. Ital. 99, 620 (1969b).

AULD, D.S., KAWAGUCHI, H., LIVINGSTON, D.M., VALLEE, B.L.: RNA-dependent DNA polymerase (reverse transcriptase) from avial myeloblastosis virus: a zinc metalloenzyme. Proc. Nat. Acad. Sci. (Wash.) 71, 2091 (1974).

BADER, J.P.: The role of DNA in the synthesis of RSV. J. Virol. 22, 462 (1964).

BADER, J.P.: Metabolic requirements for injection by RSV. I. The transient requirement for DNA synthesis. J. Virol. 29, 44 (1966).

BALDUZZI, P., MORGAN, H.R.: The mechanism of oncogenic transformation by Rouse Sarcom Virus. I. Intracellular inactivation of cell-transforming ability of RSV by 5-bromodeoxyuridine and light. J. Virol. 5, 470 (1970).

BALTIMORE, D.: RNA-dependent DNA polymerase in virions of RNA tumor viruses. Nature (Lond.) 226, 1209 (1970).

BALTIMORE, D., HUANG, A.S., STAMPFER, S.M.: RNA synthesis of vesicular stomatitis virus. II. RNA-polymerase in the virion. Proc. Nat. Acad. Sci. (Wash.) 66, 572 (1970).
BALTIMORE, D., SMOLER, D.: In: Nucleic acid protein interactions and nucleic acid synthesis in viral infection. Miama Winter Symposium (Eds. D.W. RIBBON, J.F. WOESSNER, J. SCHULTZ). Vol. 2, p. 328. Amsterdam: North-Holland Publ. 1971a.
BALTIMORE, D., SMOLER, D.: Primer requirement and template specificity of the DNA polymerase of RNA tumor viruses. Proc. Nat. Acad. Sci. (Wash.) 68, 1507 (1971b).
BALTIMORE, D., SMOLER, D.F.: Association of an endoribonuclease with AMV DNA-polymerase. J. Biol. Chem. 247, 7282 (1972).
BAUER, H., SCHÄFER, W.: Group-specific antigen of avian leukosis virus. Virology 29, 494 (1966).
BISHOP, J.M., FARAS, A.J., GARAPIN, A.C., GOODMAN, H.M., LEVINSON, W.E., STAVNEZER, J., TAYLOR, J.M., VARMUS, H.E.: Characteristics of the transcription of RNA by the DNA polymerase of Rouse Sarcoma Virus. In: DNA Synthesis in Vitro. (Eds. R.D. WILLS, R.B. INMAN), p. 341. Lancaster: Eng. Med. and Tech. Publ. Co., Ltd. 1973
BOETTIGER, D., TEMIN, H.M.: Light inactivation of focus formation by chicken embryo fibroblasts infected with avian sarcoma virus in the presence of 5-bromodeoxyuridine. Nature (Lond.) 228, 622 (1970).
BROCKMAN, W.W., CARTER, W., LI, L.H., REUSSER, F., NICHOL, F.R.: The streptovaricins inhibit RNA-dependent DNA polymerase present in an oncogenic RNA virus. Nature (Lond.) 230, 249 (1971).
CARTER, W., BROCKMAN, W.W., BORDEN, E.: Streptovaricins inhibit focus formation by MSV (MLV) complex. Nature New Biology 232, 212 (1971).
CARTER, W., BORDEN, E., BROCKMAN, W., BYRD, D., LIGON, W., ANTOSZ, F., RINEHART, K.L.: Reported at XXV Symp. on Fundamental Cancer Res., Huston, Texas (1972).
CASAZZA, A.M., FIORETTI, A., SANFILIPPO, A., VERINI, M.A.: 7th Int. Cong. Chemother., Prague, Aug 23-28, Abst A-5/41, Vol I (1971).
CHANDRA, P.: Molecular approaches for designing antiviral and antitumor compounds. In: Topics in Current Chemistry, vol. 52, p. 99. Berlin-Heidelberg-New York: Springer 1974a.
CHANDRA, P.: Inhibition of DNA-polymerases from oncornaviruses by modified nucleic acids. In: Modern Trends in Human Leukemia (Eds R. NETH, R.C. GALLO, S. SPIEGELMAN, F. STOHLMAN), p. 216. Munich: J.F. Lehmanns Verlag 1974b.
CHANDRA, P.: The relationship of chemical structure to biochemical activity of daunomycin, adriamycin and some structural analogues: Effects on viral DNA polymerases from ancorna viruses. In: Ergebnisse der Adriamycin-Therapie (Eds. M. GHIONE, J. FETZER, H. MAIER), p. 40. Berlin-Heidelberg-New York: Springer 1975.
CHANDRA, P., BARDOS, T.J.: Inhibition of DNA polymerases from RNA tumor viruses by novel template analogues: partially thiolated polycytidylic acid. Res. Commun. Chem. Path. and Pharmacol. 4, 615 (1972).
CHANDRA, P., DI MARCO, A., ZUNINO, F., CASAZZA, A.M., GERICKE, D., GUILIANI, F., SORANZO, C., THORBECK, R., GÖTZ, A., ARCAMONE, F., GHIONE, M.: The role of molecular structure on the inhibition of DNA-polymerases from RNA tumor viruses, viral multiplication and tumor growth by some antitumor antibiotics. Naturwissenschaften 59, 448 (1972a).

CHANDRA, P., EBENER, U., BARDOS, T.J., CHAKRABARTI, P., HO, Y.K., MIKULSKI, A.J., ZSINDELY, A.: Polynucleotides containing 5-mercapto substituted pyrimidines. Ann. N.Y. Acad. Sci. (in press) (1974a).
CHANDRA, P., GERICKE, D., ZUNINO, F., THORBECK, R.: Molekular Grundlagen viral bedingter Tumoren. Klin. Wschr. 51, 781 (1973).
CHANDRA, P., GÖTZ, A.: Inhibition of viral DNA polymerases by heparin. Manuscript in preparation (1974).
CHANDRA, P., GÖTZ, A., WACKER, A., VERINI, M.A., CASAZZA, A.M., FIORETTI, A., ARCAMONE, F., GHIONE, M.: Some structural requirements for the antibiotic action of distamycins. FEBS Letters 16, 249 (1971).
CHANDRA, P., GÖTZ, A., WACKER, A., VERINI, M.A., CASAZZA, A.M., FIORETTI, A., ARCAMONE, F., GHIONE, M.: Some structural requirements for the antibiotic action of distamycins. II. Structural modification of the side chains in distamycin A molecule. FEBS Letters 19, 327 (1972b).
CHANDRA, P., GÖTZ, A., WACKER, A., ZUNINO, F., DI MARCO, A., VERINI, M.A., CASAZZA, A.M., FIORETTI, A., ARCAMONE, F., GHIONE, M.: Soem structural requirements for the antibiotic action of distamycins. III. Possible interaction of formyl group of distamycin side chain with adenine. Hoppe-Seyler's Z. Physiol. Chem. 353, 393 (1972c).
CHANDRA, P., HAUSMANN, S., METZ, A.: Inhibition of DNA and RNA-polymerases of oncornaviruses and of bacteria by N-formyl-glycyl and N-formyl-alanyl derivatives of distamycin A. Pharmacol. Res. Commun. 6, 311 (1974b).
CHANDRA, P., KORNHUBER, B., GERICKE, D., GÖTZ, A., EBENER, U.: Hemmung der viralen Reverse-transkriptase und Leukämogenese mit modifizierten Nukleinsäuren. Z. Krebsforschg. Klin. Onkol. (in press) (1974c).
CHANDRA, P., WILL, G., GERICKE, D., GÖTZ, A.: Inhibition of DNA polymerases from RNA tumor viruses by tilorone and cogeners: Site of action. Biochem. Pharmacol. 23, 3259 (1974d).
CHANDRA, P., WOLTERSDORF, M.: Influence of tilorone and cogeners on the secondary structure and template activity of DNA. FEBS Letters 41, 169 (1974).
CHANDRA, P., ZIMMER, C.H., THRUM, H.: Effect of distamycin A on the structure and template activity of DNA in RNA-polymerase system. FEBS Letters 7, 90 (1970).
CHANDRA, P., ZUNINO, F., GAUR, V.P., ZACCARA, A.: Mode of tilorone hydrochloride interaction to DNA and polydeoxyribonucleotides. FEBS Letters 28, 5 (1972).
CHANDRA, P., ZUNINO, F., GÖTZ, A.: Bis-DEAE-fluorene: a specific inhibitor of DNA polymerases from RNA tumor viruses. FEBS Letters 22, 161 (1972e).
CHANDRA, P., ZUNINO, F., GÖTZ, A., GERICKE, D., DI MARCO, A.: Specific inhibition of DNA polymerases from RNA tumor viruses by some new daunomycin derivatives. FEBS Letters 21, 264 (1972f).
CHANDRA, P., ZUNINO, F., GÖTZ, A., WACKER, A., GERICKE, D., DI MARCO, A., CASAZZA, A.M., GIULIANI, F.: Template specific inhibition of DNA polymerases from RNA tumor viruses by distamycin A and its structural analogues. FEBS Letters 21, 154 (1972g).
CHANDRA, P., ZUNINO, F., ZACCARA, A.: Influence of tilorone hydrochloride on the secondary structure and template activity of DNA. FEBS Letters 23, 145 (1972h).

CHIRIGOS, M.A., PEARSON, J.W., PAPAS, T.S., WOODS, W.A., WOOD, H.B., SPAHN, G.: Effect of streptonigrin (NSC-45383) and analogs on oncornavirus replication and DNA polymerase activity. Cancer Chemo. Rep. 57, 305 (1973).

COFFIN, J.M., TEMIN, H.M.: RNase-sensitive DNA-polymerase activity in uninfected rat cells and rat cells infected with RSV. J. Virol. 8, 630 (1971).

COFFIN, J.M., TEMIN, H.M.: Hybridization of RSV DNA-polymerase products and RNA from chicken and rat cells infected with RSV. J. Virol. 9, 766 (1972).

DARNELL, J.E.: Ribonucleic acids from animal cells. Bact. Rev. 32, 262 (1968).

DE BOER, C.P., MEULMAN, P.A., WNUK, R.J., PETERSON, D.H.: Geldanamycin, a new antibiotic with cytotoxic activity. J. Antibiotic 23, 442 (1970).

DI MARCO, A., GAETANI, M., OREZZI, P., SCOTTI, T., ARCANIBE, F.: Experimental studies on distamycin A: A new antibiotic with cytotoxic activity. Cancer Chemo. Rep. 18, 15 (1962).

DI MARCO, A., GHIONE, M., MIGLIACCI, A., MORVILLO, E., SANFILIPPO, A.: Studi sul meccanismo dell'aszione antifagica dell'antibiotico distamyciny. Giorn. Microbiol. 11, 87 (1963a).

DI MARCO, A., GHIONE, M., SANFILIPPO, A., MORVILLO, E.: Selective inhibition of the multiplication of phage T_1 in *E. coli* K12. Experimentia 19, 134 (1963b).

DRAY, S., BELL, C.: Synthesis of allogeneic immunoglobulins by rabbit lymphoid cells after interaction *in vitro* or *in vivo* with RNA extracts of allogeneic lymphoid tissue. In: Possible Episomes in Eukaryotes (Ed. L.G. SILVESTRI), p. 287. Amsterdam: North-Holland Publ. Co. 1973.

DUESBERG, P.H., CANAANI, E.: Complimentarity between RSV-RNA and the *in vitro* synthesized DNA of the virus associated DNA-polymerase. Virology 42, 783 (1970).

ENGLUND, P.T.: The initial step of *in vitro* synthesis of DNA by the T_4 DNA-polymerase. J. Biol. Chem. 246, 5684 (1971).

ERICKSON, R.J., GROSCH, J.C.: The inhibition of Avian Myeloblastosis virus DNA polymerase by synthetic polynucleotides. Biochemistry 13, 1987 (1974).

FANSHIER, L., GARAPIN, A.C., McDONNELL, J., LEVINSON, W., BISHOP, J.M.: DNA polymerase associated with Avian tumor viruses. Secondary structure of the DNA product. J. Virol. 7, 76 (1971).

FARAS, A.J., TAYLOR, J.M., McDONNELL, J.P., LEVINSON, W.E., BISHOP, J.M.: Purification and characterization of the DNA-polymerase associated with RSV. Biochemistry 11, 2334 (1972).

FRIDLENDER, B., WEISSBACH, A.: DNA polymerase of tumor viruses: specific effect of ethidium bromide on the use of different synthetic templates. Proc. Nat. Acad. Sci. (Wash.) 68, 3116 (1971).

FUJINAGA, K., PARSONS, J.T., BEARD, J.W., BEARD, D., GREEN, M.: Mechanism of carcinogenesis by RNA tumor viruses. III. Formation of RNA-DNA complex and duplex DNA by DNA polymerase(s) of AMV. Proc. Nat. Acad. Sci. (Wash.) 67, 1432 (1970).

GALLO, R.C.: Reverse-transcriptase, the DNA polymerase of oncogenic RNA viruses. Nature 234, 194 (1971).

GALLO, R.C.: RNA-dependent DNA polymerases and cells. Blood 39, 117 (1972).

GALLO, R.C., ABRELL, J.W., ROBERT, M.S., YANG, S.S., SMITH, R.G.: Reverse transcriptases from Mason-Pfizer monkey tumor virus, AMV, and RLV and its responses to rifampicin derivatives. J. Nat. Cancer Inst. 48, 1185 (1972a).

GALLO, R., SARIN, P., SMITH, R., BOBROW, S., SARNGADHARAN, M., REITZ, M., ABRELL, J.: RNA-directed DNA synthesis in tumor viruses and human lymphocytes. In: DNA Synthesis *in vitro* (Eds. WELLS and INMAN), p. 251. Baltimore: Univ. Park Press 1973.

GALLO, R., SMITH, R.G., WHENG-PENG, J., TING, R.C., YANG, S.S., ABRELL, J.: RNA tumor viruses, DNA polymerases and oncogenesis: some selective effects of rifampicin derivatives. Medicine 51, 159 (1972b)

GALLO, R.C., TING, R.C.: Cancer viruses: In: CRC Critical Reviews in Clinical Lab Sciences 403 (1972).

GALLO, R.C., YANG, S.S., TING, R.S.: RNA-dependent DNA polymerase of human acute leukemic cells. Nature (Lond.) 228, 927 (1970).

GARAPIN, A.C., FANSHIER, L., LEONG, J.A., JACKSON, J., LEVINSON, W., BISHOP, J.M.: DNA polymerase of RSV: kinetics of DNA synthesis and specificity of the products. J. Virol. 7, 227 (1971).

GARAPIN, A.C., McDONNELL, J.P., LEVINSON, W., QUINTRELL, N., FANSHIER, L., BISHOP, J.M.: DNA polymerase associated with RSV and AMV: Properties of the enzyme and its product. J. Virol. 6, 589 (1970).

GELB, L.D., AARONSON, D.A., MARTIN, M.A.: Heterogeneity of murine leukemia virus *in vitro* DNA. Science 173, 1353 (1971).

GERWIN, B.J., MILSTEIN, J.B.: An oligonucleotide affinity column for RNA-dependent DNA polymerase from RNA tumor viruses. Proc. Nat. Acad. Sci. (Wash.) 69, 2599 (1972).

GERWIN, B.I., TODARO, G.J., ZEVE, V., SCOLNICK, E.M., AARONSON, S.A.: Separation of RNA-dependent DNA polymerase activity from the murine leukemia virion. Nature (Lond.) 228, 435 (1970).

GILLESPIE, D., MARSHALL, S., GALLO, R.C.: RNA of RNA tumor viruses contains poly A. Nature New Biology, Lond. 236, 227 (1972).

GIRON, D.J., SCHMIDT, J.P., BALL, R.J., PINDAK, R.: Effect of interferon inducers and interferon on bacterila infections. Antimicrob. Agents and Chemother. 1, 80 (1972).

GOLDIN, A., JOHNSON, R.K.: Adriamycin activity in experimental tumors. In: Ergebnisse der Adriamycin-Therapie (Eds. M. GHIONE, J. FETZER, H. MAIER), p. 3. Berlin-Heidelberg-New York: Springer 1975.

GOODMAN, N.C., SPIEGELMAN, S.: Distinguishing reverse-transcriptase of an RNA tumor virus from other known DNA polymerases. Proc. Nat. Acad. Sci. (Wash.) 68, 2203 (1971).

GOULIAN, M., LUCAS, Z.J., KORNBERG, A.: Enzymatic synthesis of DNA. XXV. Purification and properties of DNA-polymerase induced by infection with T_4. J. Biol. Chem. 243, 627 (1968).

GRANBOULAN, N., HUPPERT, J., LACOUR, F.: Examen au microscope electronique du RNA du virus de la myeloblastose aviaire. J. Mol. Biol. 16, 571 (1966).

GRANDGENETT, D.P., GERARD, G.I., GREEN, M.: Ribonuclease H: a ubiquitous activity in virions of RNA tumor viruses. J. Virol. 10, 1136 (1972).

GRANDGENETT, D.P., GERARD, G.F., GREEN, M.: A single subunit from AMV with both RNA-directed DNA polymerase and ribonuclease H activity. Proc. Nat. Acad. Sci. (Wash.) 70, 230 (1973).

GREEN, M., CARTAS, M.: The genome of RNA tumor viruses contains polyadenylic acid sequences. Proc. Nat. Acad. Sci. (Wash.) 69, 791 (1972).

GREEN, M., ROKUTANDA, M., JUJINAGA, K., RAY, R.K., ROKUTANDA, H., GURGO, C.: Mechanism of carcinogenesis by RNA tumor viruses. I. An RNA-dependent DNA polymerase in murine sarcome viruses. Proc. Nat. Acad. Sci. (Wash.) 67, 385 (1970).

GREEN, M., WEST, W.L.: *In vitro* cytogenetic effects of tilorone hydrochloride. Pharmacologist 13, 260 (1971).

GURGO, C., GREEN, M., RAY, R.: Rifamycin derivatives strongly inhibiting RNA-DNA polymerase reverse transcriptase of murine sarcoma virus. J. Nat. Cancer Inst. 49, 61 (1972).

HAAPALA, D.K., JASMIN, C., SINOUSSI, F., CHERMANN, J.C., RAYNAUD, M.: Inhibition of tumor virus RNA-dependent DNA-polymerase by the heteropolyanion silicotungstate. Biomed. Express 19, 7 (1973).

HAHN, F.E.: Distamycin A and Netropsin, Review. In: Antibiotics (Eds. J.W. CORCORAN, F.E. HAHN), Vol. 3, p. 79. Berlin-Heidelberg-New York: Springer 1974.

HANAFUSA, H., BALTIMORE, D., SMOLER, D., WATSON, K.F., YANIV, A., SPIEGELMAN, S.: Absence of polymerase protein in virions of α-type RSV. Science 177, 1188 (1972).

HANAFUSA, H., HANAFUSA, T.: Non-infectious RSV deficient in DNA polymerase. Virology 43, 313 (1971).

HATANAKA, M., HUEBNER, R.J., GILDEN, R.V.: DNA-polymerase activity associated with RNA tumor viruses. Proc. Nat. Acad. Sci. (Wash.) 67, 143 (1970).

HAUSEN, P., STEIN, H.: Ribonuclease H: An enzyme degrading the RNA moiety of DNA-RNA hybrids. Europ. J. Biochem. 14, 278 (1970).

HILL, M., HILLOVA, J.: RNA and DNA forms of the genetic material of C-type viruses and the integrated state of the DNA form in the cellular chromosome. Biochim. Biophys. Acta 337, 7 (1974).

HIRSCHMAN, S.Z.: Inhibitors of viral nucleic acid transcriptases. Trans. N.Y. Acad. Sci. U.S.A. 33, 595 (1971).

HIRSCHMAN, S.Z.: Inactivation of DNA polymerase of Murine Leukemia viruses by calcium elenolate. Nature New Biology 238, 277 (1972).

HOLLDORF, A.W., FRIEBE, B., STOBER, M.: Zur Wirkung von Distamycin A auf die Induktion von Enzymen in *E. coli*. Zbl. Bakt. 212, 265 (1970).

HURWITZ, J., LEIS, J.: RNA-dependent DNA-polymerase activity of RNA tumor viruses. J. Virol. 9, 116 (1972).

JOVIN, T.M., ENGLUND, P.T., KORNBERG, A.: Enzymatic synthesis of DNA. XXII. Chemical modifications of DNA-polymerases. J. Biol. Chem. 244, 3009 (1969).

KACIAN, D.L., WATSON, K.F., BURNEY, A., SPIEGELMAN, S.: Purification of the DNA polymerase of AMV. Biochim. Biophys. Acta 246, 365 (1971).

KAKEFUDA, T., BADER, J.P.: Electron microscopic observation on the RNA of murine leukemia virus. J. Virol. 4, 460 (1969).

KANG, C.Y., TEMIN, H.M.: Endogenous RNA-directed DNA polymerase activity in uninfected chicken embryos. Proc. Nat. Acad. Sci. (Wash.) 69, 1550 (1972).

KELLER, W., CROUCH, R.: Degradation of DNA-RNA hybrids by Ribonuclease H and DNA polymerases of cellular and viral origin. Proc. Nat. Acad. Sci. (Wash.) 69, 3360 (1972).

KILEY, M.P., WAGNER, R.P.: RNA species of intracellular nucleocapsids and released virions of VSV. J. Virol. 10, 244 (1972).

KISHI, T., ASAI, M., MUROI, M., HARADA, S., MIZUTA, E., TERAO, S., MIKI, T., MIZUNO, K.: Structure of tolypomycinone. Tetrahedron Letters, p. 91 (1969).

KLEINSCHMIDT, A.K., LANG, D., JACHERTS, D., ZAHN, R.K.: Darstellung des gesamten DNS-Inhaltes von T2 Backeriophagen. Biochim. Biophys. Acta 61, 857 (1972).

KOLAKOFSKY, D., BOY DE LA TOUR, E., DELIUS, H.: J. Virol. 13, 261 (1974).

KORNBERG, A.: Active center of DNA polymerase. Science 163, 1410 (1969).

KOTLER, M., BECKER, Y.: Effect of distamycin A and congocidine on DNA synthesis by RSV reverse transcriptase. FEBS Letters 22, 222 (1972).

KREY, A.K., HAHN, F.E.: Studies on the complex of distamycin A with calf thymus DNA. FEBS Letters 10, 175 (1970).

KRUEGER, R.F., YOSHIMURA, S.: Antiviral activity of bis-DEAE-fluorenone, an oral interferon-inducer. Fed. Proc. 29, 635 (1970).

KUFE, D.W., PETERS, W.P., SPIEGELMAN, S.: Unique nuclear DNA sequences in the involved tissues of Hodgkin's and Burkitt's lymphoma. Proc. Nat. Acad. Sci. (Wash.) 70, 3810 (1973).

LAI, M.M.C., DUESBERG, P.H.: Adenylic acid-rich sequences in RNAs of Rous sarcoma virus and Rauscher mouse leukemia virus. Nature (Lond.) 235, 383 (1972).

LEIS, J.P., BERKOWER, I., HURWITZ, J.: Mechanism of action of ribonuclease H isolated from AMV and *E. coli*. Proc. Nat. Acad. Sci. (Wash.) 70, 466 (1973).

LEIS, J.P., HURWITZ, J.: RNA-dependent DNA polymerase activity of RNA tumor viruses: directing influence of RNA in the reaction. J. Virol. 9, 130 (1972).

LERMAN, L.S.: Structural considerations in the interaction of DNA and acridines. J. Mol. Biol. 3, 18 (1961).

LERMAN, L.S.: Acridine mutagens and DNA structure. J. Cell. Comp. Physiol. 64, 1 (1964).

LEVINSON, W., FARAS, A., WOODSON, B., JACKSON, J., BISHOP, J.M.: Inhibition of RNA-dependent DNA polymerase of RSV by thiosemicarbazones and several cations. Proc. Nat. Acad. Sci. (Wash.) 70, 164 (1973).

LINN, S., LEHMAN, I.R.: An endonuclease from *Neurospora crassa* specific for polynucleotides lacking an ordered structure. J. Biol. Chem. 240, 1294 (1965).

LOWRY, O.H., ROSEBROUGH, N.J., FARR, A.L., RANDALL, R.J.: Protein measurement with the Folin Phenol Reagent. J. Biol. Chem. 193, 265 (1951).

LOWY, D.R., ROWE, W.P., TEICH, N., HARTLEY, J.W.: Murine leukemia virus. High frequency activation *in vitro* by 5-iododeoxyuridine and 5-bromodeoxyuridine. Science 174, 155 (1971).

MANLY, K.F., SMOLER, D.F., BROMFELD, E., BALTIMORE, D.: Forms of DNA produced by virions of RNA tumor viruses. J. Virol. 7, 106 (1971).

MASAMUNE, Y., RICHARDSON, C.C.: A mutant of bacteriophage T_7 deficient in polynucleotide ligase. J. Biol. Chem. 246, 2692 (1971).

MAYER, G.D., FINK, B.A.: Bis-DEAE-fluorenone, an oral inducer of interferone. Fed. Proc. 29, 635 (1970).

MAYER, G.D., KRUEGER, R.F.: Tilorone hydrochloride: Mode of action. Science 169, 1214 (1970).

MAYER, G.D., KRUEGER, R.F., YOSHIMURA, S.: Protection of mice from influenza infection by tilorone hydrochloride and influenza vaccine. Bact. Proc., p. 188 (abstr.) (1972).

MAYER, R.J., SMITH, R.G., GALLO, R.C.: Reverse transcription in normal rhesus monkey placenta. Science (in press) (1974).

MIZUTANI, S., BOETTIGER, D., TEMIN, H.M.: A DNA dependent DNA polymerase and a DNA endonuclease in virions of Rous Sarcoma virus. Nature 228, 424 (1970).

MIZUTANI, S., KANG, C.Y., TEMIN, H.M.: Relationships among RNA-directed DNA polymerase activities of Avian viruses and chicken cells. Cold Spring Harbor Symp. Quant. Biol. 38, 289 (1974).

MIZUTANI, S., TEMIN, H.M.: Enzymes and nucleotides in virions of Rous Sarcoma virus. Cold Spring Harbor Symp. Quant. Biol. 35, 847 (19)O).

MIZUTANI, S., TEMIN, H.M.: Enzymes and nucleotides in virions of Rous Sarcoma virus. J. Virol. 8, 409 (1971).

MIZUNO, S., YAMAZAKI, H., NITTA, K., UMEZAWA, H.: Inhibition of initiation of DNA-dependent RNA synthesis. Biochim. Biophys. Res. Commun. 35, 127 (1968).

MÖLLING, K., BOLOGNESI, D.P., BAUER, H., BÜSEN, W., PLASSMAN, H.W., HANSEN, P.: Association of the viral reverse-transcriptase with an enzyme degrading the RNA moiety of RNA-DNA hybrids. Nature New Biology 234, 240 (1971).

MÜLLER, W.E.G., YAMAZAKI, Z.I., SÖGTROP, H.H., ZAHN, R.K.: Action of 1-β-D-Arabinofuranosylcytosine on mammalian and oncogenic viral polymerases. Europ. J. Cancer 8, 421 (1972).

MÜLLER, W.E.G., ZAHN, R.K., SEIDEL, H.J.: Inhibitors acting on nucleic acid synthesis in an oncogenic RNA virus. Nature New Biology 232, 143 (1971).

MUNSON, A.E., MUNSON, J.A., REGELSON, W., WAMPLER, G.: Effect of tilorone hydrochloride and cogeners on reticuloendothelial system, tumors, and the immunse responses. Cancer Res. 32, 1397 (1972).

OROSZLAN, S., HATANAKA, M., GILDEN, R.V., HUEBNER, R.J.: Specific inhibition of mammalian RNA C-type virus DNA polymerases by rat antisera. J. Virol. 8, 816 (1971).

PAPAS, T.S., PRY, T.W., CHIRIGOS, M.A.: Inhibition of the DNA polymerase of Avian Myeloblastosis virus (AMV) by pyran-copolymer. Am Soc. Microbiol. Ann. Meeting Abstr. V247 (1973).

PARKS, W.P., SCOLNICK, E.M., ROSS, J., TODARO, G.J., AARONSON, S.A.: Immunological relationships of reverse transcriptases from ribonucleic acid tumor viruses. J. Virol. 9, 110 (1972).

PITHA, P.M., TEICH, N.M., LOWY, D.R., PITHA, J.: Inhibition of leukemia virus replication by vinyl analogs of polynucleotides. In: DNA synthesis *in vitro* (Eds. WELLS and INMAN), p. 369. Lancaster: Medical and Technical Publ. 1973.

REITZ, M., GILLESPIE, D., SAXINGER, W.C., ROBERT, W., GALLO, W. C.: Poly(rA) tracts of tumor virus 70S RNA are not transcribed in endogenous or reconstituted reactions of viral reverse transcriptase. Biochem. and Biophys. Res. Commun. 49, 1216 (1972).

RHEINS, M.S., BASKER, A.D., WILSON, H.E.: Therapeutic effects of a series of non-viral interferon inducers on a viral-induced leukemia. Canad. J. Microbiol. 17, 1257 (1971).

RICHARDSON, C.C.: DNA-polymerase from *E. coli*. In: Producers in nucleic acid Res. (Eds. CANTONI and DAVIES), Vol. I, p. 263. New York: Harper and Row 1966.

RINEHART, K., ANTOSZ, F., SOBIECZEWSKI, W., McMILLAN, M., FREIDINGER, J., ENANOZA, R.: Reported at 165th Meeting. Am. Chem. Soc., Div. Med. Chem. (1973).

RINEHART, K.L. Jr., MATHUR, H.H., SASAKY, K., MARTIN, P.K., COVERDALE, C.E.: Chemistry of streptovaricins. J. Amer. Chem. Soc. 90, 6241 (1968).

ROBERT, M.S., SMITH, R.G., GALLO, R.C., SARIN, P.S., ABRELL, J. W.: Viral and cellular DNA polymerases: comparision of activities with synthetic and natural RNA templates. Science 176, 798 (1972).

ROKUTANDA, M., ROKUTANDA, H., GREEN, M., FUJINAGA, K., RAY, R.K., GURGO, C.: Formation of viral RNA-DNA hybrid molecules by the DNA polymerases of sarcoma-leukemia viruses. Nature (Lond.) 227, 1026 (1970).

ROSS, J., SCOLNICK, E.M., TODARO, G.J., AARONSON, S.A.: Separation of murine cellular and murine leukemia virus DNA polymerases. Nature New Biology 231, 163 (1971).

ROYE, D., RHOADS, A., MORRIS, H.P., WEST, W.L.: Effect of bis-DEAE-fluorenone (tilorone) on growth of tumors in rats. Pharmacologist 13, 260 (1971).

RUPRECHT, R.M., GOODMAN, N.C., SPIEGELMAN, S.: Determination of natural host taxonomy of RNA tumor viruses by molecular hybridization: application to RD-114, a candidate human virus. Proc. Nat. Acad. Sci. (Wash.) 70, 1437 (1973).

SANFILIPPO, A., MORVILLO, E., GHIONE, M.: Activity of distamycin A on the induction of adaptive enzymes in *E. coli*. J. Gen. Microbiol. 43, 369 (1966).

SARIN, P.S., GALLO, R.C.: Characteristics of reverse transcriptase from human acute leukemic cells and RNA tumor virus. J. Int. Res. Commun. Sys. 1, 52 (1973).

SARIN, P.S., GALLO, R.C.: RNA directed DNA polymerase. Int. Rev. Sci. 6, ch. 8 (1974).

SARKAR, N.H., MOORE, D.H.: J. Virol. 5, 230 (1970).

SARNGADHARAN, M.G., SARIN, P.S., REITZ, M.S., GALLO, R.C.: Reverse transcriptase activity of human acute leukemic cells: purification of the enzyme, response to AMV 70S RNA and characterization of the DNA product. Nature New Biol. 240, 67 (1972).

SCHLOM, J., SPIEGELMAN, S.: Simultaneous detection of reverse transcriptase and high molecular weight RNA unique to oncogenic RNA viruses. Science 174, 840 (1971).

SCOLNICK, E.M., AARONSON, S.A., TODARO, G.J.: DNA synthesis by RNA-containing tumor viruses. Proc. Nat. Acad. Sci. (Wash.) 67, 1034 (1970).

SCOLNICK, E.M., PARKS, W.P., TODARO, G.J., AARONSON, S.A.: Reverse-transcriptase of primate viruses as immunological marker. Science 177, 1119 (1972a).

SCOLNICK, E.M., PARKS, W.P., TODARO, G.J., AARONSON, S.A.: Immunological characterization of primate C-type virus reverse transcriptase. Nature New Biology 235, 35 (1972b).

SENSI, P., MAGGI, N., FURESZ, S., MAFFI, G.: Chemical modifications and biological properties of rifamycins. Antimicrob. Agents Chemother., p. 699 (1967).

SPIEGELMAN, S., AXEL, R., SCHLOM, J.: Presence in human breast cancer of RNA homologous to mouse mammary tumor virus RNA. Nature (Lond.) 235, 32 (1972).

SPIEGELMAN, S., BURNY, A., DAS, M.R., KEYDAR, J., SCHLOM, J., TRAVNICEK, M., WATSON, K.: (1) Characterization of the products of RNA-directed DNA polymerases in oncogenic RNA viruses. Nature (Lond.) 227, 563 (1970).

SUGINO, A., OKAZAKI, R.: RNA-linked DNA fragments in-vitro. Proc. Nat. Acad. Sci. (Wash.) 70, 88 (1973).

SWEET, R.W., GOODMAN, N.C., CHE, J.R., RUPRECHT, R.M., REDFIELD, R.R., SPIEGELMAN, S.: The presence of unique DNA sequences after viral induction of leukemia in mice. Proc. Nat. Acad. Sci. (Wash.) 71, 1705 (1974).

TEMIN, H.M.: (1) Nature of the provirus of Rous Sarcoma. Nat. Cancer Inst. Monogr. 17, 557 (1964a).

TEMIN, H.M.: The participation of DNA in RSV production. J. Virol. 23, 486 (1964b).

TEMIN, H.M.: Malignant transformation of cells by viruses. Presp. Biol. Med. 14, 11 (1970).

TEMIN, H.M.: The protovirus hypothesis. J. Nat. Cancer Inst. 42, 111 (1971a).

TEMIN, H.: Mechanism of cell transformation by RNA tumor viruses. Am. Rev. Microbiol. 25, 609 (1971b).

TEMIN, H.M., BALTIMORE, D.: RNA-directed DNA synthesis and RNA tumor viruses. Adv. Virus Res. 17, 129 (1972).

TEMIN, H.M., KANG, C.Y., MIZUTANI, S.: RNA-directed synthesis in viruses and cells. In: Possible Episomes in Eukaryotic Cells. (Ed. L.G. SILVESTRI), pp. 1. Amsterdam: North-Holland Publ. Co. 1973.

TEMIN, H.M., MIZUTANI, S.: RNA-dependent DNA polymerase in virions of Rous sarcoma virus. Nature (Lond.) 226, 1211 (1970).

TING, R.C., YAND, S.S., GALLO, R.C.: RNA tumor virus transformation and derivatives of rifamycin SV. Nature New Biology, Lond. 236, 163 (1972).

TOCCHINI-VALENTINI, G.P., CRIPPA, M.: On the mechanism of gene amplification. In: The Biology of Oncogenic Viruses. (Ed. L.G. SILVESTRI), p. 237. Amsterdam: North-Holland Publ. Co. 1971.

TODARO, G.J., GALLO, R.C.: Immunological relationship of DNA polymerase form human acute leukemia cells and primate and mouse leukemia virus reverse transcriptase. Nature 244, 206 (1973).

TODARO, G.J., HUEBENER, R.J.: The viral oncogene hypothesis: New evidence. Proc. Nat. Acad. Sci. (Wash.) 69, 1009 (1972).

TOUMINEN, F.W., KENNEY, F.T.: Inhibition of RNA-directed DNA-polymerases from RLV by Ara-CTP. Biochem. Biophys. Res. Commun. 48, 1469 (1972).

TRONICK, S.R., SCOLNICK, E.M., PARKS, W.P.: Reversible inactivation of the DNA polymerase of RLV. J. Virol. 10, 885 (1972).

VARMUS, H.E., LEVINSON, W.E., BISHOP, J.M.: Extent of transcription by the RNA-dependent DNA polymerase of RSV. Nature New Biology 233, 19 (1971).

WATSON, K.F., MÖLLING, K., BAUER, H.: Biochem. Biophys. Res. Commun. 51, 232 (1973).

WEBER, G.H., HEINE, U., COTTLER-FOX, M., BEAUDREAU, G.S.: Visualization of single-stranded nucleic acid of RNA tumor virus with the electron microscope. Proc. Nat. Acad. Sci. (Wash.) 71, 1887 (1974).

WEHRLI, W., STAEHELIN, M.: Actions of rifamycins. Bacteriol. Rev. 35, 290 (1971).

WEIMANN, B.J., SCHMIDT, J.: RNA-dependent DNA polymerase and ribonuclease H from Friend virions. FEBS Letters 43, 37 (1974).
WELLS, R.D., FLÜGEL, R.M., LARSON, J.E., SCHENDEL, P.F., SWEET, R.W.: Comparison of some reactions catalyzed by deoxyribonucleic acid polymerase from avian myeloblastosis virus, *Escherichia coli* and *Micrococcus luteus*. Biochemistry 11, 621 (1972).
WERNER, G.H., GANTER, P., DE RATULD, Y.: Studies on the antiviral activity of distamycin A. Chemotherapia 9, 65 (1964).
WHALLEY, J.M.: In: Bibliotheca Haematologica, No. 39: Unifying concepts of leukemia (Eds. R.M. DUTCHER, L. CHIECO-BIANCHI), p. 125. Basel: S. Karger 1973.
WU, A.M., GALLO, R.C.: Interaction between murine type-C virus RNA-directed DNA polymerase and rifamycin derivatives. Biochim. Biophys. Acta (in press) (1974).
WU, A.M., SARNGADHARN, M.G., GALLO, R.C.: Separation of ribonuclease H and RNA directed DNA polymerase (reverse transcriptase) of murine type-C RNA tumor viruses. Proc. Nat. Acad. Sci. (Wash.) 71, 1871 (1974).
WU, A.M., TING, R.C.Y., GALLO, R.C.: RNA-directed DNA-polymerase and virus-induced leukemia in mice. Proc. Nat. Acad. Sci. (Wash.) 70, 1298 (1973).
WU, A.M., TING, R.C., PARAN, M., GALLO, R.C.: Cordycepin inhibits induction of murine leukovirus production. Proc. Nat. Acad. Sci. (Wash.) 69, 3820 (1972).
YAMAMATO, K., ACTON, E.M., HENRY, D.W.: Antitumor activity of some derivatives of daunorubicin at the amino and methyl ketone functions. J. Med. Chem. 15, 872 (1972).
YANG, S.S., HERRERA, F., SMITH, R.G., REITZ, M., LANCINI, G., TING, R., GALLO, R.C.: Rifamycin antibiotics. Inhibitors of Rauscher (murine) leukemia virus reverse transcriptase and purified DNA-polymerases from human normal and leukemic lymphoblasts. J. Nat. Cancer Inst. 49, 7 (1972).
ZIMMER, C., PUSCHENDORF, B., GRUNICKE, H., CHANDRA, P., VENNER, H.: Influence of netropsin and distamycin A on the secondary structure and template activity of DNA. Europ. J. Biochem. 21, 269 (1971a).
ZIMMER, C., REINERT, K.E., LUCK, G., WÄHNERT, U., LÖBER, G., THRUM, H.: Interaction of the oligopeptide antibiotics netropsin and distamycin A with nucleic acids. J. Mol. Biol. 58, 329 (1971b).
ZUNINO, F., DI MARCO, A.: Studies on the interaction of distamycin A and its derivatives with DNA. Biochem. Pharmacol. 21, 867 (1972).

Adenosine as a Physiological Regulator of Coronary Blood Flow

Ray A. Olsson, Randolph E. Patterson

I. Introduction

A. Cardiac Oxygen Metabolism

Coronary blood flow rate depends uniquely on myocardial oxygen consumption rate, as first demonstrated by ECKENHOFF et al. (1947) and confirmed by ALLELA et al. (1955) and RAYFORD et al. (1965). The energy to support the constant work of the heart comes from aerobic metabolism; fatty acids are the primary fuel, their oxidation accounting for about 70 percent of the oxygen the heart consumes (NEELY and MORGAN, 1974). Indeed, cardiac energy production is almost exclusively aerobic. Even maximally accelerated anerobic metabolism can furnish only 10 - 15 percent of the energy required to sustain cardiac performance.

The intrinsic tone of the coronary vessels is higher than that of most other organ circulations and is augmented even further when these vessels are compressed during systole by the myocardium in which they are embedded (SABISTON and GREGG, 1957). The combination of this high resistance to flow and the heart's high rate of oxygen consumption (4 - 10 ml/min per 100 g) result in an oxygen extraction by the heart which exceeds that of any other organ. The coronary venous oxygen content is on the order of 3 - 5 ml/100 ml, which corresponds to a pO_2 of 15 - 20 mm Hg. Because oxygen extraction is already very nearly complete, the heart can only respond to increased workloads by increasing coronary flow rate. The coronary vasculature is finely tuned to the oxygen requirements of the tissue it serves (GREGG and FISHER, 1963; BERNE, 1964; BRAUNWALD, 1969); the coronary flow response to an abrupt change in cardiac effort is essentially complete within a few heartbeats (KHOURI et al., 1965; PITT and GREGG, 1968).

B. Theories of Coronary Blood Flow Regulation

How does the vascular smooth muscle of the coronary resistance vessels sense and respond to the oxygen requirements of the neighboring striated muscle cells? Although neural and mechanical factors can influence coronary vascular resistance directly, control is primarily metabolic. Ordinarily these factors alter cardiac performance, thereby eliciting a metabolic response in cardiac muscle which overshadows direct effects on the coronary vessels themselves (BERNE, 1964). Two hypotheses attempt to explain how coronary vascular resistance is coupled to the oxygen requirements of the myocardium. The first proposes that oxygen availability determines the coronary smooth muscle tone (GUYTON et al., 1964; HONIG, 1968). According to this hypothesis the oxidative metabolism of the vascular smooth muscle cells acts as the "oxygen sensor" and the contractile machinery as the effector of a feedback loop. In the alternative hypothesis myocardial metabolism "senses" oxygen lack, and releases vasodilatory metabolites which diffuse to the coronary smooth muscle cells and effect their relaxation.

Currently no convincing evidence favors one hypothesis over the other; indeed some evidence suggests that the two mechanisms operate synergistically. In any event, no large body of experimental data exists to justify an analytical comparison of the two contending hypothesis. Rather, we have chosen to discuss progress in the research supporting the "adenosine hypothesis" of coronary flow regulation, i.e.,that adenosine fulfills the criteria as one physiologic mediator of the metabolic regulation of coronary blood flow.

C. The Adenosine Hypothesis

In 1963, BERNE, recognizing the heart's dependence on aerobic energy metabolism, proposed that restricted oxygen availability would curtail the generation of ATP and lead to the accumulation of its degradation products; ADP, AMP and adenosine. Although the vasodilator effects of adenosine and the adenosine phosphates were established earlier (DRURY and SZENT-GYORGI, 1929), adenosine was thought to be the most likely candidate as a physiological vasodilator as it could diffuse out of the myocardial cell. Adenosine phosphates are ionized at physiological pH and consequently have poor lipid solubility, which restricts their diffusion across biological membranes. BERNE postulated the relative oxygen lack leads to the release of adenosine into the interstitial space of the heart, where the nucleoside diffuses to the coronary arterioles and causes them to relax, increasing coronary flow and oxygen delivery.

This hypothesis has stimulated considerable research, and current evidence suggests that adenosine may be only one of several interacting "physiological regulators" of coronary flow. Nevertheless, the evidence implicating adenosine is perhaps more detailed than for any other candidate compound and may suggest experimental approaches to evaluate other flow-regulating metabolites.

The experimental evaluation of this hypothesis can be divided into three major areas: (1) the production of adenosine by cardiac cells; (2) the regulation of its concentration in the vicinity of the coronary resistance vessels, and (3) its interaction with and relaxation of coronary vascular smooth muscle. The discussion which follows reflects the authors' view that quantitative data are required to prove or refute this hypothesis and that substantial quantitative and perhaps qualitative differences in adenosine metabolism distinguish different organs and experimental animals from each other. For example, only mammalian hearts contain the enzymes necessary for adenosine production and the tissue activity of these enzymes may vary by as much as 100-fold from one species to another (NAKATSU and DRUMMOND, 1972). Likewise, the amount of adenosine in heart muscle may be as low as 0.2 nmole/g wet weight in the dog (RUBIO, BERNE, and KATORI, 1969) and as high as 6.0 nmole/g wet weight in the rat (BERNE et al., 1971). Using data from one species or organ to predict findings in another is treacherous because of these large quantitative differences.

II. Control of Adenosine Production in Mammalian Heart

Experimental methods for stimulating myocardial adenosine production require careful definition because the morphological, metabolic and functional consequences of arterial hypoxemia differ from those of myocardial ischemia (KUBLER and SPIEKERMANN, 1970; ROVETTO, WHITMER, and NEELY, 1973). Although cardiac tissue hypoxia results from either process, reducing coronary blood flow decreases delivery of other substrates and the washout of the end products of metabolism. The accumulation of lactic acid is particularly important, as it causes tissue acidosis, which impairs anerobic glycolysis during ischemia relative to the much higher glycolytic flux observed during hypoxemia (KUBLER and SPIEKERMANN, 1970; ROVETTO, WHITMER, and NEELY, 1973). Accumulation of organic acids in the heart may also impair other enzymes, such as 5'-nucleotidase, which loses activity as pH falls below its optimum of 7 - 8 or higher (BAER, DRUMMOND, and DUNCAN, 1966). Thus, the effects of ischemia on cardiac nucleotide metabolism probably differ from those of hypoxemia.

A. Pathways of 5'-AMP Catabolism in Muscle

Adenosine is present in oxygenated heart, and accumulates rapidly when the heart is made hypoxic by coronary artery occlusion (DEUTICKE and GERLACH, 1966; RUBIO and BERNE, 1969; OLSSON, 1970). Dephosphorylation of AMP by 5'-nucleotidase produces almost all the adenosine in the heart. *De novo* synthesis also occurs, but at a rate too slow to explain the rapid production of adenosine during cardiac anoxia (ZIMMER et al., 1973). AMP degradation in muscle may proceed *via* either of two pathways, deamination to IMP and dephosphorylation to adenosine. Skeletal and cardiac muscle differ with respect to which pathway predominates, deamination being preponderant in skeletal muscle and dephosphorylation being of greater importance in cardiac muscle (GERLACH, DEUTICKE, and DREISBACH, 1963; IMAI, REILEY, and BERNE, 1966; BURGER and LOWENSTEIN, 1967; BERNE et al., 1971).

B. 5'-Nucleotidase

1. Cellular Localization

Several histochemical studies demonstrate the presence of cardiac 5'-nucleotidase in the sarcolemma, T-tubules and intercalated discs (ROSTGAARD and BEHNKE, 1964; GORDON, PRICE, and BLUMBERG, 1969; RUBIO, BERNE, and DOBSON, 1973). Each study demonstrated the enzyme by trapping one of the reaction products, phosphate, as its lead salt. The lead phosphate was found in each case on the *external* face of the plasma membrane. One must be very careful in drawing inferences about enzyme structure from histochemical evidence like this. However, it is consistent with other evidence discussed below that adenosine, too, is released into the interstitial space of the heart. Even though this technique is not able to provide a detailed description of the orientation of the enzyme in the plasma membrane, it does establish its presence in the sarcolemma unequivocally (SHNITKA and SELIGMAN, 1971).

2. *Physical and Chemical Properties*

5'-nucleotidase is a glycoprotein (EVANS and GURD, 1973; OLSSON, GENTRY, and TOWNSEND, 1973). The enzyme of mouse liver plasma membranes has a molecular weight of 140 - 150,000 daltons and is composed of two probably identical subunits (EVANS and GURD, 1973), but the molecular weight and structure of the cardiac enzyme are unknown. The enzymes obtained from different mammalian hearts resemble those of mouse liver in several properties: a Km in the range of 15 - 20 μM, a requirement for Mg^{++}, the ability to hydrolyze all nucleoside monophosphates, and susceptibility to inhibition by divalent cations and nucleoside di- and triphosphates and their phosphonate analogs (BAER, DRUMMOND, and DUNCAN, 1966; EDWARDS and MAGUIRE, 1970; SULLIVAN and ALPERS, 1971; NAKATSU and DRUMMOND, 1972; OLSSON, GENTRY, and TOWNSEND, 1973). There are marked species differences in cardiac 5'-nucleotidase activity, rat hearts containing 100 times the activity found in rabbit heart (BAER, DRUMMOND, and DUNCAN, 1966). Although many properties of the enzyme from different organs and/or species of animal are similar, the type of inhibition exerted by ATP and ADP is uncertain. ATP inhibition has been reported to be competitive (BAER, DRUMMOND, and DUNCAN, 1966), mixed (SHIMIZU and ALPERS, 1971), or noncompetitive (OLSSON, GENTRY, and TOWNSEND, 1973), while ADP inhibition is said to be competitive (EDWARDS and MAGUIRE, 1970; SULLIVAN and ALPERS, 1971; OLSSON, GENTRY, and TOWNSEND, 1973), or noncompetitive (NAKATSU and DRUMMOND, 1972). This conflict is probably due to differences in experimental conditions, particularly pH and Mg^{++} concentration. We have recently found that the type of inhibition exerted by AOPCP, an ADP analog, is strongly dependent on pH, being of the mixed type below pH 7 and competitive at pH 7 and above (GENTRY and OLSSON, unpublished results). Further, inhibition by the ATP-Mg^{++} complex is at least quantitatively different from that exerted by free ATP (SULLIVAN and ALPERS, 1971). Finally, interpretation of these conflicting results is complicated by the evidence presented by IPATA (1968) that the two classes of ATP-binding sites in the enzyme from sheep brain are interactive (allosteric). The kinetics of this type of inhibition are complex and could be mistaken for one or another of the simpler types of inhibition under certain experimental conditions.

3. *Regulation of 5'-Nucleotidase in vivo*

How adenosine production is adjusted to the oxygen requirements of the heart is central to the validation of the adenosine hypothesis. One of the major obstacles to the acceptance of this hypothesis is that except for one recent report (RUBIO, WIEDMEIER, and BERNE, 1974) hypoxia and ischemia have been the only experimental interventions used to perturb tissue adenosine levels. This constraint on the type of experiment used to test the adenosine hypothesis is imposed by the imprecision of the methods for assaying adenosine in heart-muscle extracts. These methods are simply not adequate to measure the small changes which would be expected to occur during the response of the normally oxygenated heart to physiological stresses.

One approach to understanding the regulation of adenosine production is, of course, to define how 5'-nucleotidase is regulated *in vivo*. Cardiac tissue appears to contain much more 5'-nucleotidase than is required to account for the observed rates of adenosine production in beating hearts. Dog heart homogenates contain enough of this enzyme to produce about 130 nmole adenosine/g per min (FRICK and OLSSON, unpublished), yet oxygenated hearts produce less than 1 nmole/g per min (OLSSON et al., 1972) and this figure increases only about 6-fold during coronary occlusion (OLSSON, 1970).

This disparity between the actual and potential rates of adenosine production suggest that 5'-nucleotidase is markedly inhibited *in vivo*. What are the possible factors contributing to this relatively low apparent activity, and are any of these factors the means by which adenosine production is coupled to oxygen usage?

When BAER, DRUMMOND, and DUNCAN (1966) found that ATP inhibited 5'-nucleotidase they proposed that this inhibition accounted for the *in vivo* control of the enzyme. According to this postulate, hypoxia leads to increased production of adenosine because it causes ATP levels to fall, thereby relieving inhibition of the enzyme. Further, the concurrent conversion of ATP to AMP would provide more substrate for the enzyme. Current evidence cannot validate this hypothesis. ATP levels in heart do not change for at least 15 sec after coronary artery ligation (BRAASCH et al., 1968; WOLLENBERGER and KRAUSE, 1968), whereas adenosine levels increase after as little as 5 sec of coronary artery occlusion (OLSSON, 1970). Further, a decrease in ATP necessarily implies an increase in ADP, which indeed has been observed within 15 sec after coronary occlusion (WOLLENBERGER and KRAUSE, 1968). ADP inhibits 5'-nucleotidase even more strongly than ATP (SULLIVAN and ALPERS, 1971; OLSSON, GENTRY, and TOWNSEND, 1973) which should increase rather than decrease the inhibition of the enzyme.

The ionic composition of the microenvironment of the enzyme, particularly with respect of H^+, Mg^{++} and Ca^{++} may also influence *in vivo* activity of 5'-nucleotidase. Early reports of the properties of the solubilized enzyme established an inhibitory effects for Ca^{++} and Mg^{++} (BAER, DRUMMOND, and DUNCAN, 1966;

EDWARDS and MAGUIRE, 1970), while a later report showed that in addition to this direct effect on the enzyme, Mg^{++} tended to reverse the inhibition exerted by ATP and ADP (SULLIVAN and ALPERS, 1971). Because the enzyme is located in a membrane through which there are very important (though incompletely defined) ion fluxes which ultimately determine cardiac performance and oxygen usage, it may be very susceptible to this sort of local chemical control. Hydrogen ion concentration, like that of the metal ions, has a direct effect on enzyme activity which is expressed in the activity *vs* pH curve. The pH optimum of 5-nucleotidase is rather broad and has a peak between 7 and 8, so that a direct effect of pH is probably not important in oxygenated hearts. As noted above, however, there is preliminary evidence that H^+ concentration may modify the inhibitory effects of nucleoside di- and triphosphates on 5'-nucleotidase activity to an important degree.

The histochemical evidence suggesting that 5'-nucleotidase is oriented in the sarcolemma with its active site directed outward is another possible explanation for the low activity observed *in vivo*. This is supported by the physiological experiment of BAER and DRUMMOND (1968), who injected radioactively labeled AMP into the non-recirculating perfusate of a Langendorf rat heart preparation and found that about half the radioactivity in the effluent collected over the succeeding 3 min was in the form of either adenosine or inosine. These histochemical and physiological observations suggest that the low rate of *in vivo* adenosine production may be due in part to restricted access of intracellular stubstrate to a catalytic site on the exterior surface of the cell. Confirmation of this tentative conclusion is important for two reasons. If it is correct, it would tend to discount the potential importance of ADP to *in vivo* inhibition. This nucleotide is a competitive inhibitor and would by definition exert its effect by interacting with the active site. Second, it requires an explanation of how AMP traverses the cell membrane to reach the active site. Since the traditional view holds that plasma membranes are impermeable to nucleotides, this requires a more complicated model of events at the membrane level which result in adenosine production, e.g., the participation, perhaps, of an "AMP permease".

It is also possible that adenosine production is subserved by a small, perhaps intra-sarcolemmal compartment of the cellular AMP pool, and that the rate of adenosine production is limited purely because of the small size of this hypothetical pool. At this time there is only inferential evidence for compartmentation of the cardiac AMP pool (cf. III.B2 below) and no evidence that this compartment is located in the sarcolemma.

III. Regulation of the Size of Cardiac Adenosine Pool

If adenosine regulates coronary blood flow physiologically, the mechanism controlling the concentration of this nucleotide near the coronary resistance vessels must play a critical role. The experimental support for this aspect of the adenosine hypothesis rests on fairly firm ground.

A. Adenosine Pool in Oxygenated Heart

The adenosine pool of the normally-perfused heart of an anesthetized dog is 0.2 - 0.3 nmole/g wet weight (RUBIO and BERNE, 1969; OLSSON, 1970), and is higher in the rat, 6.0 nmole/g (BERNE et al., 1971). Because heart muscle cells rapidly take up and metabolize adenosine (JACOB and BERNE, 1960), this pool is probably confined to the interstitial space. The volume of the extracellular space has variously been estimated at 0.2 to 0.4 ml/g myocardial wet weight (RUBIO and BERNE, 1969; PAGE, 1962). The average concentration of adenosine, therefore, is in the range of 0.5 - 1.5 μM in the dog and about an order of magnitude higher in the rat.

B. Adenosine Uptake by Cardiac Muscle

Adenosine uptake by cardiac cells appear to be the most important way this nucleoside is removed from its cardiac pool.

1. Adenosine Transport Mechanism

JACOB and BERNE (1960) showed that rat heart rapidly takes up adenosine and incorporates it into the cardiac nucleotide pool. There is now convincing evidence that a number of different types of cells take up nucleosides by a process called facilitated diffusion (SCHOLTISSEK, 1968; PLAGEMANN, 1971; OLIVER and PATERSON, 1971; TAUBE and BERLIN, 1972; SCHRADER, BERNE and RUBIO, 1972; PLAGEMANN, 1974). There is some evidence that this type of transport, which is carrier-mediated but, unlike active transport, non-concentrative, i.e., net transport does not occur against a concentration gradient, also accounts for adenosine uptake by cardiac muscle (PFLEGER, VOLKMER, and KOLASSA, 1969; KOLASSA, PFLEGER, and RUMMEL, 1970; HOPKINS and GOLDIE, 1971; OLSSON et al., 1972). As in other cell types, the kinetics of adenosine uptake are of the Michaelis-Menten type, i.e., saturable, and adenosine analogs can competitively inhibit its transport. This is consistent with the concept of a "carrier" containing an adenosine-binding site. However, attempts made in beating hearts to demonstrate counter-transport, another characteristic of facilitated diffusion, have not been successful (OLSSON et al., 1972). The rate of adenosine uptake in the oxygenated dog heart is estimated to be about 0.6 nmole/g per min (OLSSON et al., 1972), a rate which would cause turnover of the adenosine pool twice per minute. There appear to be species differences in the kinetics of adenosine uptake and the affinity of the putative carrier for an uptake inhibitor, dipyridamole (HOPKINS and GOLDIE, 1971). This study showed that the Km of uptake in the rat and guinea pig were 5 and 1 μM, respectively, and the corresponding Vmax values were 12 and 4.5 nmole/min per g. Dipyridamole inhibited adenosine uptake in the rat heart but not in guinea pig.

2. Incorporation of Adenosine into Cardiac Nucleotide Pool

Adenosine is transported across the cardiac cell membrane intact, and most if not all is thereupon phosphorylated to AMP (LIU and FEINBERG, 1971), even under ischemic conditions (JACOB and BERNE, 1960). If uniformly ^{14}C-labeled adenosine is infused intracoronary, the ratio of base/total ^{14}C in the nucleotides subsequently isolated is the same as that of the infusate, excluding the possibility that the base is cleaved prior to uptake and is subsequently incorporated into nucleotide (WIEDMEIER, RUBIO, and BERNE, 1972). In mammalian heart, adenosine kinase catalyzes phosphorylation of adenosine to AMP (GOLDTHWAIT, 1957). *In vitro*, at optimal concentrations of ATP and adenosine, dog heart adenosine kinase can phosphorylate about 30 nmoles adenosine/min · g wet weight. The Km of this enzyme for adenosine is very low, o.4 μM (OLSSON et al., 1972; NAMM and LEADER, 1974), or one percent of the Km of adenosine deaminase (ROCKWELL and MAGUIRE, 1966). The lower Km of adenosine kinase may explain the preferential phosphorylation of adenosine despite high tissue adenosine deaminase activity, 1.2 μmole/min g wet weight in the dog (GENTRY and OLSSON, unpublished). Studies of the fate of adenosine subsequent to its uptake using the human red cell ghost as a model indicate that phosphorylation keeps pace with adenosine uptake. Even at high rates of transport, when its concentration is as high as 50 μM, adenosine does not accumulate in these cells (SCHRADER, BERNE, and RUBIO, 1972). These authors suggest that adenosine kinase may be located in the cell membrane, and have some evidence for this in the embryonic avian heart (BERNE and RUBIO, 1974). Further work in mammalian heart cells, perhaps also in cell culture preparations in which the metabolic contribution of vascular tissue has been removed, could provide a clearer definition of this aspect of the adenosine hypothesis. At this time one can only conclude that the total activity and low Km of the enzyme account adequately for the phosphorylation of at least 70 % of the adenosine entering the cell and the possibility that a superficial localization of adenosine kinase also contributes to the preponderance of phosphorylation remains an attractive hypothesis which has not been tested in mammalian heart.

The AMP formed from adenosine after its transport into cells may accumulate in a special compartment separate from, and exchanging slowly with, other AMP pools. Evidence for this compartmentalization comes from studies of LIU and FEINBERG (1971) in which ^{14}C-adenosine was infused intracoronary in rabbit hearts for 90 min and the relative specific activities of the adenine nucleotides were compared. If a labeled precursor, A*, is introduced into a compartment containing unlabeled precursor A and its products, B. C..., N one would expect the specific activity of A>B>C>...>N until equilibration occurs. However, the specific activity of the ATP and ADP pools in these hearts always exceeded by as much as six-fold that of the AMP pool. In contrast, inosine infusions resulted in specific activities of the ADP and ATP pools which were initially lower than AMP, the ratio gradually rising with continued infusion of precursor to values similar to those seen during adenosine infusions. This finding suggests that the adenosine but not inosine which is taken up by the heart is incorporated preferentially into a small nucleotide compartment which exchanges poorly with a separate "bulk" nucleotide pool. Such a compartmentalization is similar to that of the ATP subserving cyclic AMP synthesis in mammalian brain (SHIMIZU, CREVELING, and DALY, 1970a,b).

C. Adenosine Catabolism

There is a constant efflux of inosine and hypoxanthine into coronary venous blood, suggesting that degradation of adenosine may also contribute to regulation of the size of its cardiac pool. This process is quantitatively second in importance to uptake of adenosine by the cardiac cells.

1. Inosine Formation

Myocardial inosine concentrations approximate those of adenosine, about 1 nmole/g in dog heart (OLSSON, 1970) and 6 nmole/g rat heart (RUBIO and BERNE, 1969). There are three potential sources of this inosine: (1) The 5'-nucleotidase-catalyzed dephosporylation of IMP (BURGER and LOWENSTEIN, 1967), the concentration of IMP being three times greater than AMP in oxygenated rat heart (RUBIO, BERNE, and DOBSON, 1973); (2) by the deamination of that fraction ($\leq$ 30 percent) of the adenosine transport into the cardiac cells which is not phosphorylated by adenosine kinase (OLSSON et al., 1972); and (3) deamination of adenosine released into the extracellular space by contiguous vascular structures, which are rich in adenosine deaminase (CONWAY and COOKE, 1939). Cardiac muscle is rich in AMP deaminase, 5'-nucleotidase and adenosine deaminase, so that all three pathways may contribute to inosine production. The repeated failure to find adenosine in the coronary venous effluent of oxygenated heart (RUBIO, BERNE, and KATORI, 1969; RUBIO, WIEDMEIER, and BERNE, 1974; SNOW et al., 1973) argues persuasively that deamination of adenosine after its release into the cardiac interstitial space may contribute some of the inosine found in coronary venous blood. A histochemical method for demonstrating the tissue distribution of adenosine deaminase could be very useful in defining this aspect of cardiac purine metabolism.

2. Hypoxanthine Formation

The origin of the hypoxanthine found in heart muscle and coronary venous blood seems more certain than that of inosine. Hypoxanthine is generated by the phosphorolytic cleavage of ribose from inosine which is catalyzed by purine nucleoside phosphorylase. Histochemical staining for this enzyme shows that it is located in the endothelium and pericytes of the coronary capillaries (RUBIO, WIEDMEIER, and BERNE, 1972).

3. Disposition of Cardiac Inosine and Hypoxanthine

Inosine and hypoxanthine are taken up by cardiac cells and are incorporated into the cellular purine nucleotide pool, but at a much lower rate than adenosine. Rabbit heart incorporates inosine into its nucleotide pool only 7 percent as rapidly as adenosine when these nucleosides are infused intracoronary at concentrations of 0.3 μM (LIU and FEINBERG, 1971). Rat heart assimilates inosine and hypoxanthine substantially less rapidly than adenosine over a perfusate concentration range of 0.1 - 10 μM (NAMM and LEADER, 1974). Consonant with the limited ability of myocardial cells to assimilate these purines, there is a substantial efflux into coronary venous blood. In oxygenated dog hearts this rate of escape into the coronary venous effluent is about 0.3 nmole/g min (RUBIO, BERNE, and KATORI, 1969; SNOW et al., 1973), and consists exclusively of inosine and hypoxanthine.

IV. Adenosine Relaxation of Coronary Smooth Muscle

A. The "Adenosine Receptor"

The concept that the biological effects of drugs and hormones are initiated by binding of the effector agent to a "receptor" has proved to be an extremely useful intellectual tool in pharmacology and, more recently, other branches of biology as well. Indeed, the macromolecules to which neuroeffectors such as the catecholamines or acetycholine bind have been partially purified and characterized by chemical techniques, so there is little reason to doubt the existence of receptors as concrete physical entities. Except for a few specific examples like those given, however, the experimental support for the role of receptors in most biological responses remains inferential. This is clearly the case for adenosine. The evidence for a coronary adenosine receptor consists of the blockade of adenosine coronary vasodilatory effects of some adenosine analogs. There have not been any attempts chemically to isolate a vascular adenosine receptor to our knowledge.

Because the available evidence for an adenosine receptor is pharmacological, there are inherent difficulties in its interpretation. It is not always possible to separate the direct coronary effects of adenosine from its separate effects on heart rate and myocardial contractility, which change cardiac oxygen metabolism, and, indirectly, coronary vascular resistance. It is equally difficult to separate the direct coronary-relaxing effects of adenosine analogs from the possibility that these analogs cause vasodilation indirectly by interfering with the uptake of adenosine by the myocardial cells, thus causing the accumulation of endogenous adenosine. Adenosine is a sterically and chemically complex molecule which possesses at least four functional groups. The scarcity of appropriately modified analogs has slowed the systematic analysis of its structure-activity relationships.

The administration of methylxanthines, particularly theophylline, reduces the coronary vasodilatory effects of adenosine and dipyridamole (AFONSO, 1970). The effect is specific, methylxanthines

Table 1. Coronary vasoactivity of various adenosine analogs

	COBBIN, EINSTEIN and MAGUIRE (1974)	ANGUS et al. (1971)	OLSSON et al. unpubl.
Adenosine	1.00	1.00	1.00
N-1-Substituted Adenosines			
Adenosine-1-oxide			0.049
1-Methyladenosine			0.028
C-2-Substituted Adenosines			
2-Fluoroadenosine	5.3		7.8
2-Chloroadenosine		6.6	27.0
2-Bromoadenosine	2.5		
2-Hydroxyadenosine	0.82		2.8
2-Methoxyadenosine	0.31	0.28	
2-Ethoxyadenosine	0.42		
2-Aminoadenosine	0.14		0.057
2-Methylaminoadenosine	0.07		
2-Ethylaminoadenosine	1.10	0.91	
2-S-Methylthioadenosine		0.03	
2-S-Ethylthioadenosine		0.12	
2-S-n-Propylthioadenosine		0.48	
2-S-i-Propylthioadenosine		0.37	
2-S-n-Butylthioadenosine	0.27		
2-S-i-Butylthioadenosine	2.30		
2-S-t-Butylthioadenosine	0.43		
2-S-Benzylthioadenosine	0.09		
2-Trifluoromethyladenosine	0.05		
C-6 Substituted Purine Ribosides			
Purine riboside			0.023
6-Chloropurine riboside			0.023
6-Methylaminopurine riboside		0.06	0.05
6-Dimethylaminpurine riboside		<0.02	0.0012
6-Isopentenylaminopurine riboside	0.06		
6-Hydroxylaminopurine riboside	0.06		
Inosine			0.0026
6-Thiopurine riboside			0.0065
6-Methylthiopurine riboside			0.016

Table 1 (continued)

	COBBIN, EINSTEIN and MAGUIRE (1974)	ANGUS et al. (1971)	OLSSON et al. unpubl.
C-8 Substituted Adenosines			
8-Bromoadenosine			inactive
8-Thioadenosine			inactive
C-2 and C-6 Substituted Purine Ribosides			
2-Aminopurine riboside			0.006
2-Amino-6-chloropurine riboside			0.033
2-Amino-6-p-nitrobenzylthiopurine riboside			0.033
Guanosine			0.002
2-Chloro-6-methylaminopurine riboside		0.04	
6-Methylamino-2-trifluoromethylpurine riboside	0.01		
Adenine Nucleosides			
Adenine-α-D-ribofuranoside			inactive
"Carbocyclic adenosine"	0.23		
Adenine Psicofuranoside			0.006
2'-deoxyadenosine			0.01
Adenine arabinoside	inactive		inactive
2'-O-Methyladenosine			0.007
3'-deoxyadenosine			0.11
Adenine xyloside			inactive
3'-O-Methyladenosine	0.06		0.004
Adenine-α-L-lyxoside			inactive
5'-deoxyadenosine			0.22
5'-chloro-5'-deoxyadenosine	1.9		0.86
Adenosine-5'-uronic acid			inactive
Miscellaneous Nucleosides			
Tubercidin	0.005		
Formycin	0.005		
8-Azaadenosine	0.003		
"ε-Adenosine			inactive
Cytosine			inactive
Uridine			inactive
4-Aminoimidazolecarboxamide-1-riboside			inactive

having no effect on the coronary vasodilation caused by nitroglycerin, carbochromen or isuprel. Although this has been interpreted as evidence of occupancy of a common receptor site (SCHOLTHOLT, NITZ, and SCHRAVEN, 1972), which implies an interaction at the surface of the myocyte, the possibility that either or both compounds exert their respective effects at an intracellular level has not been rigorously excluded.

Almost all the adenosine analogs which have been tested to date cause some degree of coronary vasodilation (Table 1). However, many though not all also inhibit adenosine uptake (OLSSON et al., 1972), so that the possibility of indirect effects cannot be excluded. If one accepts the existence for an adenosine receptor for the purpose of this discussion, however, the available data help define its properties. The following discussion is intended to indicate the future research which is needed adequately to establish and characterize such a receptor.

B. Determinants of Binding to the Coronary "Adenosine Receptor"

Adenosine consists of two chemically dissimilar moieties, the planar, aromatic purine base with its exocyclic 6-amino group and the ribofuranose ring which is non-planar, non-aromatic and, too, has exocyclic hydroxyl and hydroxymethyl substituents.

1. Purine Base

Only three observations imply a role of the purine base itself as a determinant of binding to the receptor: (1) purine riboside is weakly vasoactive while, (2) pyrimidine nucleosides (WOLF and BERNE, 1956) and "ε-adenosine" are inactive, (3) the imidazole ring is not absolutely essential for activity, as indicated by weak vasoactivity of formycin, tubercidin and 8-azaadenosine. The effect of systematic one-for-one substitutions of carbon and nitrogen in the pyrimidine ring has apparently not been studied. The inactivity of the tricyclic "ε-adenosine" suggests that the size of the purine base could be important. Acromaticity of the purine base confers hydrophobicity which may or may not contribute to vasoactivity. Adequate evaluation of the contribution role of spatial and electronic properties of the purine base in its binding to the putative receptor must await more information about the aza- and deaza adenosines, benzimidazole nucleosides, and the nucleosides of other heterocyclic compounds with different ring sizes.

2. Exocyclic Purine Substituents

There is substantially more information about the effect of various purine substituents on the vasoactivity of adenosine, especially for nucleosides substituted at C-6, which are relatively easy to synthesize. The data summarized in Table 1 show reasonably good agreement among laboratories, which supports the validity of these results. Current evidence is consistent with some tentative structure-activity relationships: (a) Substitution at N-1 appears to produce a substantial loss of activity. (b) Substitutions at C-2, particularly by certain halogens and alkylthiols, may enhance vasoactivity. (c) The potency of adenosine far exceeds that of other 6-substituted purine ribosides, but none are inactive.

3. Glycosidic Bond

Rotation of the purine base about the axis of the N9 and C1 bond yields two conformations of the adenosine molecule designated as *syn*, in which the base lies over the ribose, and *anti*, in which the base is oriented away from the ribose moiety (HASCHENMEYER and RICH, 1967). The energy barriers to rotation about the axis of the glycosidic bond are not large (TENG et al., 1971). However, complex steric and chemical interactions of C8 of the purine with C'2, C'3, and C'5 of the ribofuranose (BERTHOHOD and PULLMAN, 1971a, 1971b), restrict this rotation and promote a preponderance of the *anti* conformer in solution (SAENGER, 1973). The absence of coronary vasoactivity of 8-bromoadenosine could be attributed to the hindrance to rotation caused by the bulk of the bromine atom which would result in a population largely if not entirely in the *syn* conformation. This conformational factor has been invoked to explain the inability of adenosine deaminase to utilize 8-bromoadenosine as a substrate (SIMON et al., 1970). Subsequent work with adenine cyclonucleosides (HAMPTON, HARPER, and SASAKI, 1972a,b; IKEHARA and FUKUI, 1974) supports this by defining the conformation of the adenosine molecule which binds to the active site of the deaminase. Similar studies of the coronary vasoactivity of compounds which are rigidly constrained into conformations with fixed glycosidic torsion angles obviously would be of great interest.

4. Furanose

The coronary vasoactivity of adenosine depends absolutely on its sugar moiety; adenine is devoid of activity. Alterations of the sugar, particularly those affecting bulk near C'-2 and C'-3, may markedly reduce or abolish coronary vasoactivity. Reduction of either the 2'- and 3'-hydroxyls (i.e., 2'- and 3'-deoxyadenosine) decreases coronary vasoactivity less than either O-methylation or epimerization (adenine arabinoside and xyloside). These findings suggest that the disposition of space-occupying hydroxyl groups may be more important than the hydroxyls themselves. Likewise, epimerization at C'4 (adenine α-L-lyxoside), which rotates C'5 below the "plane" of the furanose ring abolishes vasoactivity. The nature of C'5-substituents appears to influence vasoactivity, but the number of compounds studied to date is insufficient to define precisely these determinants of adenosine activity.

C. Adenosine Effects on Smooth Muscle Metabolism

Adenosine initiates a poorly understood series of molecular events which culminate in coronary smooth muscle relaxation. Knowledge of smooth muscle contractile mechanisms lags behind that of striated muscle and limits understanding of the effects of any vasoactive agent.

1. *Regulation of Smooth Muscle Tone*

Vascular smooth muscle contains the same organized array of actin and myosin filaments and their modulator proteins as striated muscle. Contraction of this muscle is calcium-dependent, but smooth muscle differs from striated muscle in having a much lower myosin ATPase activity, which correlates with its lower rate of shortening. The contractile state of vascular smooth muscle is determined directly by the calcium concentration in the myoplasm, which is controlled in turn by energy-dependent ion pumps, (Ca^{++}-ATPases) which sequester calcium in the mitochondria and sarcoplasmic reticulum or pump it into the extracellular space. It appears that the calcium necessary to support rapid contractile responses comes from intracellular stores, whereas the sustained maintenance of tone is supported by calcium from the extracellular pool (BOHR, 1973, and references therein; SOMLYO, 1972).

Presumably adenosine exerts its vasodilatory effects at the level of these ion pumps in the cell membrane of the coronary myocyte, since the high basal tone characteristic of these vessels implies a dependence on extracellular Ca^{++}. The factors controlling ion fluxes in this tissue have not been defined, so that the development of this aspect of the adenosine hypothesis depends heavily on further progress in the field of vascular smooth muscle physiology.

2. *Interaction of Oxygen and Adenosine*

One of the most promising insights into the possible role of adenosine in regulating coronary vascular resistance is the observation that ambient oxygen tension and adenosine act synergistically to produce coronary relaxation. Coronary artery strips cut from 100 μ rabbit coronary arteries maintained acetylcholine-induced tone despite reducing the pO_2 of the bathing medium to values as low as 5 - 10 mm Hg (GELLAI, NORTON, and DETAR, 1973). This pO_2 is probably lower than the ambient pO_2 in the vicinity of the resistance vessels *in situ*, because these vessels contain oxygenated blood and coronary venous pO_2 is higher, 15 - 25 mm Hg, probably reflecting the average interstitial pO_2 of the heart. These authors showed that adenosine concentrations of 0.1 or 1.0 μM, which caused minimal relaxation at a pO_2 of 100 mm Hg, caused much greater relaxation at a pO_2 of $<$ 10 mm Hg. Additionally, in the presence of low (0.1 μM) concentrations of adenosine, contractile tension was directly proportional to pO_2 over a range of 10 - 40 mm Hg. MOIR and JONES (1973) have extended these observations to the open-chest dog, finding a consistent enhancement of coronary vasodilator potency of adenosine as arterial pO_2 was reduced. These results indicate the need for a more detailed examination of the oxygen-adenosine interaction in hearts from species with various myocardial levels of adenosine, and emphasize the importance of controlling pO_2 in studies of metabolic effects of adenosine in vascular smooth muscle.

D. Drugs

This review has avoided consideration of the effects of coronary vasodilator drugs. In some well-defined circumstances drugs give useful information about a biological system, but unfortunately drugs have multiple effects, and it is often difficult or impossible to trace a chain of cause and effect between administration of an agent and an observed effect. The powerful coronary vasodilator dipyridamole exemplifies this difficulty (OLSSON, 1973). Its *known* effects include inhibition of adenosine deaminase (KÜBLER et al., 1963) and of cAMP phosphodiesterase (PÖCH and KUKOVETZ, 1972), as well as inhibition of adenosine uptake by heart muscle (PFLEGER, VOLKMER, and KOLASSA, 1969; HOPKINS and GOLDIE, 1972; OLSSON et al., 1972). The last effect might indirectly effect coronary vasodilation symply by increasing the size of the adenosine pool. Further, a direct effect on isolated coronary arteries has been demonstrated by WALTER and BASSENGE (1968). It is simply not possible to ascribe the vasodilator action of this drug unequivocally to one or another of these effects. Unfortunately, the same must be said of most other drugs of this type.

V. Summary

A persuasive body of evidence has evolved during the past 15 years showing that adenosine satisfies many criteria of a physiological regulator of coronary vascular tone. Clearly, 5'-nucleotidase, the enzyme catalyzing the production of adenosine, is present in the sarcolemma and its projections into the interior of the myocardial cell. The 5'-nucleotidase activity recoverable from heart homogenates exceeds the observed rates of adenosine production *in vivo*. It is possible that the explanation of this disparity may identify the currently unknown mechanism by which adenosine production is coupled to myocardial oxygen demand. The evidence indicating that adenosine may be formed on the external surface of the cardiac membrane, together with very high levels of adenosine deaminase within the myocardial cells, argues for an adenosine pool localized only in the cardiac extracellular space. The major process removing adenosine from this pool, facilitated diffusion back into the myocardial cell, has been identified and quantitated. There is a cardiac inosine pool which presumably results from the deamination of adenosine, but other sources of this nucleoside have not been excluded. Purine nucleoside phosphorylase, the enzyme catalyzing the degradation of inosine, has been demonstrated in vascular and perivascular cells, but localization in other cell types has not been excluded. The rates of cellular uptake of inosine and hypoxanthine are substantially lower than that of adenosine and this, together with a sizeable arteriovenous difference of these compounds in oxygenated hearts, suggests that washout by the coronary circulation removes these compounds from their cardiac pools. To the extent that inosine is a degradation product of adenosine, this constitutes a secondary though quantitatively important determinant of the size of the cardiac adenosine pool. The contractile state of vascular smooth muscle depends on myoplasmic [Ca^{++}], but the processes influencing the intracellular concentration of this ion are not defined. There is some evidence for an "adenosine receptor" in the coronary myocyte and studies of vasodilatory effects of adenosine nucleosides are generating a data base for describing the determinants of binding to it. Adenosine and oxygen appear to act synergistically to regulate coronary vascular tone, but the details of this regulation await further progress in understanding the regulation of vascular smooth muscle metabolism and its influence on myoplasmic [Ca^{++}].

Acknowledgment

In conducting the research described in this report, the investigators adhered to the "Guide for Laboratory Animal Facilities and Care", as promulgated by the Committee on the Guide for Laboratory Animal Facilities and Care of the Institute of Laboratory Animal Resources, National Academy of Sciences - National Research Council.

References

AFONSO, S.: Inhibition of coronary vasodilating action of diyridamole and adenosine by aminophylline in the dog. Circ. Res. 26, 743-752 (1970).

ALLELA, A., WILLIAMS, F.L., BOLENE-WILLIAMS, C., KATZ, L.N.: Interrelation between cardiac oxygen consumption and coronary blood flow. Amer. J. Physiol. 183, 570-582 (1955).

ANGUS, J.A., COBBIN, L.B., EINSTEIN, R., MAGUIRE, M.H.: Cardiovascular actions of substituted adenosine analogs. Brit. J. Pharmacol. 41, 592-599 (1971).

BAER, H.P., DRUMMOND, G.I., DUNCAN, E.L.: Formation and deamination of adenosine by cardiac muscle enzymes. Mol. Pharmacol. 2, 67-76 (1966).

BAER, H.P., DRUMMOND, G.I.: Catabolism of adenine nucleotides by the isolates perfused rat heart. Proc. Soc. Exp. Biol. Med. 127, 33-36 (1968).

BERNE, R.M.: Cardiac nucleotides in hypoxia: Possible role in regulation of coronary blood flow. Amer. J. Physiol. 204, 317-322 (1963).

BERNE, R.M.: Regulation of coronary blood flow. Physiol. Rev. 44, 1-29 (1964).

BERNE, R.M., RUBIO, R., DOBSON, J.B., Jr., CURNISH, R.R.: Adenosine and adenine nucleotides as possible mediators of cardiac and skeletal muscle blood flow regulation. Circ. Res. 28 (Suppl I)I, 115-119 (1971).

BERNE, R.M., RUBIO, R.: Adenine nucleotide metabolism in the heart. Circ. Res. 35 (Suppl III)III, 109-118 (1974).

BERTHOD, H., PULLMAN, B.: Molecular orbital calculations on the conformation of nucleic acids and their constituents. I. Conformational energies of B-nucleosides with C(3')- and C(2')-endo sugars. Biochim. Biophys. Acta 232, 595-606 (1971a).

BERTHOD, H., PULLMAN, B.: Molecular orbital calculations on the conformation of nucleic acids and their constituents. II. Conformational energies of nucleosides with C(3')- and C(2')-exo sugars. Biochim. Biophys. Acta 246, 359-364 (1971b).

BOHR, D.F.: Vascular smooth muscle updated. Circ. Res. 32, 665-672 (1973).

BRAASCH, W., GUDBJARNASON, S., PURI, P.S., RAVENS, K.G., BING, R.J.: Early changes in energy metabolism in the myocardium following acute coronary artery occlusion in anesthetized dogs. Circ. Res. 23, 429-438 (1968).

BRAUNWALD, E.: Determinants of myocardial oxygen consumption. Physiologist 12, 65-93 (1969).

BURGER, R.M., LOWENSTEIN, J.M.: Adenylate deaminase III. Regulation of deamination pathways in extracts of rat heart and lung. J. Biol. Chem. 242, 5281-5288 (1967).

COBBIN, L.B., EINSTEIN, R., MAGUIRE, M.H.: Studies on the coronary dilator actions of some adenosine analogs. Brit. J. Pharmacol. 50, 23-33 (1974).

CONWAY, E.J., COOKE, R.: Deaminase of adenosine and adenylic acid in blood and tissues. Biochem. J. 33, 479-492 (1939).

DEUTICKE, B., GERLACH, E.: Abbau freier Nucleotide im Herz, Skeletmuskel, Gerhirn und Leber der Ratte bei Sauerstoffmangel. Pflügers Archiv 292, 239-254 (1966).

DRURY, A.M., SZENT-GYORGYI, A.: The physiological activity of the adenine compounds with reference to action on the mammalian heart. J. Physiol. (Lond.) 68, 213-237 (1929).

ECKENHOFF, J.E., HAFKENSCHIEL, J.H., LANDMESSER, C.M., HARMEL, M.: Cardiac oxygen metabolism and control of the coronary circulation. Am. J. Physiol. 149, 634-649 (1947).

EDWARDS, M.J., MAGUIRE, M.H.: Purification and properties of rat heart 5'-nucleotidase. Mol. Pharmacol. 6, 641-648 (1970).

EVANS, W.H., GURD, J.W.: Properties of a 5'-nucleotidase purified from mouse liever plasma membranes. Biochem. J. 133, 189-199 (1973).

GELLAI, M., NORTON, J.M., DETAR, R.: Evidence for direct control of coronary vascular tone by oxygen. Circ. Res. 32, 279-289 (1973).

GENTRY, M.K., OLSSON, R.A.: A simple, specific, radioisotopic assay for 5'-nucleotidase. Analyt. Biochem. 64, 624-627 (1975).

GERLACH, E., DEUTICKE, B., DREISBACH, R.H.: Der Nucleotidabbau im Herzmuskel bei Sauerstoffmangel und seine mögliche Bedeutung für die Coronardurchblutung. Naturwissenschaften 50, 228-229 (1963).

GOLDTHWAIT, D.A.: Mechanisms of synthesis of purine nucleotides in heart muscle extracts. J. Clin. Invest. 36, 1572-1578 (1957).

GORDON, G.B., PRICE, H.M., BLUMBERG, J.M.: Electron microscopic localization of phosphatase activities within striated muscle fibers. Lab. Invest. 16, 422-435 (1967).

GREGG, D.E., FISHER, L.C.: Blood supply to the heart. In: Handbook of Physiology, vol. II, pp. 1517-1584. Washington: American Physiological Society 1963.

GUYTON, A.C., ROSS, J.M., CARRIER, O., Jr., WALKER, J.R.: Evidence for tissue oxygen demand as the major factor causing autoregulation. Circ. Res. 15 (Suppl. I)I, 60-69 (1964).

HAMPTON, A., HARPER, P.J., SASAKI, T.: Substrate properties of cycloadenosines with adenosine aminohydrolase as evidence for the conformation of enzyme-bound adenosine. Biochemistry 11, 4736-4739 (1972a).

HAMPTON, A., HARPER, P.J., SASAKI, T.: Evidence for the conformation of enzyme-bound adenosine 5'-phosphate. Substrate and inhibitor properties of 8,5'-cycloadenosine 5'-phosphate with adenylate kinase, adenylate aminohydrolase, adenylosuccinate lysase, and 5'-nucleotidase. Biochemistry 11, 4965-4969 (1972b).

HASCHENMEYER, A.E.V., RICH, A.: Nucleoside conformations: analysis of steric barriers to rotation about the glycosidic bond. Mol. Biol. 27, 369-384 (1967).

HONIG, C.R.: Control of smooth muscle actomyosin by phosphate and 5'-AMP: Possible role in metabolic autoregulation. Microvasc. Res. 1, 133-146 (1968).

HOPKINS, S.V., GOLDIE, R.G.: A species difference in the uptake of adenosine by the heart. Biochem. Pharm. 20, 3359-3365 (1971).

IKEHARA, M., FUKUI, T.: Studies of nucleosides and nucleotides. LVIII. Deamination of adenosine analogs with calf intestine adenosine deaminase. Biochim. Biophys. Acta 338, 512-519 (1974).

IMAI, S., RILEY, A.L., BERNE, R.M.: Effect of ischemia on adenine nucleotides in cardiac and skeletal muscle. Circ. Res. 25, 443-450 (1964).

IPATA, P.L.: Sheep brain 5'-nucleotidase. Some enzymatic properties and allosteric inhibition by nucleoside triphosphates. Biochemistry 7, 507-515 (1968).

JACOB, M.I., BERNE, R.M.: Metabolism of purine derivatives the isolated cat heart. Am. J. Physiol. 198, 322-326 (1960).

JACOB, M.I., BERNE, R.M.: Metabolism of adenosine by the isolated anoxic cat heart. Proc. Soc. Exp. Biol. Med. 107, 738-739 (1961).

KHOURI, E.M., GREGG, D.E., RAYFORD, C.R.: Effect of exercise on cardiac output, left coronary flow and myocardial metabolism in the unanesthetized dog. Circ. Res. 17, 427-437 (1965).

KOLASSA, N., PFLEGER, K., RUMMEL, W.: Specificity of adenosine uptake in heart and inhibition by dipyridamole. Europ. J. Pharmacol. 9, 265-268 (1970).

KUBLER, W., SPIECKERMANN, P.G.: Regulation of glycolysis in the ischemic and the anoxic myocardium. J. Mol. Cellular Cardiology 1, 351-377 (1970).

LIU, M.S., FEINBERG, H.: Incorporation of adenosine-8-^{14}C and inosine-8-^{14}C into rabbit heart adenine nucleotides. Am. J. Physiol. 220, 1242-1248 (1971).

MAGUIRE, M.H., LUKAS, M.C., RETTIE, J.F.: Adenine nucleotide salvage synthesis in the rat heart: Pathways of adenosine salvage. Biochim. Biophys. Acta 262, 108-115 (1972).

MCLEAN, J.D., KHOURI, E.M., OLSSON, R.A.: Sterochemical determinants of coronary vasoactivity of adenosine (abstr). Fed. Proc. 33, 395A (1974).

MOIR, T.W., JONES, P.K.: Observations on the effect of changes in arterial oxygenation on adenosine-induced coronary vasodilation. In: Current Topics in Coronary Research (Eds. C.M. BLOOR, R.A., OLSSON), pp. 11-26. New York: Plenum Publishing Corporation 1973.

MORAN, N.C.: Pharmacological characterization of adrenergic receptors. Pharmacol. Rev. 18, 503-513 (1966).

MUSTAFA, S.J., RUBIO, R.: Membrane phosphorylation of adenosine by embryonic chick cardiac cells (abstr). Fed. Proc. 37, 388A (1973).

NAKATSU, K., DRUMMOND, G.I.: Adenylate metabolism and adenosine formation in the heart. Am. J. Physiol. 223, 1119-1127 (1972).

NAMM, D.H., LEADER, J.: A sensitive analytical method for the detection and quantitation of adenosine in biological samples. Anal. Biochem. 58, 511-524 (1974).

NEELY, J.R., MORGAN, H.E.: Relationship between carbohydrate and lipid metabolism and the energy balance of heart muscle. Ann. Rev. Physiol. 36, 413-459 (1974).

OGILVIE, K.K., SLOTIN, L., RHEAULT, P.: Novel substrate of adenosine deaminase. Biochem. Biophys. Res. Commun. 45, 297-300 (1971).

OLIVER, J.J., PATERSON, A.R.P.: Nucleoside transport: I. Mediated process in human erythrocytes. Canad. J. Biochem. 49, 262-270 (1971).

OLSSON, R.A.: Changes in content of purine nucleosides in canine myocardium during coronary occlusion. Circ. Res. 26, 301-306 (1970).

OLSSON, R.A.: Discussion in Current Topics in Coronary Research (Eds. C.M.BLOOR, R.A.OLSSON), p. 83. New York: Plenum Publishing Corp. 1973.

OLSSON, R.A., GENTRY, M.K., SNOW, J.A.: Steric requirements for binding of adenosine to a membrane carrier in canine heart. Biochim. Biophys. Acta 311, 242-250 (1973).

OLSSON, R.A., GENTRY, M.K., TOWNSEND, R.S.: Adenosine metabolism: Properties of dog heart microsomal 5'-nucleotidase. In: Current Topics in Coronary Research (Eds. C.M. BLOOR, R.A. OLSSON), pp. 27-39. New York: Plenum Publishing Corp. 1973.

OLSSON, R.A., SNOW, J.A., GENTRY, M.K., FRICK, G.P.: Adenosine uptake by canine heart. Circ. Res. 29, 767-778 (1972).

PAGE, E.: Cat heart muscle *in vitro*: III. Extracellular space. J. Gen. Physiol. 46, 201-213 (1962).

PFLEGER, K., VOLKMER, I., KOLASSA, N.: Hemmung der Aufnahme von Adenosin und Verstärkung seiner Wirkung am isolierten Warmblüterherzen durch coronarwirksame Substanzen. Arzneim.-Forsch. 19, 1972-1974 (1969).

PITT, B., GREGG, D.E.: Coronary hemodynamic effects of increasing ventricular rate in the unanesthetized dog. Circ. Res. 22, 753-761 (1968).

PLAGEMANN, P.G.W.: Nucleoside transport by Novikoff rat hepatoma cells growing in suspension culture: Specificity and mechanism of transport reactions and relationship to nucleoside incorporation into nucleic acids. Biochim. Biophys. Acta 233, 688-701 (1971).

PLAGEMANN, P.G.W., RICHEY, D.P.: Transport of nucleosides, nucleic acid bases, choline and glucose by animal cells in culture. Biochim. Biophys. Acta 344, 263-305 (1974).

PÖCH, G., KUKOVETZ, W.R.: Studies on the possible role of cyclic AMP in drug-induced coronary vasodialtion. In: Advances in Cyclic Nucleotide Research (Eds. P. GREENGARD, G.A. ROBINSON), vol. I, pp. 195-221. New York: Raven Press 1972.

RAYFORD, C.R., KHOURI, E.M., GREGG, D.E.: Effect of excitement on coronary and systemic energetics in unanesthetized dogs. Am. J. Physiol. 209, 680-688 (1965).

ROCKWELL, M., MAGUIRE, M.H.: Studies on adenosine deaminase. I. Purification and properties of ox heart adenosine deaminase. Mol. Pharmacol. 2, 574-584 (1966).

ROSTGAARD, J., BEHNKE, O.: Fine structural localization of adenine nucleoside phosphatase activity in the sarcoplasmic reticulum and the T system of rat myocardium. Ultrastructure Res. 12, 579-591 (1965).

ROVETTO, M.J., WHITMER, J.T., NEELY, J.R.: Comparison of the effects of anoxia and whole heart ischemia on carbohydrate utilization in isolated working rat hearts. Circ. Res. 32, 699-711 (1973).

RUBIO, R., BERNE, R.M.: Release of adenosine by the normal myocardium in dogs and its relationship to the regulation of coronary resistance. Circ. Res. 25, 407-415 (1969).

RUBIO, R., BERNE, R.M., DOBSON, J.G., Jr.: Sites of adenosine production in cardiac and skeletal muscle. Am. J. Physiol. 225, 938-953 (1973).

RUBIO, R., BERNE, R.M., KATORI, M.: Release of adenosine in reactive hyperemia of the dog heart. Am. J. Physiol. 216, 56-62 (1969).

RUBIO, R., WIEDMEIER, V.T., BERNE, R.M.: Relationship between coronary flow and adenosine production and release. J. Molec. Cell. Cardiol. 6, 561-566 (1974).

SABISTON, D.C., Jr., GREGG, D.E.: Effect of cardiac contraction on coronary blood flow. Circulation 15, 14-20 (1957).

SAENGER, W.: Structure and function of nucleosides and nucleotides. Angew. Chem. (Int. Ed.) 12, 591-601 (1973).

SATTIN, A., RALL, T.W.: The effect of adenosine and adenine nucleotides on the cyclic adenosine 3'-5'-phosphate content of guinea pig cerebral cortex slices. Mol. Pharmacol. 6, 13-23 (1970).

SCHOLTHOLT, J., NITZ, R.E., SCHRAVEN, E.: On the mechanism of the antagonistic action of xanthine derivatives against adenosine and coronary vasodilators. Arzneimittel-Forsch. 22, 1255-1259 (1972).

SCHOLTISSEK, C.: Studies on the uptake of nucleic acid precursors into cells in tissue culture. Biochim. Biophys. Acta 158, 434-447 (1968).

SCHRADER, J., BERNE, R.M., RUBIO, R.: Uptake and metabolism of adenosine by human erythrocyte ghosts. Am. J. Physiol. 223, 159-166 (1972).

SHIMIZU, H., CREVELING, C.R., DALY, J.W.: Cyclic adenosine 3', 5'-monophosphate formation in brain slices: Stimulation by batrachotoxin. ouabain, veratridine and potassium ions. Mol. Pharmacol. 6, 184-188 (1970a).

SHIMIZU, H., CREVELING, C.R., DALY, J.W.: Stimulated formation of adenosine 3',5'-cyclic phosphate in cerebral cortex: Synergism between electrical activity and biogenic amines. Proc. Nat. Acad. Sci. (Wash.) 65, 1033-1040 (1970b).

SHNITKA, T.K., SELIGMAN, A.M.: Ultrastructural localization of enzymes. Ann. Rev. Biochem. 41, 375-396 (1971).

SIMON, L.N., BAUER, R.J., TOLMAN, R.L., ROBINS, R.K.: Calf intestine adenosine deaminase. Substrate specificity. Biochemistry 9, 573-577 (1970).

SNOW, J.A., OLSSON, R.A., GENTRY, M.K.: Myocardial: blood purine nucleoside concentration ratios in canine myocardium. In: Current Topics in Coronary Research (Eds. C.M. BLOOR, R.A. OLSSON), pp. 41-54. New York: Plenum Publishing Corporation 1973.

SOMLYO, A.P.: Excitation-contraction coupling in vertebrate smooth muscle: Correlation of ultrastructure with function. Physiologist 15, 338-348 (1972).

SULLIVAN, J.M., ALPERS, J.B.: *In vitro* regulation of rat heart 5'-nucleotidase by adenine nucleotides and magnesium. J. Biol. Chem. 246, 3057-3063 (1971).

TADA, M., KIRCHBERGER, M.A., CRIBBIN, K.M., REPKE, D.I., KATZ, A.M.: Stimulation of calcium-activated ATPase of dog cardiac microsomes by cyclic AM-dependent protein kinase (abstr). Fed. Proc. 33, 711A (1974).

TAUBE, R.A., BERLIN, R.D.: Membrane transport of nucleosides in rabbit polymorphonuclear leukocytes. Biochim. Biophys. Acta 255, 6-18 (1972).

TAVALE, S.S., SOBELL, H.M.: Crystal and molecular structure of 8-bromoguanosine and 8-cromoadenosine, two purine nucleosides in the *syn* conformation. J. Mo. Biol. 48, 109-123 (1970).

TENG, N.N.H., ITZKOWITZ, M.D., TINOCO, I., Jr.: Calculation of the rotational strengths of mononucleosides. J. Am. Chem. Soc. 93, 6257-6264 (1971).

VAN BELLE, H.: Uptake and deamination of adenosine by blood: Species differences effect of pH ions, temperature and metabolic inhibitors. Biochim. Biophys. Acta 192, 124-132 (1969).

WALTER, P., BASSENGE, E.: Wirkung von ATP, A-3, 5-MP, Adenosin und Dipyridamol an Streifenpräparaten der A. coronaria, A. renalis und der V. portae. Pflügers Arch. Ges. Physiol. 229, 52-65 (1968).

WIEDMEIER, V.T., RUBIO, R., BERNE, R.M.: Incorporation and turnover of adenosine-U-^{14}C in perfused guinea pig myocardium. Am. J. Physiol. 223, 51-54 (1972).

WILLIAMSON, J.R.: Glycolytic control mechanisms. II. Kinetics of intermediate changes during the aeorbic-anoxic transition in perfused rat heart. J. Biol. Chem. 241, 5026-5036 (1966).

WOLF, M.M., BERNE, R.M.: Coronary vasodilator properties of purine and pyrimidine derivatives. Circ. Res. 4, 343-348 (1956).

WOLLENBERGER, A., KRAUSE, E.G.: Metabolic control characteristics of the acutely ischemic myocardium. Am. J. Cardiol. 22, 349-359 (1968).

WRAY, H.L., GRAY, R.R., OLSSON, R.A.: Cyclic adenosine 3',-5'-monophosphate stimulation Ca^{2+}, Mg^{2+}ATPase in cardiac microsomes (abstr). Circulation 48 (Suppl.IV)IV, 232A (1973).

ZIMMER, H.G., TRENDELENBURG, C., KAMMERMEIER, H., GERLACH, E.: De novo synthesis of myocardial adenine nucleotides in the rat: Acceleration during recovery from oxygen deficiency. Cir. Res. 28, 635-642 (1973).

Subject Index

Progress in Molecular and Subcellular Biology

Volume 1

By B.W. Agranoff, J. Davies, F.E. Hahn, H.G. Mandel, N.S. Scott, R.M. Smillie, C.R. Woese
32 figures. VII, 237 pages
1969

Volume 1 contains contributions by Woese, Davies and Mandel who are concerned either theoretically or experimentally with the nature of the genetic code and its correct translation. Smillie and Scott write about the biosynthesis of chloroplasts and the photoregulation of the underlying basic processes. Agranoff reviews critically current ideas concerning the molecular basis of higher nervous activities. The managing editor, Hahn, makes a critical analysis of the origin, conceptual content and probable future developments in molecular biology.

Volume 2

Proceedings of the Research Symposium on Complexes of Biologically Active Substances with Nucleic Acids and Their Modes of Action

Held at the Walter Reed Army Institute of Research, Washington, 16-19 March, 1970. 158 figures. IX
IX, 400 pages. 1971

These 28 contributions are a definitive account of the most recent research results on selected aspects of the title topic. The major topical subdivisions are: Antibiotics and Nucleic Acids – Antimalarials and Nucleic Acids – Alkaloids, Natural Polyamines and DNA – Intercalation into Supercoiled DNA – Synthetic Drugs and Dyes Binding to Nucleic Acids – Antimutagens – Carcinogens and Nucleic Acids – The Natural State of DNA. Among the contributors are D.M. Crothers, G.F. Gause, W. and H. Kersten, L.S. Lerman, H.R. Mahler, P.O.P. Ts'o, J. Vinograd, and M. Waring.
The editor, F.E. Hahn, introduces Volume 2 with a review of the history, current state and perspectives of the field of nucleic acid complexes. The book will interest research scientists active in the study of nucleic acid complexes and molecular biologists, molecular pharmacologists, biochemists and biophysicists seeking comprehensive and up-to-date information in this specialized field.

Volume 3

By A.S. Braverman, D.J. Brenner, B.P. Doctor, A.B. Edmundson, K.R. Ely, M.J. Fournier, F.E. Hahn, A. Kaji, C.A. Paoletti, G. Riou, M. Schiffer, M.K. Wood
58 figures. VII, 251 pages
1973

The volume is concerned with the transcription and translation of genetic information and with products of both processes in prokaryotic and eukaryotic organisms, Reverse transcription is reviewed with respect to its theoretical importance to molecular biology and its practical importance to cancer research. Translation, i.e., protein biosynthesis, is analyzed in the light of knowledge of numerous inhibitors of individual reaction steps in the overall sequence. The transcription of transfer RNA is reviewed in detail as one aspect of transcription of genetic information. Thalassemia and Bence-Jones proteins have been treated as selected excamples of molecular pathology in eukaryotes. The nature of mitochondrial DNA in neoplastic cells is reviewed as one contribution to the molecular biology of cancer. The book signifies the advancement of molecular biology from the study of bacteria and their viruses to the consideration of events and entities in eukaryotic organisms.

Springer-Verlag
Berlin Heidelberg New York

Molecular Biology, Biochemistry and Biophysics

Editors: A. Kleinzeller, G.F. Springer, H.G. Wittmann

Vol. 1: J.H. van't Hoff
Imagination in Science
Translated into English with notes and a general introduction by G.F. Springer
1 portrait. VI, 18 pp. 1967

Vol. 2: K. Freudenberg, A.C. Neish
Constitution and Biosynthesis of Lignin
10 figs. IX, 129 pp. 1968

Vol. 3: T. Robinson
The Biochemistry of Alkaloids
37 figs. X, 149 pp. 1968

Vol. 5: B. Jirgensons
Optical Activity of Proteins and Other Macromolecules
2nd revised and enlarged edition. 71 figs. IX, 199 pp. 1973

Vol. 6: F. Egami, K. Nakamura
Microbial Ribonucleases
5 figs. IX, 90 pp. 1969

Vol. 8: **Protein Sequence Determination**
A Sourcebook of Methods and Techniques
Edited by S.B. Needleman
2nd revised and enlarged edition. 80 figs.
XVIII, 393 pp. 1975

Vol. 9: R. Grubb
The Genetic Markers of Human Immunoglobulins
8 figs. XII, 152 pp. 1970

Vol. 10: R.J. Lukens
Chemistry of Fungicidal Action
8 figs. XIII, 136 pp. 1971

Vol. 11: P. Reeves
The Bacteriocins
9 figs. XI, 142 pp. 1972

Vol. 12: T. Ando, M. Yamasaki, K. Suzuki
Protamines
Isolation, Characterization, Structure and Function
24 figs. 17 tables
IX, 144 pp. 1973

Vol. 13: P. Jolles, A. Paraf
Chemical and Biological Basis of Adjuvants
24 figs. 41 tables.
VIII, 153 pp. 1973

Vol. 14: **Micromethods in Molecular Biology**
Edited by V. Neuhoff
With contributions by numerous experts.
275 figs. (2 in color).
23 tables. XV, 428 pp. 1973

Vol. 15: M. Weissbluth
Hemoglobin
Cooperativity and Electronic Properties. 50 figs.
VIII, 175 pp. 1974

Vol. 16: S. Shulman
Tissue Specificity and Autoimmunity
32 figs. XI, 196 pp. 1974

Vol. 17: Y.A. Vinnikov
Sensory Reception
Cytology, Molecular Mechanisms and Evolution
Translated from the Russian by W.L. Gray and B.M. Crook
124 figs. (173 separate ill.)
IX, 392 pp. 1974

Vol. 18: H. Kersten, W. Kersten
Inhibitors of Nucleic Acid Synthesis
Biophysical and Biochemical Aspects
73 figs. IX, 184 pp. 1974

Vol. 19: M.B. Mathews
Connective Tissue
Macromolecular Structure and Evolution
31 figs. XII, 318 pp. 1975

Vol. 20: M.A. Lauffer
Entropy-Driven Processes in Biology
Polymerization of Tobacco Mosaic Virus Protein and Similar Reactions
90 figs. X, 264 pp. 1975

Vol. 21: R.C. Burns, R.W.F. Hardy
Nitrogen Fixation in Bacteria and Higher Plants
27 figs. X, 189 pp. 1975

Vol. 22: H. Fromm
Initial Rate Enzyme Kinetics
88 figs. 19 tables
X, 321 pp. 1975

**Springer-Verlag
Berlin
Heidelberg
New York**